DESIGN RELIABILITY

Fundamentals and Applications

DESIGN RELIABILITY

Fundamentals and Applications

B. S. Dhillon

Department of Mechanical Engineering
University of Ottawa
Ottawa, Ontario, Canada

CRC Press is an imprint of the
Taylor & Francis Group, an informa business

CRC Press
Taylor & Francis Group
6000 Broken Sound Parkway NW, Suite 300
Boca Raton, FL 33487-2742

First issued in paperback 2021

ISBN-13: 978-0-8493-1465-0 (hbk)
ISBN-13: 978-1-03-218021-2 (pbk)
DOI: 10.1201/9780367802400

Publisher's Note

The publisher has gone to great lengths to ensure the quality of this reprint but points out that some imperfections in the original copies may be apparent.

Library of Congress Cataloging-in-Publication Data

Dhillon, B. S.
Design reliability : fundamentals and applications / B.S. Dhillon.
p. cm.
Includes bibliographical references (p.
ISBN 0-8493-1465-8 (alk. paper)
1. Engineering design. 2. Reliability (Engineering). I. Title.
TA174.D4929 1999
620'.0042--dc21 99-28211
CIP

Visit the Taylor & Francis Web site at
http://www.taylorandfrancis.com

and the CRC Press Web site at
http://www.crcpress.com

Dedication

This book is affectionately dedicate to my mother
Udham Kaur.

Preface

Today at the dawn of the twenty-first century, we are designing and building new complex and sophisticated engineering systems for use not only on Earth but to explore other heavenly bodies. Time is not that far off when humankind will permanently reside on other planets and explore stars outside our solar system. Needless to say, the reliability of systems being used for space explorations and for use on Earth is becoming increasingly important because of factors such as cost, competition, public demand, and usage of untried new technology. The only effective way to ensure reliability of engineering systems is to consider the reliability factor seriously during their design.

Over the years as engineering systems have become more complex and sophisticated, the knowledge in reliability engineering has also increased tremendously and it has specialized into areas such as human reliability, software reliability, mechanical reliability, robot reliability, medical device reliability, and reliability and maintainability management.

Even though there are a large number of texts already directly or indirectly related to reliability engineering, there is still a need for an up-to-date book emphasizing design reliability with respect to specialized, application, and related areas. Thus, the main objective of writing this book is to include new findings as well as tailor the text in such a manner so that it effectively satisfies the needs of modern design reliability. Therefore, this book is written to meet this challenge and its emphasis is on the structure of concepts rather than on mathematical rigor and minute details. However, the topics are treated in such a manner that the reader needs no previous knowledge to understand them. Also, the source of most of the material presented in the book is given in references if the reader wishes to delve deeper into particular topics. The book contains a large number of examples along with their solutions, and at the end of each chapter there are numerous problems to test reader comprehension.

The book is composed of 17 chapters. Chapter 1 presents the historical aspect of reliability engineering, the need of reliability in engineering design, reliability in the product design process, and important terms and definitions, and information sources. Chapter 2 reviews mathematics essential to understanding subsequent chapters. Fundamental aspects of engineering design and reliability management are presented in Chapter 3. As the failure data collection and analysis are considered the backbone of reliability engineering, Chapter 4 presents many associated aspects. Chapter 5 presents many basic reliability evaluation and allocation methods. However, the emphasis of the chapter is on the basic reliability evaluation methods. Chapters 6 and 7 describe in detail two most widely used methods (i.e., failure modes and effect analysis and fault tree analysis, respectively) to evaluate engineering design with respect to reliability. Chapter 8 presents two important topics of reliability engineering: common-cause failures and three state devices. Two specialized areas of reliability, mechanical reliability and human reliability, are discussed in Chapters 9 and 10, respectively.

Chapter 11 presents the topics of reliability testing and growth essential in the design phase of an engineering system. Chapters 12 through 14 present three application areas of reliability, i.e., reliability in computer systems, robot reliability, and medical device reliability, respectively. In particular, the emphasis of Chapter 12 is on computer software reliability. Chapter 15 presents important aspects of design maintainability and reliability centered maintenance. Chapters 16 and 17 describe three topics directly or indirectly related to design reliability: total quality management, risk assessment, and life cycle costing.

The book is intended primarily for senior level undergraduate and graduate students, professional engineers, college and university level teachers, short reliability course instructors and students, researchers, and engineering system design managers.

The author is deeply indebted to many individuals including colleagues, students, friends, and reliability and maintainability professionals for their invisible inputs and encouragement at the moment of need. I thank my children Jasmine and Mark for their patience and intermittent disturbances leading to desirable coffee and other breaks. And last, but not least, I thank my boss, friend, and wife, Rosy, for typing various portions of this book and other related materials, and for her timely help in proofreading.

B.S. Dhillon
Ottawa, Ontario

The Author

B.S. Dhillon, Ph.D., is a professor of Mechanical Engineering at the University of Ottawa. He has served as a Chairman/Director of Mechanical Engineering Department/Engineering Management Program for 11 years at the same institution. He has published over 260 articles on Reliability and Maintainability Engineering and on related areas. In addition, he has written 20 books on various aspects of system reliability, safety, human factors, design, and engineering management published by Wiley (1981), Van Nostrand (1982), Butterworth (1983), Marcel Dekker (1984), and Pergamon (1986) among others. His books on Reliability have been translated into several languages including Russian, Chinese, and German. He has served as General Chairman of two international conferences on reliability and quality control held in Los Angeles and Paris in 1987. Also, he is or has been on the editorial board of five international journals.

Dr. Dhillon is a recipient of the American Society of Quality Control Austin J. Bonis Reliability Award, the Society of Reliability Engineers' Merit Award, the Gold Medal of Honor (American Biographical Institute), and Faculty of Engineering Glinski Award for excellence in Reliability Engineering Research. He is a registered Professional Engineer in Ontario and is listed in the *American Men and Women of Science*, *Men of Achievements*, *International Dictionary of Biography*, *Who's Who in International Intellectuals*, and *Who's Who in Technology*.

At the University of Ottawa, he has been teaching reliability and maintainability for over 18 years. Dr. Dhillon attended the University of Wales where he received a B.S. in electrical and electronic engineering and an M.S. in mechanical engineering. He received a Ph.D. in industrial engineering from the University of Windsor.

Table of Contents

CHAPTER 9: Mechanical Reliability

CHAPTER 10: Human Reliability in Engineering Systems

CHAPTER 13: Robot Reliability and Safety

CHAPTER 14: Medical Device Reliability

1 Introduction

1.1 RELIABILITY HISTORY

The history of the reliability discipline goes back to the early 1930s when probability concepts were applied to electric power generation related problems [1-3]. During World War II, Germans applied the basic reliability concepts to improve reliability of their V1 and V2 rockets. In 1947, Aeronautical Radio, Inc. and Cornell University conducted a reliability study of over 100,000 electronic tubes. In 1950, an *ad hoc* committee on reliability was established by the United States Department of Defense and in 1952 it was transformed to a permanent body: Advisory Group on the Reliability of Electronic Equipment (AGREE) [4].

In 1951, Weibull published a statistical function that subsequently became known as the Weibull distribution [5]. In 1952, exponential distribution received a distinct edge after the publication of a paper, presenting failure data and the results of various goodness-of-fit tests for competing failure distribution, by Davis [6].

In 1954, a National Symposium on Reliability and Quality Control was held for the first time in the United States and in the following year, the Institute of Electrical and Electronic Engineers (IEEE) formed an organization called the Reliability and Quality Control Society. During the following two years, three important documents concerning reliability appeared:

- 1956: a book entitled *Reliability Factors for Ground Electronic Equipment* [7],
- 1957: AGREE report [8],
- 1957: first military reliability specification: MIL-R-25717 (USAF): Reliability Assurance Program for Electronic Equipment [9].

In 1962, the Air Force Institute of Technology of the United States Air Force (USAF), Dayton, Ohio, started the first master's degree program in system reliability engineering. Nonetheless, ever since the inception of the reliability field many individuals have contributed to it and hundreds of publications on the topic have appeared [10,11].

This chapter discusses various introductory aspects of the reliability discipline.

1.2 NEED OF RELIABILITY IN PRODUCT DESIGN

There have been many factors responsible for the consideration of reliability in product design including product complexity, insertion of reliability related-clauses in design specifications, competition, awareness of cost effectiveness, public demand, and the past system failures. Some of these factors are described below in detail.

Even if we consider the increase in the product complexity with respect to parts alone, there has been a phenomenal growth of some products. For example, today a typical Boeing 747 jumbo jet airplane is made up of approximately 4.5 million parts, including fasteners. Even for relatively simpler products, there has been a significant increase in complexity with respect to parts. For example, in 1935 a farm tractor was made up of 1200 critical parts and in 1990 the number increased to around 2900.

With respect to cost effectiveness, many studies have indicated that the most effective for profit contribution is the involvement of reliability professionals with product designers. In fact, according to some experts, if it would cost $1 to rectify a design defect prior to the initial drafting release, the cost would increase to $10 after the final release, $100 at the prototype stage, $1000 at the pre-production stage, and $10,000 at the production stage. Nonetheless, various studies have revealed that design-related problems are generally the greatest causes for product failures. For example, a study performed by the U.S. Navy concerning electronic equipment failure causes attributed 43% of the failures to design, 30% to operation and maintenance, 20% to manufacturing, and 7% to miscellaneous factors [12].

Well-publicized system failures such as those listed below may have also contributed to more serious consideration of reliability in product design [13-15]:

- **Space Shuttle Challenger Disaster:** This debacle occurred in 1986, in which all crew members lost their lives. The main reason for this disaster was design defects.
- **Chernobyl Nuclear Reactor Explosion:** This disaster also occurred in 1986, in the former Soviet Union, in which 31 lives were lost. This debacle was also the result of design defects.
- **Point Pleasant Bridge Disaster:** This bridge located on the West Virginia/Ohio border collapsed in 1967. The disaster resulted in the loss of 46 lives and its basic cause was the metal fatigue of a critical eye bar.

1.3 RELIABILITY IN THE PRODUCT DESIGN PROCESS

Reliability of the design, to a large extent, is determined by the reliability tasks performed during the product design. These reliability tasks include: establishing reliability requirements definition, using reliability design standards/guides/checklists, allocating reliability, predicting reliability, reliability modeling, monitoring subcontractor/supplier reliability activities, performing failure modes effects and criticality analysis, monitoring reliability growth, assessing software reliability, environmental stress screening, preparing critical items list, and performing electronic parts/circuits tolerance analysis [16].

Reliability tasks such as those listed above, if performed effectively, will contribute tremendously to the product design.

1.4 RELIABILITY SPECIALIZED AND APPLICATION AREAS

Ever since the inception of the reliability field, the reliability discipline has branched into many specialized and application areas such as those listed below [10,11]:

- **Mechanical reliability.** This is concerned with the reliability of mechanical items. Many textbooks and other publications have appeared on this topic. A comprehensive list of publications on mechanical reliability is given in Reference 17.
- **Software reliability.** This is an important emerging area of reliability as the use of computers is increasing at an alarming rate. Many books have been written on this topic alone. A comprehensive list of publications on software reliability may be found in References 10 and 18.
- **Human reliability.** In the past, many times systems have failed not due to technical faults but due to human error. The first book on the topic appeared in 1986 [18]. A comprehensive list of publications on human reliability is given in Reference 19.
- **Reliability optimization.** This is concerned with the reliability optimization of engineering systems. So far, at least one book has been written on the topic and a list of publications on reliability optimization is given in References 20 and 21.
- **Reliability growth.** This is basically concerned with monitoring reliability growth of engineering systems during their design and development. A comprehensive list of publications on the topic is available in Reference 10.
- **Structural reliability.** This is concerned with the reliability of engineering structures, in particular civil engineering. A large number of publications including books have appeared on the subject [11].
- **Reliability general.** This includes developments on reliability of a general nature. Usually, mathematicians and related professionals contribute to the area [10].
- **Power system reliability.** This is a well-developed area and is basically concerned with the application of reliability principles to conventional power system related problems. Many books on the subject have appeared over the years including a vast number of other publications [22].
- **Robot reliability and safety.** This is an emerging new area of the application of basic reliability and safety principles to robot associated problems. Over the years many publications on the subject have appeared including one textbook [23].
- **Life cycle costing.** This is an important subject that is directly related to reliability. In particular, when estimating the ownership cost of the product, the knowledge regarding its failure rate is essential. In the past, many publications on life cycle costing have appeared including several books [24].
- **Maintainability.** This is closely coupled to reliability and is concerned with the maintaining aspect of the product. Over the years, a vast number of publications on the subject have appeared including some books [10].

1.5 TERMS AND DEFINITIONS

There are many terms and definitions used in reliability engineering. Some of the frequently used terms and definitions are presented below [25-28].

- **Reliability.** This is the probability that an item will carry out its assigned mission satisfactorily for the stated time period when used under the specified conditions.
- **Failure.** This is the inability of an item to function within the initially defined guidelines.
- **Downtime.** This is the time period during which the item is not in a condition to carry out its stated mission.
- **Maintainability.** This is the probability that a failed item will be repaired to its satisfactory working state.
- **Redundancy.** This is the existence of more than one means for accomplishing a defined function.
- **Active redundancy.** This is a type of redundancy when all redundant items are operating simultaneously.
- **Availability.** This is the probability that an item is available for application or use when needed.
- **Mean time to failure (exponential distribution).** This is the sum of the operating time of given items divided by the total number of failures.
- **Useful life.** This is the length of time an item operates within an acceptable level of failure rate.
- **Mission time.** This is the time during which the item is performing its specified mission.
- **Human error.** This is the failure to perform a given task (or the performance of a forbidden action) that could lead to disruption of scheduled operations or result in damage to property/equipment.
- **Human reliability.** This is the probability of completing a job/task successfully by humans at any required stage in the system operation within a defined minimum time limit (if the time requirement is specified).

1.6 RELIABILITY INFORMATION SOURCES

There are many different sources for obtaining reliability related information. They may be classified into various different categories. Three such categories are as follows:

1. Journals
    - *Quality and Reliability Engineering*, published by John Wiley & Sons four times a year.
    - *IEEE Transactions on Reliability*, jointly published by the Institute of Electrical and Electronic Engineers (IEEE) and the American Society for Quality Control (ASQC) five times a year.
    - *Microelectronics Reliability*, published by Pergamon Press 12 times a year.
    - *Reliability, Quality, and Safety Engineering*, published by the World Scientific Publishing Company four times a year.
    - *Quality and Reliability Management*, published by MCB University Press several times a year.

- *Reliability Engineering and System Safety*, published by Elsevier Science Publishers several times a year.
- *Reliability Review*, published by the Reliability Division of ASQC four times a year.

2. Conference Proceedings
   - *Proceedings of the Annual Reliability and Maintainability Symposium* (U.S.)
   - *Proceedings of the Annual Reliability Engineering Conference for the Electric Power Industry* (U.S.)
   - *Proceedings of the European Conference on Safety and Reliability* (Europe)
   - *Proceedings of the Symposium on Reliability in Electronics* (Hungary)
   - *Proceedings of the International Conference on Reliability, Maintainability, and Safety* (China)
   - *Proceedings of the International Conference on Reliability and Exploitation of Computer Systems* (Poland)
3. Agencies
   - Reliability Analysis Center
Rome Air Development Center (RADC)
Griffis Air Force Base
New York, NY 13441-5700
   - Government Industry Data Exchange Program (GIDEP)
GIDEP Operations Center
U.S. Department of Navy
Naval Weapons Station
Seal Beach
Corona, CA 91720
   - National Aeronautics and Space Administration (NASA) Parts Reliability Information Center
George C. Marshall Space Flight Center
Huntsville, AL 35812
   - National Technical Information Center (NTIS)
5285 Port Royal Road
Springfield, VA 22151
   - Defense Technical Information Center
DTIC-FDAC
8725 John J. Kingman Road, Suite 0944
Fort Belvoir, VA 22060-6218
   - American National Standards Institute (ANSI)
11 W. 42nd St.
New York, NY 10036
   - Technical Services Department
American Society for Quality Control
611 W. Wisconsin Ave., P.O. Box 3005
Milwaukee, WI 53201-3005

- Space Documentation Service
  European Space Agency
  Via Galileo Galilei
  Frascati 00044, Italy
- System Reliability Service
  Safety and Reliability Directorate
  UKAEA
  Wigshaw Lane, Culcheth
  Warrington, WA3 4NE, U.K.

1.7 MILITARY AND OTHER RELIABILITY DOCUMENTS

Over the years many military and other documents concerning reliability have been developed. Such documents can serve as a useful tool in conducting practically inclined reliability studies and related tasks. Some of these documents are as follows:

- MIL-HDBK-338, Electronics Reliability Design Handbook, Department of Defense, Washington, D.C.
- MIL-HDBK-251, Reliability/Design Thermal Applications, Department of Defense, Washington, D.C.
- MIL-HDBK-217, Reliability Prediction of Electronic Equipment, Department of Defense, Washington, D.C.
- MIL-STD-785, Reliability Program for Systems and Equipment, Development and Production, Department of Defense, Washington, D.C.
- MIL-HDBK-189, Reliability Growth Management, Department of Defense, Washington, D.C.
- MIL-STD-1556, Government Industry Data Exchange Program (GIDEP), Department of Defense, Washington, D.C.
- MIL-STD-756, Reliability Modeling and Prediction, Department of Defense, Washington, D.C.
- MIL-STD-721, Definitions of Terms for Reliability and Maintainability, Department of Defense, Washington, D.C.
- MIL-STD-52779, Software Quality Assurance Program Requirements, Department of Defense, Washington, D.C.
- MIL-STD-2167, Defense System Software Development, Department of Defense, Washington, D.C.
- MIL-STD-1629, Procedures for Performing a Failure Mode, Effects and Criticality Analysis, Department of Defense, Washington, D.C.
- MIL-STD-2155, Failure Reporting, Analysis and Corrective Action System (FRACAS), Department of Defense, Washington, D.C.
- MIL-STD-790, Reliability Assurance Program for Electronic Parts Specifications, Department of Defense, Washington, D.C.
- ANSI/AIAA R-013, Recommended Practice for Software Reliability, American National Standards Institute, New York.
- MIL-STD-337, Design to Cost, Department of Defense, Washington, D.C.

- MIL-STD-1908, Definitions of Human Factors Terms, Department of Defense, Washington, D.C.
- MIL-STD-1472, Human Engineering Design Criteria for Military Systems, Equipment and Facilities, Department of Defense, Washington, D.C.
- MIL-HDBK-H108, Sampling Procedures and Tables for Life and Reliability Testing (Based on Exponential Distribution), Department of Defense, Washington, D.C.
- MIL-STD-781, Reliability Design, Qualification and Production Acceptance Tests: Exponential Distribution, Department of Defense, Washington, D.C.
- MIL-STD-2074, Failure Classification for Reliability Testing, Department of Defense, Washington, D.C.
- MIL-STD-690, Failure Rate Sampling Plans and Procedures, Department of Defense, Washington, D.C.
- SAE ARD 50010, Recommended Reliability, Maintainability, Supportability (RMS) Terms and Parameters, Society of Automotive Engineers (SAE), Warrendale, PA.
- MIL-STD-472, Maintainability Prediction, Department of Defense, Washington, D.C.

1.8 PROBLEMS

1. Write an essay on the history of reliability.
2. Discuss the need for reliability in product design.
3. Discuss at least three well-publicized system failures and the reasons for their occurrence.
4. What are the reliability related tasks performed during the product design process?
5. Describe the following two specialized areas of reliability:
   - Human reliability
   - Software reliability
6. What are the differences between reliability and maintainability?
7. Define the following terms:
   - Failure rate
   - Redundancy
   - Failure
8. Describe the following two agencies for obtaining reliability related information:
   - Reliability Analysis Center
   - GIDEP
9. In your opinion, what are the five U.S. military reliability related documents that are the most important during the design process? Give reasons for your selection.
10. Make comparisons between mechanical reliability and power system reliability.

1.9 REFERENCES

1. Lyman, W.J., Fundamental consideration in preparing a master system plan, *Electrical World*, 101, 778-792, 1933.
2. Smith, S.A., Service reliability measured by probabilities of outage, *Electrical World*, 103, 371-374, 1934.
3. Dhillon, B.S., *Power System Reliability, Safety and Management*, Ann Arbor Science Publishers, Ann Arbor, MI, 1983.
4. Coppola, A., Reliability engineering of electronic equipment: a historical perspective, *IEEE Trans. Reliability*, 33, 29-35, 1984.
5. Weibull, W., A statistical distribution function of wide applicability, *J. Appl. Mech.*, 18, 293-297, 1951.
6. Davis, D.J., An analysis of some failure data, *J. Am. Statist. Assoc.*, 47, 113-150, 1952.
7. Henney, K., Ed., *Reliability Factors for Ground Electronic Equipment*, McGraw-Hill, New York, 1956.
8. AGREE Report, Advisory Group on Reliability of Electronic Equipment (AGREE), Reliability of Military Electronic Equipment, Office of the Assistant Secretary of Defense (Research and Engineering), Department of Defense, Washington, D.C., 1957.
9. MIL-R-25717 (USAF), Reliability Assurance Program for Electronic Equipment, Department of Defense, Washington, D.C.
10. Dhillon, B.S., *Reliability and Quality Control: Bibliography on General and Specialized Areas*, Beta Publishers, Gloucester, Ontario, 1992.
11. Dhillon, B.S., *Reliability Engineering Application: Bibliography on Important Application Areas*, Beta Publishers, Gloucester, Ontario, 1992.
12. Niebel, B.W., *Engineering Maintenance Management*, Marcel Dekker, New York, 1994.
13. Dhillon, B.S., *Engineering Design: A Modern Approach*, Richard D. Irwin, Chicago, IL, 1996.
14. Elsayed, E.A., *Reliability Engineering*, Addison Wesley Longman, Reading, MA, 1996.
15. Dhillon, B.S., *Advanced Design Concepts for Engineers*, Technomic Publishing Company, Lancaster, PA, 1998.
16. Reliability, Maintainability, and Supportability Guidebook, SAE G-11 RMS Committee Report, Society of Automotive Engineers (SAE), Warrendale, PA, 1990.
17. Dhillon, B.S., *Mechanical Reliability: Theory, Models and Applications*, American Institute of Aeronautics and Astronautics, Washington, D.C., 1988.
18. Dhillon, B.S., *Reliability in Computer System Design*, Ablex Publishing, Norwood, NJ, 1987.
19. Dhillon, B.S., *Human Reliability with Human Factors*, Pergamon Press, New York, 1986.
20. Tillman, F.A., Hwang, C.L., and Kuo, W., *Optimization of Systems Reliability*, Marcel Dekker, New York, 1980.
21. Tillman, F.A., Hwang, C.L., and Kuo, W., Optimization techniques for system reliability with redundancy: A review, *IEEE Trans. Reliability*, 26, 148-155, 1977.
22. Dhillon, B.S., *Power System Reliability, Safety and Management*, Ann Arbor Science, Ann Arbor, MI, 1983.
23. Dhillon, B.S., *Robot Reliability and Safety*, Springer-Verlag, New York, 1991.
24. Dhillon, B.S., *Life Cycle Costing: Techniques, Models, and Applications*, Gordon and Breach Science Publishers, New York, 1989.

25. MIL-STD-721, Definitions of Effectiveness Terms for Reliability, Maintainability, Human Factors and Safety, Department of Defense, Washington, D.C.
26. Omdahl, T.P., Ed., *Reliability, Availability, and Maintainability (RAM) Dictionary,* ASQC Quality Press, Milwaukee, WI, 1988.
27. Von Alven, Ed., *Reliability Engineering,* Prentice-Hall, Englewood Cliffs, NJ, 1964.
28. Naresky, J.J., Reliability definitions, *IEEE Trans. Reliability,* 19, 198-200, 1970.

2 Design Reliability Mathematics

2.1 INTRODUCTION

The history of our current number symbols often referred to as the "Hindu-Arabic numeral system" may be traced back over 2000 years to the stone columns erected by the Scythian Indian emperor Asoka. Some of these columns erected around 250 B.C. contain the present number symbols [1]. It appears the Hindus invented the numeral system and the Arabs transmitted it to western Europe after their invasion of the Spanish peninsula in 711 A.D. For example, the numeral symbols are found in a tenth-century Spanish manuscript [1]. However, zero, the Hindu **Sunya** meaning "empty" or "void" and Arabic **Sifr** were only introduced into Germany in the thirteenth century. Nonetheless, around 2000 B.C. Babylonians solved quadratic equations successfully.

Even though the traces of the "differentiation" may be found back to the ancient Greeks, the idea of "modern differentiation" comes from Pierre Fermat (1601–1665). Laplace transforms often used to find solutions to differential equations were one of the works of Pierre-Simon Laplace (1749–1827).

The history of probability may be traced back to the writing of a gambler's manual by Girolamo Cardano (1501–1576), in which he considered some interesting issues on probability [1, 2]. However, Pierre Fermat and Blaise Pascal (1623–1662) solved independently and correctly the problem of dividing the winnings in a game of chance. The first formal treatise on probability based on the Pascal–Fermat correspondence was written by Christiaan Huygens (1629–1695) in 1657. Boolean algebra, which plays a significant role in modern probability theory, is named after the mathematician George Boole (1815–1864), who published a pamphlet entitled "The Mathematical Analysis of Logic, Being an Essay towards a Calculus of Deductive Reasoning" in 1847 [1, 3].

This chapter presents various mathematical concepts related to design reliability.

2.2 BOOLEAN ALGEBRA LAWS

As Boolean algebra plays on important role in probability theory and reliability studies, some of its associated laws are presented below [3-5]:

Commutative Law:

$$X \cdot Y = Y \cdot X \tag{2.1}$$

where

X is an arbitrary set or event.
Y is an arbitrary set or event.
Dot (·) denotes the intersection of sets. Sometime Equation (2.1) is written without the dot but it still conveys the same meaning.

$$X+Y=Y+X \tag{2.2}$$

where

+ denotes the union of sets.

Associative Law:

$$(XY)Z=X(YZ) \tag{2.3}$$

where

Z is an arbitrary set or event.

$$(X+Y)+Z=X+(Y+Z) \tag{2.4}$$

Distributive Law:

$$X(Y+Z)=XY+XZ \tag{2.5}$$

$$X+YZ=(X+Y)(X+Z) \tag{2.6}$$

Absorption Law

$$X+(XY)=X \tag{2.7}$$

$$X(X+Y)=X \tag{2.8}$$

Idempotent Law

$$XX=X \tag{2.9}$$

$$X+X=X \tag{2.10}$$

2.3 PROBABILITY PROPERTIES

As the basis for reliability theory is probability, this section presents basic properties of probability. Some of these properties are as follows [6,7]:

- The probability of occurrence of event, say A, is

$$0 \leq P(A) \leq 1 \quad (2.11)$$

- Probability of the sample space S is

$$P(S) = 1 \quad (2.12)$$

- Probability of the negation of the sample space S is

$$P(\bar{S}) = 1 \quad (2.13)$$

where

$\bar{S}$ is the negation of the sample space S.

- The probability of the union of n independent events is

$$P(A_1 + A_2 + A_3 + \ldots + A_n) = 1 - \prod_{i=1}^{n} (1 - P(A_i)) \quad (2.14)$$

where

A_i is the *ith* event; for i = 1, 2, ..., n.
P (A_i) is the probability of occurrence of event A_i; for i = 1, 2, ..., n.

For n = 2, Equation (2.14) reduces to

$$P(A_1 + A_2) = P(A_1) + P(A_2) - P(A_1)P(A_2) \quad (2.15)$$

- The probability of the union of n mutually exclusive events is

$$P(A_1 + A_2 + A_3 + \ldots + A_n) = \sum_{i=1}^{n} P(A_i) \quad (2.16)$$

- The probability of an intersection of n independent events is

$$P(A_1 A_2 --- A_n) = P(A_1)P(A_2)---P(A_n) \tag{2.17}$$

- The probability of occurrence and nonoccurrence of an event, say A, is

$$P(A) + P(\bar{A}) = 1 \tag{2.18}$$

where

P (A) is the probability of occurrence of A.
P ($\bar{A}$) is the probability of nonoccurrence of A.

2.4 USEFUL DEFINITIONS

This section presents some definitions that are considered to be useful in performing various design reliability studies.

2.4.1 Definition of Probability

This is expressed as [6]:

$$P(X) = \lim_{n \to \infty} (N/n) \tag{2.19}$$

where

P (X) is the probability of occurrence of event X.
N is the number of times that X occurs in the n repeated experiments.

2.4.2 Cumulative Distribution Function

For continuous random variables, this is defined by [6]:

$$F(t) = \int_{-\infty}^{t} f(x)dx \tag{2.20}$$

where

t is time.
F (t) is the cumulative distribution function.
f (t) is the probability density function.

For $t = \infty$, Equation (2.20) becomes

$$F(\infty) = \int_{-\infty}^{\infty} f(x)dx \tag{2.21}$$
$$= 1$$

It means the total area under the probability density curve is equal to unity.

2.4.3 Probability Density Function

This is given by [6, 8]:

$$\frac{dF(t)}{dt} = \frac{d\left(\int_{-\infty}^{t} f(x)dx\right)}{dt} \tag{2.22}$$
$$= f(t)$$

2.4.4 Reliability Function

This is expressed by:

$$R(t) = 1 - F(t)$$
$$= 1 - \int_{-\infty}^{t} f(x)dx \tag{2.23}$$

where

$R(t)$ is the reliability function.

2.4.5 Hazard Rate Function

This is defined by:

$$\lambda(t) = f(t)/R(t)$$
$$= f(t)/(1 - F(t)) \tag{2.24}$$

where

$\lambda(t)$ is the hazard rate function.

2.4.6 Laplace Transform Definition

The Laplace transform of the function f (t) is defined by:

$$f(s) = \int_0^\infty f(t) e^{-st} \, dt \tag{2.25}$$

where

s is the Laplace transform variable.
t is the time variable.
f (s) is the Laplace transform of f (t).

Example 2.1

If f (t) = $e^{-\lambda t}$, obtain the Laplace transform of this function.

Thus, substituting the above function into Equation (2.25), we get

$$\begin{aligned} f(s) &= \int_0^\infty e^{-\lambda t} e^{-st} \, dt \\ &= \int_0^\infty e^{-(s+\lambda)t} \, dt \\ &= \frac{e^{-(s+\lambda)t}}{-(s+\lambda)} \Big|_0^\infty \\ &= \frac{1}{s+\lambda} \end{aligned} \tag{2.26}$$

Example 2.2

Obtain the Laplace transform of the function f (t) = 1.

By inserting the above function into Equation (2.25), we get

$$\begin{aligned} f(s) &= \int_0^\infty 1 . e^{-st} \, dt \\ &= \frac{e^{-st}}{-s} \Big|_0^\infty \\ &= \frac{1}{s} \end{aligned} \tag{2.27}$$

Table 2.1 presents Laplace transforms of various functions considered useful for performing mathematical design reliability analysis [9].

TABLE 2.1
Laplace Transforms of Some Commonly Occurring Functions in Reliability Work

f (t)	f (s)
$e^{-\lambda t}$	$1/(s+\lambda)$
c, a constant	c/s
$d\,f(t)/d\,t$	$s\,f(s) - f(0)$
$\int_0^t f(t)\,d(t)$	$f(s)/s$
$t\,f(t)$	$-d\,f(s)/d\,s$
$\theta_1 f_1(t) + \theta_2 f_2(t)$	$\theta_1 f_1(s) + \theta_2 f_2(s)$
t^n, for n = 0, 1, 2, 3, ...	$n!/s^{n+1}$
$t\,e^{-\lambda t}$	$1/(s+\lambda)^2$

2.4.7 Laplace Transform: Final-Value Theorem

If the following limits exist, then the final-value theorem may be expressed as

$$\lim_{t\to\infty} f(t) = \lim_{s\to 0}\left[s\,f(s)\right] \tag{2.28}$$

2.4.8 Expected Value

The expected value, E(t), of a continuous random variable is expressed by

$$E(t) = \mu = \int_{-\infty}^{\infty} t\,f(t)\,dt \tag{2.29}$$

where

t is a continuous random variable.
f (t) is the probability density function.
μ is the mean value.

Similarly, the expected value, E(t), of a discrete random variable t is defined by

$$E(t) = \sum_{i=1}^{m} t_i\,f(t_i) \tag{2.30}$$

where

m is the number of discrete values of the random variable t.

2.5 PROBABILITY DISTRIBUTIONS

In mathematical reliability analysis, various types of probability distributions are used; for example, Binomial, Poisson, exponential, Rayleigh, Weibull, general, and normal. All these distributions are described below [10].

2.5.1 Binomial Distribution

This distribution has applications in many combinational-type reliability problems and sometimes it is also called a Bernoulli (i.e., after Jakob Bernoulli (1654–1705)) distribution [1].

This distribution becomes very useful in situations where one is concerned with the probabilities of outcome such as the total number of failures in a sequence of k trials. However, it is to be noted that for this distribution, each trial has two possible outcomes (e.g., success and failure), but the probability of each trial remains constant or unchanged.

The binomial probability density function, f (y), is defined by

$$f(y) = \binom{k}{i} p^{y} q^{k-y}, \quad \text{for} \quad y = 0, 1, 2, \ldots, k \tag{2.31}$$

where

$\binom{k}{i} = k!/i!(k-i)!$
p is the single trial probability of success.
q is the single trial probability of failure.
y is the number of failures in k trials.

It is to be noted that $p + q = 1$ always. The cumulative distribution function is given by

$$F(y) = \sum_{i=0}^{y} \binom{k}{i} p^{i} q^{k-i} \tag{2.32}$$

where

F(y) is the cumulative distribution function or the probability of y or less failures in k trials.

2.5.2 Poisson Distribution

This is another distribution that finds applications in reliability analysis when one is interested in the occurrence of a number of events that are of the same type. Each event's occurrence is denoted as a point on a time scale and each event represents a failure. The Poisson density function is expressed by

$$f(m) = \frac{(\lambda t)^{m} e^{-\lambda t}}{m!}, \quad \text{for} \quad m = 0, 1, 2, \ldots \tag{2.33}$$

where

λ is the constant failure or arrival rate.
t is the time.

The cumulative distribution function, F, is

$$F = \sum_{i=0}^{m} \left[(\lambda t)^i \, e^{-\lambda t}\right] / i! \tag{2.34}$$

2.5.3 Exponential Distribution

This is probably the most widely used distribution in reliability engineering because many engineering items exhibit constant hazard rate during their useful life [11]. Also, it is relatively easy to handle in performing reliability analysis.

The probability density function of the distribution is defined by

$$f(t) = \lambda e^{-\lambda t} \qquad t \geq 0, \lambda > 0 \tag{2.35}$$

where

f (t) is the probability density function.
t is time.
λ is the distribution parameter. In reliability studies, it is known as the constant failure rate.

Using Equations (2.20) and (2.35), we get the following expression for the cumulative distribution function:

$$F(t) = 1 - e^{-\lambda t} \tag{2.36}$$

With the aid of Equations (2.29) and (2.35), the following expression for the expected value of t was obtained:

$$E(t) = 1/\lambda \tag{2.37}$$

2.5.4 Rayleigh Distribution

This distribution is often used in reliability work and in the theory of sound and is named after John Rayleigh (1842–1919), its originator [1]. The probability density function of the distribution is defined as

$$f(t) = \left(\frac{2}{\theta^2}\right) t e^{-\left(\frac{t}{\theta}\right)^2} \qquad t \geq 0, \;\; \theta > 0 \tag{2.38}$$

where

θ is the distribution parameter.

Inserting Equation (2.38) into Equation (2.20), we get the following equation for the cumulative distribution function:

$$F(t) = 1 - e^{-\left(\frac{t}{\theta}\right)^2} \tag{2.39}$$

Using Equations (2.29) and (2.38), we get the following expression for the expected value of t:

$$E(t) = \theta\, \Gamma\,(3/2) \tag{2.40}$$

where

$\Gamma\,(\cdot)$ is the gamma function and is defined by

$$\Gamma(x) = \int_0^\infty t^{x-1}\, e^{-t}\, dt, \quad \text{for } x > 0 \tag{2.41}$$

2.5.5 Weibull Distribution

This distribution can be used to represent many different physical phenomena and it was developed by Weibull in the early 1950s [12]. The probability density function for the distribution is defined by

$$f(t) = \frac{b\, t^{b-1}}{\theta^b}\, e^{-\left(\frac{t}{\theta}\right)^b} \quad t \geq 0,\ \theta > 0,\ b > 0 \tag{2.42}$$

where

b and θ are the shape and scale parameters, respectively.

Using Equations (2.20) and (2.42), we get the following cumulative distribution function:

$$F(t) = 1 - e^{-\left(\frac{t}{\theta}\right)^b} \tag{2.43}$$

Substituting Equation (2.42) into Equation (2.29), we get the following equation for the expected value of t:

$$E(t) = \theta\, \Gamma\left(1 + \frac{1}{b}\right) \tag{2.44}$$

For b = 1 and 2, the exponential and Rayleigh distributions are the special cases of this distribution, respectively.

2.5.6 General Distribution

This distribution can be used to represent the entire bathtub shape hazard rate curve [13]. Its probability density function is defined by:

$$f(t) = \left[k\lambda u t^{u-1} + (1-k) b t^{b-1} \gamma e^{\gamma t^b}\right]\left[\exp\left[-k\lambda t^u - (1-k)\left(e^{\gamma t^b} - 1\right)\right]\right] \quad (2.45)$$

$$\text{for } 0 \le k \le 1 \text{ and } \lambda, u, b, \gamma > 0$$

where

u and b are the shape parameters.
λ and γ are the scale parameters.
t is time.

Substituting Equation (2.45) into Equation (2.20) yields

$$F(t) = 1 - \exp\left[-k\lambda t^u - (1-k)\left(e^{\gamma t^b} - 1\right)\right] \quad (2.46)$$

The following distributions are the special cases of this distribution:

- Bathtub hazard rate curve; for b = 1, u = 0.5
- Weibull; for k = 1
- Extreme value; for k = 0, b = 1
- Makeham; for u = 1, b = 1
- Rayleigh; for k = 1, u = 2
- Exponential; for k = 1, u = 1

2.5.7 Normal Distribution

This is one of the most widely known distributions and sometimes it is called the Gaussian distribution after Carl Friedrich Gauss (1777–1855), the German mathematician. Nonetheless, its use in reliability studies is rather limited. The probability density function of the distribution is defined by:

$$f(t) = \frac{1}{\sigma\sqrt{2\pi}} \exp\left[-\frac{(t-\mu_n)^2}{2\sigma^2}\right]; \quad -\infty < t < +\infty \quad (2.47)$$

where

μ_n and σ are the distribution parameters (i.e., mean and standard deviation, respectively).

Using Equations (2.20) and (2.47), we get the following cumulative distribution function:

$$F(t)=\frac{1}{\sigma\sqrt{2\pi}}\int_{-\infty}^{t}\exp\left[-\frac{(x-\mu_n)^2}{2\sigma^2}\right]dx \tag{2.48}$$

Substituting Equation (2.47) into Equation (2.29) yields:

$$E(t)=\mu_n \tag{2.49}$$

2.6 HAZARD RATE MODELS

In reliability work, the term "hazard rate" is also often used. In particular, when the hazard rate of an item is constant, it is simply referred to as the item's failure rate. Nonetheless, generally the hazard rate of an item varies with the time to failure distribution. This section presents hazard rates for exponential, Rayleigh, Weibull, general, and normal distributions.

2.6.1 Exponential Distribution

By substituting Equations (2.35) and (2.36) into Equation (2.24), we get the following hazard rate function for the exponential distribution:

$$\begin{aligned}\lambda(t)&=\frac{\lambda e^{-\lambda t}}{e^{-\lambda t}}\\&=\lambda\end{aligned} \tag{2.50}$$

It means the hazard rate of an exponential distribution is constant and it is simply called the failure rate.

2.6.2 Rayleigh Distribution

Using Equations (2.24), (2.38), and (2.39), the following hazard rate expression is for the Rayleigh distribution:

$$\begin{aligned}\lambda(t)&=\left[\frac{2}{\theta^2}te^{-(t/\theta)^2}\right]\Big/e^{-(t/\theta)^2}\\&=2t/\theta^2\end{aligned} \tag{2.51}$$

Equation (2.51) indicates that the hazard rate of the Rayleigh distribution increases linearly with time.

2.6.3 Weibull Distribution

Inserting Equations (2.42) and (2.43) into Equation (2.24) yields:

$$\lambda(t) = \left[\frac{b}{\theta^b} t^{b-1} e^{-(t/\theta)^b}\right] \Big/ e^{-(t/\theta)^b} = \frac{b}{\theta^b} t^{b-1} \tag{2.52}$$

Equation (2.52) indicates that the Weibull hazard rate is a function of time t and at b = 1 and b = 2, it reduces to the hazard rates for exponential and Rayleigh distributions, respectively.

2.6.4 General Distribution

With the aid of Equations (2.24), (2.45), and (2.46), we get the following expression for the general distribution hazard rate:

$$\begin{aligned}\lambda(t) &= \left[k\lambda u t^{u-1} + (1-k) b t^{b-1} \gamma e^{\gamma t^b}\right]\left[\exp\left[-k\lambda t^u - (1-k)\left(e^{\gamma t^b} - 1\right)\right]\right] \\ &\Big/ \exp\left[-k\lambda t^u - (1-k)\left(e^{\gamma t^b} - 1\right)\right] \\ &= k\lambda u t^{u-1} + (1-k) b t^{b-1} \gamma e^{\gamma t^b}\end{aligned} \tag{2.53}$$

Obviously, in Equation (2.53) the hazard rate, λ (t), varies with time and its special case hazard rates are for Weibull, bathtub, extreme value, Makeham, Rayleigh, and exponential distributions.

2.6.5 Normal Distribution

Using Equations (2.24), (2.47), and (2.48), we get the following hazard rate expression for the normal distribution hazard rate:

$$\lambda(t) = \left[\frac{1}{\sigma\sqrt{2\pi}} \exp\left[-\frac{(t-\mu_n)^2}{2\sigma^2}\right]\right] \Big/ \left[\frac{1}{\sigma\sqrt{2\pi}} \int_t^\infty \exp\left[-\frac{(x-\mu_n)^2}{2\sigma^2}\right] dx\right] \tag{2.54}$$

2.7 PARTIAL FRACTION METHOD AND EQUATION ROOTS

2.7.1 Partial Fraction Method

This is a useful approach to determine inverse Laplace transforms of a rational function such as M(s)/Q(s), where M(s) and Q(s) are polynomials, and the degree of M(s) is less than that of Q(s). Consequently, the ratio of M(s)/Q(s) may be expressed as the sum of rational functions or partial fractions as follows:

- $A/(as+b)^m$, for m = 1, 2, 3, ... (2.55)

- $(Bs+C)/(as^2+bs+c)^m$, for m = 1, 2, 3, ... (2.56)

where

A, B, C, a, b, and c are the constants.
s is the Laplace transform variable.

Heaviside Theorem

This theorem is due to Oliver Heaviside (1850–1925) [1] and is used to obtain partial fractions and the inverse of a rational function, M(s)/Q(s). Thus, the inverse of M(s)/Q(s) is expressed as follows:

$$\mathcal{L}^{-1}\left[\frac{M(s)}{Q(s)}\right]=\sum_{i=1}^{m}\frac{M(\alpha_i)}{Q'(\alpha_i)}e^{\alpha_i t} \tag{2.57}$$

where

the prime on Q denotes differentiation with respect to s.
m denotes the total number of distinct zeros of Q(s).
α_i denotes zero i; for i = 1, 2, 3, ..., m.
t is time.

Example 2.3

Suppose

$$\frac{M(s)}{Q(s)}=\frac{s}{(s-5)(s-7)}. \tag{2.58}$$

Find its inverse Laplace transforms by applying the Heaviside theorem.

Hence,

$$M(s)=s,\ Q(s)=(s-5)(s-7)=s^2-12s+35,\ Q'(s)$$
$$=2s-12,\ m=2,\ \alpha_1=5,\ \text{and}\ \alpha_2=7.$$

Using the above values in the right-hand side of Equation (2.57) yields:

$$
\begin{aligned}
&= \frac{M(\alpha_1)}{Q'(\alpha_1)} e^{\alpha_1 t} + \frac{M(\alpha_2)}{Q'(\alpha_2)} e^{\alpha_2 t} \\
&= \frac{M(5)}{Q'(5)} e^{5t} + \frac{M(7)}{Q'(7)} e^{7t} \\
&= -\frac{5}{2} e^{5t} + \frac{7}{2} e^{7t} \\
&= \frac{7}{2} e^{7t} - \frac{5}{2} e^{5t}
\end{aligned}
\tag{2.59}
$$

Thus, the inverse Laplace transform of Equation (2.58) is given by Equation (2.59).

2.7.2 Equation Roots

In reliability work, sometimes it may involve finding roots of algebraic equations. A root is a value of variable say, y, that when substituted into the polynomial equation results in the value of the equation equal to zero. A polynomial equation is considered solved when all its roots have been found.

Even though around 2,000 B.C. Babylonians solved quadratic equations and the ancient Indian mathematicians such as Brahmagupta recognized negative roots, in the western society prior to the 17th century, the theory of equations was handicapped by the failure of mathematicians to recognize negative or complex numbers as roots of equations [1]. Nonetheless, formulas to obtain roots of quadratic and cubic equations are presented below [14, 15].

Quadratic Equation

This is defined by:

$$ay^2 + by + c = 0 \tag{2.60}$$

where

a, b, and c are the constants.

Solutions to Equation (2.60) are as follows:

$$y_1, y_2 = \left[-b \pm W^{1/2}\right] / 2a \tag{2.61}$$

where

$$W \equiv b^2 - 4ac \tag{2.62}$$

If y_1 and y_2 are the roots, then

$$y_1 + y_2 = -b/a \tag{2.63}$$

and

$$y_1 y_2 = c/a \tag{2.64}$$

If a, b, and c are real, then the roots are as follows:

- Real and equal, if $W = 0$
- Real and unequal, if $W > 0$
- Complex conjugate if $W < 0$

Cubic Equation

$$y^3 + a_2 y^2 + a_1 y + a_0 = 0 \tag{2.65}$$

Let

$$M = \frac{1}{3} a_1 - \frac{1}{9} a_2^2 \tag{2.66}$$

$$N = \frac{1}{6}(a_1 a_2 - 3 a_0) - \frac{1}{27} a_2^3 \tag{2.67}$$

If y_1, y_2, and y_3 are the roots of Equation (2.65), then we have:

$$y_1 y_2 y_3 = -a_0 \tag{2.68}$$

$$y_1 + y_2 + y_3 = -a_2 \tag{2.69}$$

$$y_1 y_2 + y_1 y_3 + y_2 y_3 = a_1 \tag{2.70}$$

The following types of roots are associated with Equation (2.65):

- All roots real (irreducible case) if $M^3 + N^2 < 0$.
- All roots real and at least two equal if $M^3 + N^2 = 0$.
- One real root and one pair of complex conjugate roots if $M^3 + N^2 > 0$.

If we let

$$g_1 = \left[N + \left(M^3 + N^2\right)^{1/2}\right]^{1/2} \tag{2.71}$$

and

$$g_1 = \left[N - \left(M^3 + N^2\right)^{1/2}\right]^{1/2} \tag{2.72}$$

then we have the following roots:

$$y_1 = \left(g_1 + g_2\right) - a_2/3 \tag{2.73}$$

$$y_2 = -\frac{1}{2}\left(g_1 + g_2\right) - \frac{a_2}{3} + \frac{i\sqrt{3}}{2}\left(g_1 - g_2\right) \tag{2.74}$$

$$y_3 = -\frac{1}{2}\left(g_1 + g_2\right) - \frac{a_2}{3} - \frac{i\sqrt{3}}{2}\left(g_1 - g_2\right) \tag{2.75}$$

2.8 DIFFERENTIAL EQUATIONS

In reliability work, the use of the Markov method results in a system of linear first-order differential equations. Thus, in order to obtain state probabilities at time t, the solution to these differential equations has to be found. For this purpose, there are various methods that can be used, but the Laplace transform approach appears to be the most attractive.

Finding solutions to a system of differential equations is demonstrated through the following example:

Example 2.4

Assume that an engineering system can either be in two states: operating or failed. The following two differential equations describe the system:

$$\frac{d\,P_0(t)}{dt} + \lambda P_0(t) = 0 \tag{2.76}$$

$$\frac{d\,P_1(t)}{dt} - P_0(t)\,\lambda = 0 \tag{2.77}$$

where

> $P_i(t)$ is the probability that the engineering system is in state i at time t; for i = 0 (operating), i = 1 (failed).
> λ is the system failure rate.

At time t = 0, $P_0(0) = 1$, and $P_1(0) = 0$.

Find solutions to Equations (2.76) and (2.77).

In this case, we use the Laplace transform method. Thus, taking Laplace transforms of Equations (2.76) and (2.77) and using the given initial conditions, we get

$$P_0(s) = \frac{1}{s+\lambda} \tag{2.78}$$

$$P_1(s) = \lambda / s(s+\lambda) \tag{2.79}$$

The inverse Laplace transforms of Equations (2.78) and (2.79) are as follows:

$$P_0(t) = e^{-\lambda t} \tag{2.80}$$

$$P_1(t) = 1 - e^{-\lambda t} \tag{2.81}$$

Thus, Equations (2.80) and (2.81) represent solutions to differential Equations (2.76) and (2.77).

2.9 PROBLEMS

1. Write an essay on the history of mathematics including probability theory.
2. Prove the following Boolean algebra expression:

$$X + YZ = (X+Y)(X+Z) \tag{2.82}$$

3. Prove that

$$R(t) + F(t) = 1 \tag{2.83}$$

where

$R(t)$ is the item reliability at time t.
$F(t)$ is the item failure probability at time t.

4. Obtain the Laplace transform of the following function:

$$f(t) = \frac{t^{k-1}\, e^{at}}{(k-1)!}, \quad \text{for } k = 1, 2, 3, \ldots \tag{2.84}$$

5. Take the inverse Laplace transform of the following:

$$f(s) = \frac{s}{(s+4)(s+3)} \tag{2.85}$$

6. Obtain an expression for the expected value of the general distribution. Comment on your result.
7. Discuss hazard rates of exponential and Rayleigh distributions.
8. Prove that the expected value of the Weibull distribution is given by

$$E(t) = \theta\, \Gamma\left(1 + \frac{1}{b}\right) \tag{2.86}$$

where

θ and b are the scale and shape parameters, respectively.

9. What are the special case distributions of the Weibull distribution?
10. What are the special case distributions of the general distribution?

2.10 REFERENCES

1. Eves, H., *An Introduction to the History of Mathematics*, Holt, Rinehart, and Winston, New York, 1976.
2. Owen, D.B., Ed., *On the History of Statistics and Probability*, Marcel Dekker, New York, 1976.
3. Lipschutz, S., *Set Theory*, McGraw-Hill, New York, 1964.
4. Fault Tree Handbook, Report No. NUREG-0492, U.S. Nuclear Regulatory Commission, Washington, D.C., 1981.
5. Ramakumar, R., *Engineering Reliability: Fundamentals and Applications*, Prentice-Hall, Englewood Cliffs, NJ, 1993.
6. Mann, N.R., Schafer, R.E., and Singpurwalla, N.D., *Methods for Statistical Analysis of Reliability and Life Data*, John Wiley & Sons, New York, 1974.
7. Lipschutz, S., *Probability*, McGraw-Hill, New York, 1965.
8. Shooman, M.L., *Probabilistic Reliability: An Engineering Approach*, McGraw-Hill, New York, 1968.
9. Oberhettinger, F. and Badic, L., *Tables of Laplace Transforms*, Springer-Verlag, New York, 1973.
10. Patel, J.K., Kapadia, C.H., and Owen, D.B., *Handbook of Statistical Distributions*, Marcel Dekker, New York, 1976.
11. Davis, D.J., An analysis of some failure data, *J. Am. Stat. Assoc.*, June, 113- 150, 1952.
12. Weibull, W., A statistical distribution function of wide applicability, *J. Appl. Mech.*, 18, 293-297, 1951.
13. Dhillon, B.S., A hazard rate model, *IEEE Trans. Reliability*, 29, 150, 1979.
14. Abramowitz, M. and Stegun, I.A., Eds., *Handbook of Mathematical Functions with Formulas, Graphs, and Mathematical Tables*, National Bureau of Standards, Washington, D.C., 1972.
15. Spiegel, M.R., *Mathematical Handbook of Formulas and Tables*, McGraw-Hill, New York, 1968.

3 Engineering Design and Reliability Management

3.1 INTRODUCTION

For large engineering systems, management of design and reliability becomes an important issue. Thus, this chapter is devoted to engineering design and reliability management.

In engineering, the team "design/designer" may convey different meanings to different people [1]. For example, to some a design is the creation of a sophisticated and complex item such as a space satellite, computer, or aircraft and to others a designer is simply a person who uses drawing tools to draw the details of an item. Nonetheless, it is the design that distinguishes the engineer from the scientist as well as it is the design that ties engineers to markets, which give vent to the creativity of the engineering profession.

Imhotep, who constructed the first known Egyptian pyramid, Saqqara, in 2650 BC may be called the first design engineer. As the engineering drawings are closely tied to engineering design, their history could be traced back to 4000 BC when the Chaldean engineer Gudea engraved upon a stone tablet the plan view of a fortress [2]. The written evidence of the use of technical drawings only goes back to 30 BC when Vitruvius, a Roman architect, wrote a treatise on architecture. In 1849, in modern context, an American named William Minifie was the first person to write a book on engineering drawings entitled *Geometrical Drawings*.

Today, there are a vast number of publications on both engineering design and engineering drawings [3]. An engineering design is only good if it is effective and reliable.

The history of reliability management is much shorter than that of the engineering design and it can only be traced back to the 1950s. Probably the first prime evidence related to reliability management is the formation of an *ad hoc* committee in 1958 for Guided Missile Reliability under the Office of the Assistant Secretary of Defense for Research and Engineering [4]. In 1959, a reliability program document (exhibit 58-10) was developed by the Ballistic Missile Division of the U.S. Air Force. The military specification, MIL-R-27542, was the result of Department of Defense efforts to develop requirements for an organized contractor reliability program. In 1985, a book entitled *Reliability and Maintainability Management* was published [5] and a comprehensive list of publications on the subject is given in Reference 6.

This chapter discusses engineering design and reliability management separately.

3.2 ENGINEERING DESIGN

Nowadays, design is increasingly seen as the key to industrial success in all markets. In simple terms, design is all about producing the right products for the right markets at the right price, the right time, the right quality and reliability to the right specification and performance [7]. Nonetheless, it may be said that even though a good design may not guarantee success, it is an extremely powerful contributor to it. Some of the important characteristics of a good design include long useful life, high reliability, low cost and maintenance, high accuracy, and attractive appearance.

There would be many reasons for designing engineering products: reducing hazard, developing a new way, lowering the cost, meeting social changes, meeting competition, etc. The subsequent sections address the different aspects of engineering design.

3.2.1 Design Failures and Their Common Reasons

History has witnessed many engineering design failures, in particular, in the immediate past two well-publicized design failures were the Chernobl nuclear reactors and the space shuttle Challenger. Some of the other important design failures are as follows [8, 9]:

- In 1937, a high school in Texas converted from burning methane city gas to less expensive natural gas. Subsequently an explosion occurred and 455 persons died. An investigation of the accident indicated that the area through which the pipes ran was poorly ventilated. A recommendation was made for the installation of malodorants to all natural gas to detect any possible leaks.
- In 1963, a U.S. Navy nuclear submarine, the U.S.S. Thresher, slipped beneath the Atlantic Ocean surface by exceeding its designed maximum test depth and imploded.
- In 1979, a DC-10 airplane lost an engine during a flight and subsequently crashed. A follow up investigation indicated that the normal engine service operation was the cause of the problem. More specifically, as the engines were periodically dismounted, serviced, and remounted, the mounting holes became elongated during the servicing process and subsequently led to the disaster.
- In 1980, in the North Sea an offshore oil rig named Alexander L. Kielland broke up under normal weather conditions. A subsequent study of the disaster revealed that a 3-in. crack in a part close to a weld joint was the basic cause of the disaster.
- In 1988, a Boeing 737-200 lost its cabin roof during a flight. A preliminary investigation of the disaster indicated that various metal fatigue cracks were emanating from rivet holes in the aluminum skin.

Over the years many professionals have studied various design failures and concluded that there could be many different reasons for product failures ranging from a disaster to simple malfunction. Most of the common reasons are given in Table 3.1

TABLE 3.1
Common Reasons for Product Failures

No.	Reason
1	Wrong usage by the consumer
2	Incorrect manufacturing
3	Faulty reasoning
4	Poor understanding of the problem to be solved
5	Unsatisfactory data collection
6	Incorrect storage
7	Incorrectly stated problem with respect to basic principles
8	Wrong or over-extended assumptions
9	Erroneous data

3.2.2 The Design Process and Functions

The design process is an imaginative integration of scientific information, engineering technology, and marketing for developing a useful and profitable item. The process may be described in as little as from 5 to 25 steps; for example, in 6 steps by Dieter [10], 8 steps by Vidosic [11], and 12 steps by Hill [12]. Nonetheless, for our purpose we will describe it in six steps:

1. **Need recognition.** This is concerned with understanding and identifying the needs of the user.
2. **Problem definition.** This is concerned with developing concise problem statement, identifying problem related needs and limitations, and obtaining related information.
3. **Information gathering.** This is concerned with collecting various types of information from sources including specifications and codes, patent gazette, Internet, handbooks, technical experts, vendor catalogs, and journals.
4. **Conceptualization.** The results of conceptualization may take the form of sketches or free-hand drawings and at this stage, the design process may be regarded as a creative and innovative activity that generates various possible alternative solutions to the identified goal. Some of the idea generation techniques that design professionals can use include group brainstorming, attribute listing, synectics, and morphological chart.
5. **Evaluation.** This is concerned with deciding on the best solution out of all the potential solutions. Some of the evaluations include evaluation based on feasibility judgment, evaluation based on technology-readiness assessment, and evaluation based on GO/NO-GO screening.
6. **Communication of design.** The final solution to the engineering problem leads to final documents representing the product or the product itself, for the purpose of communicating design to others. The design documents include items such as engineering drawings, operation and maintenance instructions, information on quality assurance, patent applications, and bills of materials.

There are basically five functions involved in engineering design: (1) research, (2) engineering, (3) manufacturing, (4) quality assurance, and (5) commercial. The research function activities include conducting basic and applied research, developing specifications for quality testing procedures, and preparing process specifications for the testing of highly stressed parts. The engineering functions are the subcomponents of the design activity and include activities such as developing new design concepts, estimating cost, developing production design, making provisions of maintenance instructions, and analyzing field problems. The manufacturing functions include activities such as assembly, manufacturing planning, determining tooling needs, and purchasing materials. The main objective of the quality assurance function is to assure the quality of the end product. Some of the activities covered by the quality assurance function are designing quality related methods and procedures, and setting up design auditing. The commercial function is concerned with the relationship of various clients and its activities include conducting market surveys and tendering, advertising, managing contracts, and arranging delivery and payments.

3.2.3 The Design Team and Member Responsibilities

The design engineer is not the only one person who usually develops design. In real life, there are many other individuals who also participate. For example, representatives from areas such as manufacturing, field services, and quality control are involved. The nature of the product dictates the degree of their participation. The design team members' responsibilities may be grouped into four areas: (1) design, (2) manufacturing, (3) quality, and (4) field service [7]. The design related responsibilities include initial design concept, functional integrity, prototype build, test procedures, documentation issue, specifications, cost estimates, materials, performance goals, bill of materials, and coordination with reliability group, safety group, standards group, planning group, and model shop.

Some of the responsibilities belonging to the manufacturing area are tooling, model build co-ordination, assembly procedures, test and set up procedures, vendor involvement, pre-production builds, procurement schedule, and co-ordination with quality control, product planning, materials management, and industrial engineering.

The quality related responsibilities include vendor training, quality audits, field reliability, total quality management, quality circles, supplier approval, and product verification.

The field service related responsibilities are service procedures, service tools, service training, customer training, spares list, service documentation, and so on.

3.2.4 Design Reviews

The design reviews are an important factor during the design phase of an engineering product. The basic purpose of such reviews is to assure the application of correct design principles, in addition to determining whether the design effort is progressing according to specifications, plans, and so on.

Various types of reviews are performed during the design phase of a product and the total cost of performing such reviews varies between 1 and 2% of the total engineering cost of a project [12].

Different writers and practitioners categorize design reviews differently. For our purpose, we have divided them into three major categories: (1) preliminary design review, (2) intermediate design review, and (3) critical design review [12, 13]. Each of these three category is discussed below.

1. **Preliminary design review.** This design review is usually conducted prior to the formulation of the initial design. The primary aim of this review is to evaluate each and every design specification requirement with respect to accuracy, validity, and completeness. The areas that could be reviewed during this design review include cost objective, design alternatives, present/future availability of materials, applicable legislations, potentional users/customers, schedule imposed requirements, required functions, customer needs, and critical parts/components.
2. **Intermediate design review.** This review is conducted before starting the detailed production drawings. The main objective of this review is to make comparison of each specification requirement with the design under development. It is to be noted that prior to starting this review, the design selection process and preliminary layout drawings are complete.
3. **Critical design review.** This is also known as the final design review and is conducted after the completion of production drawings. The emphasis of this review is on areas such as design producibility, value engineering, review of analysis results, and so on.

Design Review Team

Professionals from various areas participate in design reviews and their number and type may vary from one project to another. For an effective performance, the size of the team should not exceed 12 participants [12]. A typical design review team is made up of professionals such as design engineer, senior design engineer, design review board chairman, manufacturing engineer, tooling engineer, procurement engineer, customer representative(s) (if any), test engineer, reliability engineer, materials engineer, and quality control engineer. Each of these individuals evaluates design from his/her perspective.

Design Review Related Information and Topics

In order to conduct effective reviews, the design review team members must have access to various information items, as appropriate, including parts list, specifications and schematic diagrams, acceleration and shock data, circuit descriptions, results of reliability prediction studies, list of inputs/outputs, results of failure modes and effect analysis (FMEA), and vibration and thermal tests data.

Usually during design reviews, topics such as mechanical design (i.e., results of tests, thermal analysis, balance, etc.), specifications (i.e., adherence to specifications, correctness of specifications, etc.), human factors (i.e., glare, control and display, labeling and marking, etc.), reproducibility (i.e., reliance of products on a single part supplier, economical assembly of product in the production shop, etc.), electrical design (i.e., design simplification, electrical interference, performance, results of circuit analysis, etc.), standardization, value engineering, safety, maintainability (i.e., interchangeability, maintenance philosophy, etc.), and reliability (i.e., reliability predictions, reliability allocations, the results of failure modes and effect analysis (FMEA), etc.) are discussed.

3.2.5 Design Engineer and Design Review Board Chairman

Broadly speaking, a design engineer is basically concerned with developing a product or solving a problem. More specifically, some of the tasks performed by this individual includes designing the product, keeping the design within specified constraints, keeping abreast of changing environments and technology, answering questions concerning the design under consideration, participating in design reviews, and optimizing the design.

The important qualities of a typical design engineer include creative and innovative mind, good conceptual ability, excellent scientific knowledge, ability to think logically, good communication skills, and good personal qualities.

The design review board chairman is a key member of the design review team. Usually, this person belongs to the engineering department and is not in direct line of authority to the designer whose design is under consideration. Nonetheless, a configuration manager often is appointed by the design review board chairman [14].

Some of the important qualities of the design review board chairman are general technical competence, understanding of the problem under consideration, free of bias for or against the proposed design, appropriate skills to lead a technical meeting, a pleasant personality, and a high degree of tact and discretion.

Some of the functions performed by the design review board chairman are determining the type of review to be conducted, scheduling the design reviews, chairing the meetings of the design review board, establishing the procedure for selecting particular items for review, evaluating comments from the design review meetings, supervising the publication of minutes, and providing appropriate assistance to the design organization with respect to preparation of necessary design review data.

3.2.6 Designing for Reliability and Design Reliability Check List Items

A design is only useful if it is reliable. Thus, reliability is a crucial factor in engineering design. Nonetheless, initial design concepts influenced by the overall reliability requirement will depend upon factors such as an item's/product's functional complexity, the cost target, item's/product's usage requirements, the size envelope, and the state of development and technical understanding available with new materials/parts/technologies considered for use.

The design process aspects [7, 15] that can have significant impact on product reliability include design simplicity, use of parts with known history, failure mode analysis, understanding of the requirements and the rules, allowance for possible test house approval requirements in design of electrical, mechanical, safety, and material selection aspects, procuring outside components with care, making allowance for rough handling, use abuse, and transportation, simplicity in manufacturability, and confirmation of mechanical design choices mathematically as much as possible, e.g., bearings, load deflections, etc.

In order to consider various aspects of reliability during the design process effectively, it is useful to prepare a checklist of reliability related items. Some of these items are as follows [13]:

- Minimization of adjustments
- Establishment of requirements for performance/reliability/environment/life/signals for each and every concerned component, interface, and structure
- Adequacy of mechanical support structures
- Efficiency of heat transfer devices or designs
- Effective usage of failure-indicating devices
- Specification of the particular design criteria for reliability and safety of each and every product component
- Identification of components requiring special procurement, testing, and handling
- Provisions for improvements to eradicate any design deficiencies observed during testing
- Establishment of maintenance requirements and maintainability goals
- Adequate protection against insertion of similar plugs and connectors in wrong sockets
- Establishment of satisfactory reliability assurance, acceptance, qualification, and sampling tests
- Identification of limited life parts
- Installation of self-monitoring or self-calibration devices in major products where feasible
- Selection of components, interfaces, and structures to meet reliability objectives
- Determination of shelf life of parts chosen for final design
- Usage of appropriate safety factors in the application of parts/interfaces/structures
- Establishment of stability requirements of all parts and structures associated with each and every adjustment
- Utilization of best available methods/techniques for lowering the adverse effects of operational environments on critical structures
- Classification of all critical adjustments with respect to factory, pre-operational, or operator types
- Performance of studies considering variability and degradation of part/structure parameters
- Effort expended to make the developmental model as close as the production model
- Emphasis of the design to eliminate shock mounts and vibration isolators where it is feasible
- Design of packaging and mechanical layout to facilitate effective maintenance.

3.2.7 Design Liability

This is a very important factor in product design and can simply be stated as the liability of designers for injury, loss, or damage caused by defects in products designed by them. According to Roche [16], approximately 40% of causes of product liability suits are due to defects in design.

In determining the defectiveness of a product, factors such as the marketing of product, stated instructions, stated warnings, time of sale, and foreseeable application and intended use are taken into consideration. In situations where injury results, the important points (i.e., in U.K.) affecting design liability include [17]:

- The designer cannot be liable when the risk of injury was not reasonably foreseeable.
- The designer cannot be totally liable when the risk of injury was foreseeable to him/her and he/she took reasonable design precautions.
- The designer is liable when the risk of injury was reasonably foreseeable to him/her and was not obvious to the product user but the designer took no reasonable measures to warn.
- The designer cannot be totally liable when the risk of injury was reasonably foreseeable and obvious to the user.

All in all, there is no simple answer as to how to reduce design-related risks, but the design reviews are an appropriate vehicle for identifying and evaluating potential safety hazards.

3.3 RELIABILITY MANAGEMENT

In the 1990s, there has been a major shift in the perception of reliability engineering by government, manufacturer, and consumer sectors. For example, in the 1960s and 1970s, the reliability engineering function was a discrete unit of a matrix organization; in today's environment it is an integral component of an engineering design team.

Over the years, as the engineering projects have become more complex and sophisticated, the application of reliability principles has also become a trying management task. This has led to the development of various reliability management related areas [18].

3.3.1 General Management Reliability Program Related Responsibilities and Guiding Force Associated Facts for an Effective Reliability Program

In order to have an effective reliability program, there are various factors in which the responsibilities of management [19] lie:

- Establishing a program to access with respect to reliability of the current performance of company operations and the product it manufactures;
- Establishing certain reliability objectives or goals;

- Establishing an effective program to fulfill set reliability goals and eradicating current deficiencies. An absolutely effective program should be able to pay in return many times its establishing cost;
- Providing necessary program related authority, funds, manpower, and time schedule;
- Monitoring the program on a regular basis and modifying associated policies, procedures, organization, and so on, to the most desirable level.

Facts such as the following will be a guiding force for the general management to have an effective reliability program [20]:

- Reliability is an important factor in the management, planning, and design of an engineering product.
- Changes in maintenance, manufacturing, storage and shipping, testing and usage in the field of the engineering product tend to lower the reliability of the design.
- Planned programs are needed for application in design, manufacturing, testing and field phases of the engineering product to control reliability.
- Reliability is established by the basic design.
- It is during the early phases of the design and evaluation testing programs when high levels of reliability can be achieved most economically.
- Improvement in reliability can be through design changes only.
- Human error degrades the reliability of the design.
- In achieving the desired reliability in a mature engineering product in a timely manner, deficiency data collection, analysis, and feedback are important.

3.3.2 Military Specification Guidelines for Developing Reliability Programs and a Procedure to Establish Reliability Goals

For the development of reliability programs, the military specification, MIL-R-27542, presents the following guidelines [21]:

- Assign reliability associated goals for system in question.
- To obtain maximum "inherent equipment reliability," put in as much effort as possible during the design phase.
- Evaluate reliability margins.
- Perform specification review.
- Perform design and procedure reviews.
- Establish an appropriate testing program.
- Conduct a review of changes in specification and drawings with respect to reliability.
- Establish and maintain control during production through inspection, sample testing, and effective production control.
- Develop a closed-loop system for failure reporting, analysis, and feedback to engineering specialists for corrective measures to stop re-occurrence.

- Assign responsibility for reliability to a single group.
- Make sure that the reliability group is reporting to an appropriate authority so that the reliability work can be performed effectively.
- Develop an on-the-job training facility and program.

When working from preestablished requirements, developing reliability goals are essential and the reliability goal setting basically consists of reducing the overall requirements to a series of subgoals. A general procedure to establish reliability goals is as follows [22]:

- Clarify requirements.
- Review the organizational objective and the organizational unit's mission prior to developing the unit's goals/purposes/missions to pursue the main goal.
- Highlight desired important result areas.
- Determine likely highest payoff areas.
- Choose most promising result areas to pursue.
- Choose goal candidates.
- Review resource requirements to pursue each candidate goal to achieve it successfully.
- Highlight any problem area in achieving the goal and proposed solutions.
- Rank all candidate goals with respect to ease of payment and the degree of payoff.
- Examine goal interdependencies and make adjustments in goal candidates for maximum coordination.
- Review goals with respect to factors such as attainability, compatibility, relevance, acceptability, measurability, and supportability.
- Make final selection of goals and develop milestones for their successful achievement. Develop parameters to measure success.
- Establish action plans for the goal achievement by considering provisions for motivational initiatives/supervisions and management support.
- Communicate goals in written form to all concerned bodies and review their progress periodically.
- Make adjustments as appropriate.

3.3.3 Reliability and Maintainability Management Tasks in the Product Life Cycle

To obtain required reliability of an engineering system in the field, the management tasks with respect to reliability and maintainability has to be coordinated throughout its life cycle. The life cycle of a system may be divided into four major distinct phases: (1) the concept and definition phase, (2) the acquisition phase, (3) the operation and maintenance phase, and (4) the disposal phase [5]. Reliability and maintainability management is involved in all four phases. Its tasks in each of these phases are presented below.

The Concept and Definition Phase

In this phase, system requirements are established and the basic characteristics are defined. There are various reliability and maintainability related management tasks involved in this phase. Some of them are as follows:

- Defining all terms used, the system capability requirements, management controls, and parts control requirements.
- Defining a failure.
- Defining the reliability and maintainability goals for the system in quantitative terms.
- Defining hardware and software standard documents to be used to fulfill reliability and maintainability requirements.
- Defining methods to be followed during the design and manufacturing phase.
- Defining constraints proven to be harmful to reliability.
- Defining system safety requirements and the basic maintenance philosophy.
- Defining data collection and analysis needs during the system life cycle.
- Defining management control for documentation and providing necessary documentation.
- Defining system environmental factors during its life cycle.

The Acquisition Phase

This is the second phase of the system life cycle, and it is concerned with activities associated with system acquisition and installation as well as planning for the eventual support of the system. During this phase, some of the reliability and maintainability related management tasks are as follows:

- Define the system technical requirements, major design and development methods to be used, documents required as part of the final system, type of evaluation methods to assess the system, and demonstration requirements.
- Define the reliability and maintainability needs that must be fulfilled.
- Define the meaning of a failure or a degradation.
- Define the types of reviews to be conducted.
- Define the type of data to be supplied by the manufacturer to the customer.
- Define the life-cycle cost information to be developed.
- Define the kind of field studies to be performed, if any.
- Define the kind of logistic support required.

The Operation and Maintenance Phase

This is the third phase of the system life cycle and is concerned with tasks related to the maintenance, management of the engineering and the support of the system over its operational life period. Some of the reliability and maintainability related management tasks involved during the operation and maintenance phase are as follows:

- Developing failure data banks.
- Analyzing reliability and maintainability data.
- Providing adequate tools for maintenance.
- Providing appropriately trained manpower.
- Managing and predicting spare parts.
- Developing engineering change proposals.
- Preparing maintenance documents.
- Reviewing documents in light of any engineering change.

The Disposal Phase

This is the last phase of the system life cycle and it is concerned with activities needed to remove the system and its nonessential supporting parts. Two of the reliability and maintainability management tasks involved in this phase are calculation of the final life-cycle cost and the final reliability and maintainability values of the system in question. The life-cycle cost includes the disposal action cost or income. The final reliability and maintainability values are calculated for the buyer of the used system and for use in purchasing of similar systems.

3.3.4 Documents and Tools for Reliability Management

Reliability management makes use of various kinds of documents and tools. Some of the examples of documents used by the reliability management are the in-house reliability manual; national and international standards, specification or publications; documents explaining policy and procedures; instructions and plans; reports; and drawings and contract documents. Similarly, the reliability management makes use of various kinds of management tools. Some of them are value engineering, configuration management and critical path method or program evaluation and review technique. Some of these items are discussed in the following.

The reliability manual is the backbone of any reliability organization. Its existence is vital for any organization irrespective of its size. An effective reliability manual covers topics such as:

- Organizational setup and individual responsibilities
- Company reliability policy and procedures
- Applicable reliability models, reliability prediction techniques, etc.
- Procedures for failure data collection and analysis
- Methods to be used in equipment reliability testing

Value engineering is concerned with the orderly application of established methods to identify the functions of a product or a service, and also with providing those identified functions at the minimum cost. Historical evidence indicates that the application of value engineering concept has returned between $15 to $30 for each dollar spent [23].

Hundreds of engineering changes are associated with the development of a complex system. Configuration management assures both the customer and the manufacturer that the resulting system hardware and software fully meet the contract specification. Configuration management is defined as the management of technical requirements which define the engineering system as well as changes thereto.

Both the critical path method (CPM) and the program evaluation and review technique (PERT) were developed in the late 1950s to manage engineering projects effectively. Nowadays, both these methods are used quite frequently. The application of both these techniques have been found effective in areas such as planning and smoothing resource utilization, obtaining cost estimates for proposed projects, scheduling projects, and evaluating existing schedules and cost vs. budget, as the work on the project progresses.

3.3.5 Reliability Auditing and Pitfalls in Reliability Program Management

The audits are important in reliability work to indicate its strong, weak, and acceptable areas. There are various guidelines in conducting reliability audits. Some of them are maintain auditing schedule, record the audit results, conduct audits without prior announcements, perform audits using checklists, choose unbiased persons for auditing, take follow-up actions, and avoid appointing someone permanently for the auditing purpose.

Reliability auditing has many advantages. For example, it helps problem areas to surface, determines if the customer specifications are met satisfactorily, helps to reduce the complaints received from customers, is useful in predicting the reaction of the buyer toward reliability, and finds out if the company reliability objectives are fulfilled.

Pitfalls in reliability program management are the causes for many reliability program uncertainties and problems. These pitfalls are associated with areas such as follows [24]:

- reliability organization
- reliability testing
- programming
- manufacturing

Two important examples of reliability organization pitfalls are having several organizational tiers and a number of individuals with authority to make commitments without internal dialog and coordination. Furthermore, an important example of the reliability testing pitfalls is the delayed start of the reliability demonstration test. This kind of delay may compound the problem of incorporating reliability-related changes into deliverable product to the customer. Programming pitfalls are also important and must be given full attention. One example of these pitfalls associated with program requirements is the assumption that each concerned individual with the program understands the reliability requirements.

There are also pitfalls that occur during the manufacturing phase. For example, when the parts acquisition lead time is incompatible with the system manufacturing schedule, the parts buyers or even the design engineers will (under certain circumstances) authorize substitute parts without paying much attention to their effect on reliability. These kinds of pitfalls can only be avoided by an effective reliability management team.

3.3.6 Reliability and Maintainability Engineering Departments and their Responsibilities

Probably the most crucial factor in the success or failure of reliability and maintainability programs in an organization is the attitude and the thinking philosophy of top level management toward reliability and maintainability. More clearly, without the support of the top management, reliability and maintainability programs in a company will not be effective at all. Once the high-up management's positive and effective attitude is generated, then appropriate reliability and maintainability organizations are to be formed. Two distinct departments, i.e., reliability and maintainability, will be dictated by the need of the company in question. Sometimes the reliability and maintainability functions may be combined within a single department or assigned as the responsibility of the quality assurance department. As to the reporting structure of the reliability and maintainability departments, there is no iron-clad rule as to how it should be. However, in order to command respect for these programs their heads should be high enough on the organizational ladder with necessary authority [25].

Reliability Engineering Department Responsibilities

A reliability engineering department may have various kinds of responsibilities. However, the major ones are as follows:

- Establishing reliability policy, plans and procedures
- Reliability allocation
- Reliability prediction
- Specification and design reviews with respect to reliability
- Reliability growth monitoring
- Providing reliability related inputs to design specifications and proposals
- Reliability demonstration
- Training reliability manpower and performing reliability-related research and development work
- Monitoring the reliability activities of subcontractors, if any
- Auditing the reliability activities
- Failure data collection and reporting
- Failure data analysis
- Consulting

Maintainability Engineering Department Responsibilities

Similarly, as for the reliability engineering department, the maintainability engineering department has many responsibilities as well. Some of them are as follows:

- Establishing maintainability policies and procedures
- Training and staffing maintainability manpower
- Developing plans for and conducting maintainability demonstration
- Collecting and analyzing failure data
- Taking part in design and specification reviews
- Maintainability assessment
- Maintainability allocation
- Conducting maintainability tradeoff studies
- Monitoring maintainability activities of subcontractors
- Providing consulting services to others
- Documenting maintainability related information

3.3.7 Reliability Manpower

The reliability group is composed of people who have specialties in various branches of reliability engineering. However, according to some experts [21], the personnel involved in reliability work should have experience and background in areas such as quality control, probability theory, project management, system engineering, operations research, environmental testing, components design, manufacturing methods, data analysis and collection, developing specifications, and test planning. Furthermore, it is to be noted that professions such as engineering physics, mechanical engineering, statistics, metallurgy, and electronics engineering are related to reliability. Their relationships are briefly discussed in Reference 26. Some areas related to reliability personnel are described below.

Rules for Reliability Professionals

These rules are useful for the effective implementation of reliability programs in an organization by the reliability manager or engineer as well as the concerned professional who can expect some advancement in their careers [27]. Some of these rules are presented below.

- Speak as briefly as possible. In other words, get to the core of the problem by presenting a summary and recommendations as soon as possible.
- When talking to others, make sure that you converse in a language they understand. Avoid using terms and abbreviations with which they are uncomfortable.
- Avoid using statistical jargon as much as possible especially when dealing with top-level management.
- Make sure that all reliability functions are included in the reliability program plan.
- Contribute to the solution by getting directly involved with the problem investigation effort and work out a group of corrective action options.
- Develop for yourself how to be just as comfortable reporting successes as reporting failures with respect to reliability.
- Avoid, whenever it is possible, paralysis of analysis.

- Develop in yourself the belief that the reliability engineer is a pessimist, the designer is an optimist, and the combination is a success.
- Avoid assuming that the reliability function is responsible for accomplishing reliability goals.
- Learn to believe that new reliability techniques are not the final result, but are only the processes that allow one to achieve product and business goals.

Tasks of a Reliability Engineer

During planning, design and development, manufacturing, and improvement phases of a system, the reliability engineer performs various tasks. Many of these tasks are as follows:

- Performing analysis of a proposed design
- Securing resources for an effective reliability program during the planning stage
- Analyzing customer complaints with reliability
- Investigating field failures
- Running tests on the system, subsystems, and parts
- Examining difficulties associated with maintaining a system
- Developing tests on the system, subsystems, and components
- Budgeting the tolerable system failure down to the component level
- Developing a reliability program plan
- Keeping track of parts
- Determining reliability of alternative designs
- Providing information to designers regarding improving the ease of maintenance and system life
- Providing relevant reliability information to management
- Ensuring that the relevant information flows to all concerned in an effective manner
- Ensuring that the end results of all reliability tasks are organized systematically and effectively
- Monitoring subcontractor's reliability performance
- Participating in developing requirements for new systems
- Participating in evaluating requests for proposals
- Ensuring that enough consideration is given to reliability in new contracts
- Developing reliability prediction models and techniques
- Participating in design reviews

Training of Reliability Manpower

This is an important area concerning every reliability manager because for an effective performance training, developing manpower is a never-ending process. Need for training has to be planned well in advance in order to avoid any difficulty at the moment of need. Some of the major objectives [28] of reliability training indoctrination are as follows:

- Make all personnel in manufacturing, engineering, reliability, purchasing, quality control, etc. aware of the specific effect of tasks performed by them on product reliability;
- Assure that the persons involved in reliability work are in a position to carry out their tasks in an effective and efficient manner;
- Enhance reliability consciousness in people associated with the project under consideration;
- Focus on those activity areas deemed to be specifically amenable to a reliability improvement effort.

There are many useful guidelines with respect to training of reliability manpower including identifing the weak areas, forecasting the manpower needed, planning in advance to meet future needs of the organization, making presentations on the latest reliability state of the art, sponsoring personnel involved with reliability work to conferences, courses, and seminars, and joining professional reliability societies [27].

Difficulties in Motivating Design Engineers for Reliability

In recent times, during the design phase, reliability has been receiving wider attention than ever before. However, there still remain several difficulties in motivating design engineers to accept reliability as one of the design parameters. Some of these difficulties are as follows [29]:

- Understanding product users
- Requirement for reliability specifications
- Choosing the right person for the job
- Understanding an engineering design recipe
- Educating educators (i.e., design engineering professors) on reliability
- Assisting instead of controlling
- Comprehending business considerations

3.4 PROBLEMS

1. Write an essay on important design failures that have occurred over the past 50 years.
2. List at least five of the most important common reasons for product failures.
3. Describe a typical design process practiced in the industrial sector.
4. What are the advantages of conducting design review?
5. Discuss the following design reviews:
    - Preliminary design review
    - Intermediate design review
    - Critical design review
6. Describe the functions of the following two professionals:
    - Design engineer
    - Reliability engineer

7. What are the design process aspects that have significant impact on product reliability?
8. List the military specification (i.e., MIL-R-27542) guidelines for developing reliability programs.
9. Discuss the documents used by the reliability management.
10. Briefly discuss the tools used by the reliability management.
11. Describe the reliability program management related pitfalls.
12. What are the typical responsibilities of a reliability engineering department?
13. List the difficulties experienced in motivating design engineers to accept reliability as one of the design parameters.

3.5 REFERENCES

1. Shigley, J.D. and Mitchell, L.D., *Mechanical Engineering Design*, McGraw-Hill, New York, 1983.
2. Farr, M., *Design Management*, Cambridge University Press, London, 1955.
3. Dhillon, B.S., *Engineering Design: A Modern Approach*, Richard D. Irwin, Chicago, IL, 1996.
4. Austin-Davis, W., Reliability management: A challenge, *IEEE Trans. Reliability*, R-12, 6-9, 1963.
5. Dhillon, B.S. and Reiche, H., *Reliability and Maintainability Management*, Van Nostrand Reinhold Company, New York, 1985.
6. Dhillon, B.S., *Reliability and Quality Control: Bibliography on General and Specialized Areas*, Beta Publishers, Gloucester, Ontario, 1992.
7. Hurricks, P.L., *Handbook of Electromechanical Product Design*, Longman Scientific and Technical, Longman Group UK Limited, London, 1994.
8. Walton, J.W., *Engineering Design*, West Publishing Company, New York, 1991.
9. Elsayed, E.A., *Reliability Engineering*, Addison Wesley Longman, Reading, MA, 1996.
10. Dieter, G.E., *Engineering Design*, McGraw-Hill, New York, 1983.
11. Vidosic, J.P., *Elements of Design Engineering*, The Ronald Press Co., New York, 1969.
12. Hill, P.H., *The Science of Engineering Design*, Holt, Rhinehart, and Winston, New York, 1970.
13. Pecht, M., Ed., *Product Reliability, Maintainability, and Supportability Handbook*, CRC Press, Boca Raton, FL, 1995.
14. AMCP 706-196, *Engineering Design Handbook, Part II: Design for Reliability*, 1976. Prepared by Headquarters, U.S. Army Material Command, Alexandria, VA.
15. Carter, A.D.S., *Mechanical Reliability*, MacMillan, London, 1986.
16. Roche, J.G., Design implications of product liability, *Int. J. Quality and Reliability Manage.*, 2, 1988.
17. Abbot, H., *Safer by Design*, The Design Council, London, 1987.
18. Dhillon, B.S., Engineering reliability management, *IEEE J. Selected Areas Comm.*, 4, 1015-1020, 1986.
19. Heyel, C., *The Encyclopedia of Management*, Van Nostrand Reinhold Company, New York, 1979.
20. Finch, W.L., Reliability: A technical management challenge, *Proc. Am. Soc. Quality Control Annu. Conf.*, 851-856, 1981.

21. Karger, D.W. and Murdick, R.G., *Managing Engineering and Research,* Industrial Press, New York, 1980.
22. Grant Ireson, W., Coombs, C.F., and Moss, R.Y., Eds., *Handbook of Reliability Engineering and Management,* McGraw-Hill, New York, 1996.
23. Demarle, D.J. and Shillito, M.L., Value Engineering, in *Handbook of Industrial Engineering,* Salvendy, Ed., John Wiley & Sons, New York, 1982, 7.3.1–7.3.20.
24. Thomas, E.F., Pitfalls in reliability program management, *Proc. Annu. Reliability Maintainability Symp.,* 369-373, 1976.
25. Jennings, J.A., Reliability of aerospace electrical equipment: How much does it cost?, *IEEE Trans. Aerospace,* 1, 38-40, 1963.
26. *Reliability Engineering: A Career for You,* Booklet published jointly by the American Society for Quality Control (ASQC) and the Institute of Electrical and Electronics Engineers (IEEE). Available from the Director of Technical Programs, ASQC, 230 West Wells St., Milwakee, WI 53203.
27. Ekings, J.D., Ten rules for the reliability professional, *Proc. Annu. Am. Soc. Quality Control Conf.,* 343-351, 1982.
28. McClure, J.Y., Organizing for Reliability and Quality Control, in *Reliability Handbook,* Ireson, G., Ed., McGraw-Hill, New York, 16-37, 1966.
29. Bajaria, H.J., Motivating design engineers for reliability, *Proc. Annu. Conf. Am. Soc. Quality Control,* 767-773, 1979.

4 Failure Data Collection and Analysis

4.1 INTRODUCTION

Failure data are the backbone of reliability studies because they provide invaluable information to concerned professionals such as reliability engineers, design engineers, and managers. In fact, the failure data of a product are the final proof of reliability related efforts expended during its design and manufacture. It would be impossible to have an effective reliability program without the collection, analysis, and use of information acquired through the testing and operation of products used in industrial, military, and consumer sectors.

It may be stated, more specifically, that the basic purpose of developing a formal system of collecting, analyzing, and retrieving information acquired from past experiences is to allow design and development of better and more reliable products without repeating the previous efforts concerning research, design, and testing to achieve the current product reliability. Thus, the fundamental goal of a failure data collection and analysis system is to convert the relevant information accumulated in various sources into an effectively organized form so that it can efficiently be used by individuals with confidence in conducting their assigned reliability related tasks. There are many different ways to store and present this organized form of information to the end users. As each such way has advantages and disadvantages, a careful consideration must be given in their selection.

In particular, the nature of the products and their intended applications are important factors in deciding the extensiveness, form, etc. of a failure data collection and analysis system. For example, a company stamping out stainless steel flatware will require a simple data system as opposed to a complex data system for a company manufacturing aircraft. All in all, prior to deciding the size, type, etc. of a failure data system, consider factors such as individuals requiring information, information required by the individuals to meet their needs, and frequency of their need.

An extensive list of publications on failure data is given in Reference 1. This chapter presents different aspects of failure data collection and analysis.

4.2 FAILURE DATA USES

There are many areas for the uses of failure data including conceptual design, preliminary design, test planning, design reviews, manufacturing, quality assurance and inspection planning, logistic support planning inventory management, procurement, top management budgeting, and field service planning [2]. Nonetheless, some of the more specific uses of failure data are estimating hazard rate of an item, performing life cycle cost studies, making decisions regarding introduction of redundancy, predicting an item's reliability/availability, performing equipment replacement studies, recommending design changes for improving the product's reliability, performing cost vs. reliability studies, determining the maintenance needs of a new item, and conducting preventive maintenance studies [3–5].

4.3 FAILURE DATA COLLECTION SOURCES IN EQUIPMENT LIFE CYCLE AND QUALITY CONTROL DATA

There are many sources of collecting failure related data during an equipment life cycle. According to Reference 6, eight of these sources are as follows:

1. Warranty claims
2. Previous experience with similar or identical equipment
3. Repair facility records
4. Factory acceptance testing
5. Records generated during the development phase
6. Customers' failure reporting systems
7. Tests: field demonstration, environmental qualification, and field installation
8. Inspection records generated by quality control/manufacturing groups

In order to maintain the required quality standards, the quality control groups within organizations regularly perform tests/inspections on products/equipment. As the result of such actions, valuable data are generated that can be very useful in design reviews, part vendor selection, providing feedback to designers on the problems related to production, and so on. The important components of the quality control data are incoming inspection and test results, quality audit records and final test results, in-process quality control data, results of machine and process capability studies, and calibration records of measuring instruments [2].

4.4 FAILURE REPORTING AND DOCUMENTATION SYSTEM DESIGN GUIDELINES AND FAILURE DATA COLLECTION FORMS

A failure reporting and documentation system could be very effective if a careful consideration is given during its design. Some of the associated design guidelines are simplicity and clarity of the failure reporting form, effective and efficient documentation of the information on failures, clarity in terms and definitions, efficiency in conducting recorded failure data analysis, user involvements, flexibility, cost effectiveness, feasibility of prerecording any static information on the data form, elimination of the needs for memorizing codes, and feedback of the analyzed data to involved personnel in an effective manner [7].

The backbone of a failure data system is the failure data collection form. Such forms must be designed so that they require minimum effort to collect failure data and provide maximum benefits. Also, these forms must be tailored to meet the

TABLE 4.1
Selective Standard Documents for Failure Data Collection

No.	Document classification No.	Document title	Developed by:
1	MIL-STD-2155	Failure Reporting, Analysis and Corrective Action System (FRACAS)	U.S. Department of Defense
2	IEC 706 PT3	Guide on Maintainability of Equipment: Verification and Collection, Analysis and Presentation of Data	International Electrotechnical Commission (IEC)
3	MODUKDSTAN 00-44	Reliability and Maintainability Data Collection and Classification	British Defense Standards (U.K. Department of Defense)
4	IEEE 500	Guide to the Collection and Presentation of Electrical, Electronic, Sensing Component, and Mechanical Equipment Reliability Data for Nuclear Power Generating Stations	Institute of Electrical and Electronics Engineers (IEEE)
5	IEC 362	Guide for the Collection of Reliability, Availability, and Maintainability Data from Field Performance of Electronic Items	International Electrotechnical Commission (IEC)
6	—	A Reliability Guide to Failure Reporting, Analysis, and Corrective Action Systems	American Society for Quality Control (ASQC)

requirements of a specific organization or program effectively. Nonetheless, such forms are generally designed to include information such as form number, item description, location of the hardware, serial number of the failed item and manufacturer's name, description of the failure, form completion date, dates on which failure occurred, was detected and corrected, operating hours from the previous failure, time taken for repair, repair cost, serial number and manufacturer of the replaced part, repair-person's name and appropriate signatures [8].

Table 4.1 presents useful standard documents for failure data collection.

4.5 EXTERNAL FAILURE DATA SOURCES

There are a large number of organizations, data banks, and documents to obtain failure data related information. Some of these are as follows [2, 5, 9]:

- **Organizations**
  - **Government Industry Data Exchange Program (GIDEP)** [10]. It is probably the most comprehensive source for obtaining data and is operated and funded by the U.S. government. A company must formally

become a full or partial participating member of the GIDEP before making use of it. This means that the company must also submit its own data to GIDEP in order to receive data from it. The system is managed by the GIDEP Operations Center, Fleet Missile Systems, Analysis and Evaluation Group, Department of Defense, Corona, California 91720.

- **National Technical Information Service (NTIS).** This is also a good source for obtaining various types of failure data and the service is managed by the U.S. Department of Commerce. The address of the NTIS is 5285 Port Royal Road, Springfield, Virginia 22161.
- **Reliability Analysis Center (RAC).** RAC provides various types of information concerning reliability and is managed by the U.S. Air Force (USAF). In particular, the Nonelectronic Parts Reliability Data (NPRD) reports are released periodically by the center. The address of the RAC is Rome Air Development Center (RADC), Griffiss Air Force Base, Department of Defense, Rome, New York 13441.
- **Defense Technical Information Center.** This center is a useful source for obtaining various types of failure data related information, particularly concerning the defense equipment. The address of the center is Defense Logistics Agency, Cameron Station, Alexandria, Virginia 22314.
- **Parts Reliability Information Center (PRINCE).** This center is quite useful to obtain various types of data related to space systems. The address of the center is Reliability Office, George C. Marshall Space Flight Center, National Aeronautics and Space Administration (NASA), Huntsville, Alabama 35812.
- **Institute of Electrical and Electronics Engineers (IEEE).** This is the largest professional body in the world and provides failure data concerning various electrical related items. The address of this organization is 345 East 47th Street, New York, New York 10017.
- **National Electric Reliability Council (NERC).** NERC publishes various types of failure data collected annually from the U.S. power plants. This organization is located in New York City.
- **Insurance Information Institute (III).** This organization provides data on disaster impacts and related matters. III is located at 110 William Street, New York, New York 10017.

- **Data Banks**
  - **Nuclear Plant Reliability Data System (NPRDS).** This system provides various types of failure data on equipment used in nuclear power generation. NPRDS is managed by the South West Research Institute, San Antonio, Texas.
  - **Equipment Reliability Information System (ERIS).** This system provides failure data on equipment used in electric power generation. The system is managed by the Canadian Electrical Association (CEA), Montreal, Quebec.
  - **SYREL: Reliability Data Bank.** This data bank provides failure data on equipment used in power generation. The data bank is managed by

the Systems Reliability Service, Safety and Reliability Directorate, United Kingdom Atomic Energy Authority (UKAEA), Wigshaw Lane, Culcheth, Warrington, Lancashire WA3 4NE, England.

- **ESA Electronic Component Data Bank.** This data bank provides data on various types of electronic parts, especially used in space environment. Its address is Space Documentation Service, European Space Agency (ESA), Via Galileo, 00044 Frascati, Italy.

- **Documents**
  - MIL-HDBK-217, Reliability Prediction of Electronic Equipment, Department of Defense, Washington, D.C.
  - Component Parts Failure Data Compendium, Engineering Dept., Electronic Industries Association, 11 W. 42nd St., New York, New York 10036.
  - RADC-TR-85-194, RADC Nonelectronic Reliability Notebook, Reliability Analysis Center, Rome Air Development Center (RADC), Griffiss Air Force Base, Rome, New York 13441-5700, 1985.
  - IEEE-STD-500-1977, IEEE Nuclear Reliability Data Manual, John Wiley & Sons, New York, 1977.
  - WASH-1400 (NUREG 75/014), Appendix 3 and 4, Reactor Safety Study: An Assessment of Accident Risks in U.S. Commercial Nuclear Power Plants, U.S. Nuclear Regulatory Commission, Washington, D.C., 1975.
  - Dhillon, B.S., *Mechanical Reliability: Theory, Models, and Applications,* American Institute of Aeronautics and Astronautics, Washington, D.C., 1988 (Chapter 9 contains failure data on various mechanical parts).
  - RAC EMD 1 and 2, Electronic Equipment Reliability Data, Reliability Analysis Center (RAC), Rome Air Development Center, Griffiss Air Force Base, Rome, New York.
  - NUREG/CR-1278, Handbook of Human Reliability Analysis with Emphasis on Nuclear Power Plant Applications, U.S. Nuclear Regulatory Commission, Washington, D.C.

4.6 FAILURE DATA FOR SELECTIVE ITEMS AND TASKS

For many decades, failure data on engineering items and various tasks performed by humans have been collected and analyzed. In order to demonstrate the type of analyzed and available failure data, Tables 4.2 [11] and 4.3 [5, 12–15] present failure rates for some selected electronic and mechanical items, respectively, and Table 4.4 [16–19] contains human error data for a few selective tasks.

4.7 HAZARD PLOTTING METHOD

This is a powerful graphical approach to perform failure data analysis. The method is popular among engineers because of its advantages such as fits the data to a straight line, results obtained through plotting are convincing, easy to understand, straightforward to visualize the theoretical distribution that fits the field data, and equally

TABLE 4.2
Failure Rates for Some Electronic Items

No.	Item description	Failure rate (failures/10^6 h)
1	Neon lamp	0.2
2	Weld connection	0.000015[a]
3	Fuse (cartridge class H or instrument type)	0.010[a]
4	Fiber optic cable (single fiber types only)	0.1 (per fiber km)
5	Solid state relay (commercial grade)	0.0551[a]
6	Vibrator (MIL-V-95): 60-cycle	15
7	Terminal block connection	0.062[a]
8	Microwave ferrite devices: isolators and circulatory (≤100 watts)	0.10[a]
9	Solid state relay (military specification grade)	0.029[a]
10	Single fiber optic connector	0.1
11	Crimp connection	0.00026[a]
12	Spring contact connection	0.17[a]
13	Hybrid relay (commercial grade)	0.0551[a]
14	Vibrator (MIL-V-95): 120-cycle	20
15	Fuse: current limiter type (aircraft)	0.01[a]

[a] Use environment: ground, benign.

applicable to analyze data with complete and incomplete observations [20–24]. Another important advantage of the approach is that it indicates if the given data set belongs to the tried distribution. If so, then it estimates the associated parameters.

The complete observations may be described as data in which the times to failure of all units of a sample are known. On the other hand, when the running times of unfailed units and the failure times of failed units are given, the data are called incomplete or censored data/observations. In turn, the running times of unfailed units are called the censoring times. In the event of having different censoring times of the unfailed units, the data are known as the multi-censored data. In contrast, when the censoring times of all the unfailed units are the same and greater than the failed units' failure times, the data are said to be singly censored. Nonetheless, the multi-censored data occur due to factors such as follows:

- Removing items/units or terminating their use prior to their failure.
- Some extraneous causes were responsible for the units'/items' failure.
- Collecting data from active units/items.

4.7.1 Hazard Plotting Mathematical Theory

The hazard function of a statistical distribution is the basis for this theory. Thus, the hazard rate is defined by

TABLE 4.3
Failure Rates for Some Mechanical Items

No.	Item description	Failure rate (failures/10^6 h)
1	Pivot	1
2	Heat exchanger	6.11–244.3
3	Relief valve	0.5–10
4	Conveyor belt (heavy load)	20–140
5	Heavy duty ball bearing	20
6	Piston	1
7	Washer	0.5
8	Nut or bolt	0.02
9	Rotating seal	4.4
10	Pipe	0.2
11	Spring (torsion)	14.296[a]
12	Gearbox (reduction)	18.755[b]
13	Bellow	5
14	Crankshaft (general)	33.292[b]
15	Cylinder	0.1
16	Knob (general)	2.081[a]
17	Flexible coupling	9.987[b]
18	Axle (general)	9.539[b]
19	Hair spring	1
20	Slip ring (general)	0.667[a]

[a] Use environment: ground, fixed.
[b] Use environment: ground, mobile.

TABLE 4.4
Human Error Rates for Some Tasks

No.	Error description	Error rate
1	Improper servicing/reassembly by the maintenance person	0.0153[a]
2	Closing valve incorrectly	1800[b]
3	Incorrect adjustment by the maintenance person	0.0134[a]
4	Procedural error in reading given instructions	64500[b]
5	Reading gauge incorrectly	5000[a]
6	Misunderstanding of requirements by the operator	0.0076[a]
7	Installation error	0.0401[a]

[a] Errors per plant month (for pressurized water reactors).
[b] Errors per million operations.

$$z(t) = \frac{f(t)}{R(t)} = \frac{f(t)}{1 - F(t)} \tag{4.1}$$

where

$z(t)$ is the hazard rate or instantaneous failure rate.
$R(t)$ is the reliability at time t.
$f(t)$ is the distribution probability density function.
$F(t)$ is the cumulative distribution function.

The cumulative distribution function is defined by

$$F(t) = \int_0^t f(t)\,dt \tag{4.2}$$

The cumulative hazard function, $z_c(t)$, is defined by

$$z_c(t) = \int_0^t z(t)\,dt \tag{4.3}$$

Equations (4.1) through (4.3) are used to obtain a straight line expression for estimating graphically the parameters of failure distributions such as the following [20, 25]:

Exponential Distribution

This is a single parameter distribution and its probability density function is defined by

$$f(t) = \lambda e^{-\lambda t} \qquad t \geq 0 \tag{4.4}$$

where

λ is the parameter known as the constant failure rate.
t is time.

Inserting Equation (4.4) into Equation (4.2) yields

$$F(t) = 1 - e^{-\lambda t} \tag{4.5}$$

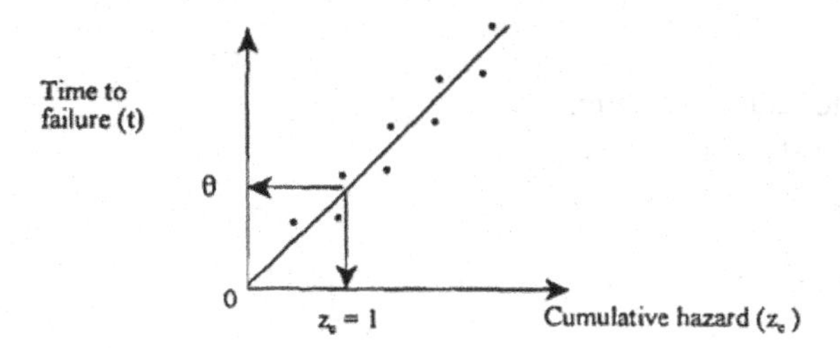

FIGURE 4.1 A hypothetical plot of time to failure, t, against cumulative hazard, z_c.

By substituting Equations (4.4) and (4.5) into Equation (4.1), we get

$$z(t) = \lambda \tag{4.6}$$

By inserting Equation (4.6) into Equation (4.3), we obtain the following expression for the cumulative hazard function:

$$z_c(t) = \lambda t \tag{4.7}$$

By letting $\lambda = 1/\theta$, where θ is the mean time to failure and rearranging Equation (4.7) to express the time to failure, t, as a function of z_c, we get

$$t(z_c) = \theta z_c \tag{4.8}$$

Equation (4.8) is the equation of a straight line passing through the origin. In order to estimate the value of θ, the time to failure t against the cumulative hazard z_c is plotted. If the plotted field data points fall roughly on a straight line, then a line is drawn to estimate θ. At $z_c = 1$ on this plot, the corresponding value of t is the estimated value of θ. The value of the θ also equals the slope of the straight line. When the plotted data points do not approximately fall along a straight line, this indicates that the failure data do not belong to the exponential distribution and another distribution should be tried. A hypothetical plot of Equation (4.8) is shown in Figure 4.1.

Bathtub Hazard Rate Distribution

This distribution can represent a wide range of failure data representing increasing, decreasing, and bathtub hazard rates. The probability density function of the distribution is defined by [26, 27]:

$$f(t) = \theta \gamma (\gamma t)^{\theta - 1} e^{-\left[e^{(\gamma t)^{\theta}} - (\gamma t)^{\theta} - 1\right]} \tag{4.9}$$

for $\theta, \gamma > 0$, and $t \geq 0$

where

γ is the scale parameter.
θ is the shape parameter.
t is time.

Substituting Equation (4.9) into Equation (4.2) yields

$$F(t) = 1 - e^{-\left\{e^{(\gamma t)^{\theta}} - 1\right\}} \tag{4.10}$$

The following are the special cases of this distribution:

At $\theta = 1$: extreme value distribution
$\theta = 0.5$: bathtub hazard rate curve

By inserting Equations (4.9) and (4.10) into Equation (4.1), we obtain

$$z(t) = \theta\, \gamma\, (\gamma t)^{\theta - 1}\, e^{(\gamma t)^{\theta}}. \tag{4.11}$$

Substituting Equation (4.11) into Equation (4.3) yields

$$z_c(t) = e^{(\gamma t)^{\theta}} - 1 \tag{4.12}$$

Rearranging (4.12) and twice taking natural logarithms of $(z_c + 1)$, we get

$$\ln t = \frac{1}{\theta} \ln x - \ln \gamma \tag{4.13}$$

where

$x \equiv \ln(z_c + 1)$

The plot of ln t against ln x gives a straight line; thus, the slope of the line and its intercept are equal to $1/\theta$ and $(-\ln\gamma)$, respectively. If the plotted field data points fall roughly on a straight line, then a line can be drawn to estimate the values of θ and γ. On the other hand, when the plotted data do not approximately fall along a straight line, this obviously indicates that the failure data under consideration do not belong to the bathtub hazard rate distribution; thus, try another distribution.

Weibull Distribution

This is a widely used distribution in reliability studies and it can represent a wide range of failure data. The probability density function of the distribution is defined by

$$f(t)=\frac{\theta}{\beta^{\theta}}t^{\theta-1}\,e^{-(t/\beta)^{\theta}},\quad \text{for } t\geq 0 \tag{4.14}$$

where

t is time.
β is the scale parameter.
θ is the shape parameter.

Inserting Equation (4.14) into Equation (4.2) yields

$$F(t)=1-e^{-(t/\beta)^{\theta}} \tag{4.15}$$

At $\theta = 1$ and 2, the Weibull distribution becomes exponential and Rayleigh distributions, respectively.

By substituting Equations (4.14) and (4.15) into Equation (4.1), we get

$$z(t)=\frac{\theta}{\beta^{\theta}}\,t^{\theta-1} \tag{4.16}$$

Inserting Equation (4.16) into Equation (4.3) leads to

$$z_c(t)=(t/\beta)^{\theta} \tag{4.17}$$

In order to express the time to failure t as a function of z_c, we rearrange Equation (4.17) as follows:

$$t(z_c)=\beta(z_c)^{1/\theta} \tag{4.18}$$

Taking natural logarithms of Equation (4.18) results in

$$\ln t=\frac{1}{\theta}\ln z_c+\ln\beta \tag{4.19}$$

In Equation (4.19), $\ln t$ against $\ln z_c$ plots linearly or in a straight line with slope $1/\theta$ and the intercept $\ln\beta$. Also, at $z_c = 1$, the corresponding time $\ln t$ equals $\ln\beta$. Nonetheless, graphically, this fact is used to estimate the value of the scale parameter β.

Nonetheless, from the graphical standpoint, if the plotted field data points fall approximately along a straight line, then a line is drawn to estimate the values of the parameters θ and β. In the event of such points not falling along the straight line, try another distribution.

Similarly, the straight line expressions for other failure distributions can be obtained [20, 25].

TABLE 4.5
Failure and Running (Censoring) Times for 16 Identical Units

Unit No.	Failure and running (censoring) times (h)
1	14,092[a]
2	3,973
3	2,037
4	360
5	628
6	3,279
7	2,965
8	837[a]
9	79
10	13,890[a]
11	184
12	13,906[a]
13	499
14	134
15	10,012
16	3,439[a]

[a] running (censoring) time

Example 4.1

Assume that 16 identical units, at time t = 0, were put on test and Table 4.5 presents failure and running times of these units. Determine the statistical distribution fit to the given data and estimate values for its parameters by using the hazard plotting steps described below.

Hazard Plotting Steps

The following eight steps are used to construct a hazard plot:

1. Order the data containing m times from smallest to largest without making any distinction whether these data are running (censoring) or failure times of units. Use an asterisk (*) or other means to identify the running or censoring times. In the event of having some running and failure times equal, mix such times well on the ordered list of smallest to largest times. For Example 4.1 data given in Table 4.5, in Column 1 of Table 4.6, the failure and running times of 16 units are ranked from smallest to largest with the censoring times identified by a superscript "a".
2. Label the ordered times with reverse ranks. For example, label the first time with m, the second with (m-1), the third with (m-2), …, and the mth with 1.

   Column 2 of Table 4.6 presents reverse rank labels for the 16 ordered times of Column 1.

TABLE 4.6
Ordered Times and Hazard Values

Col. 1 Ordered times t in hours (smallest to largest)	Col. 2 Reverse rank labels	Col. 3 Hazard value (100/Col. 2 value)	Col. 4 Cumulative hazard (z_c)
79	16	6.25	6.25
134	15	6.67	12.92
184	14	7.14	20.06
360	13	7.69	27.75
499	12	8.33	36.08
628	11	9.09	45.17
837[a]	10	—	—
2,037	9	11.11	56.28
2,965	8	12.5	68.78
3,279	7	14.29	83.07
3,439[a]	6	—	—
3,973	5	20	103.07
10,012	4	25	128.07
13,890[a]	3	—	—
13,906[a]	2	—	—
14,092[a]	1	—	—

[a] running (censoring) time

3. Compute a hazard value for each *failure* using 100/m, 100/(m-1), 100/(m-2), etc. The hazard value may be described as the conditional probability of failure time. In other words, it is the observed instantaneous failure rate at a certain failure time.

   For example, as per Columns 1 and 2 of Table 4.6, the first failure occurred after 79 h out of 16 units put on test at time t = 0. Thus, the hazard value is

$$\frac{100}{16} = 6.25\%$$

   After 79 h, only 15 units were operating and 1 failed after 134 h. Similarly, the hazard value is

$$\frac{100}{15} = 6.67\%$$

   In a similar manner, the hazard values at other failure times were computed as shown in Column 3 of Table 4.6. These values are simply 100 divided by the corresponding number of Column 2.
4. Obtain the cumulative hazard value for each *failure* by adding its hazard value to the cumulative hazard value of the preceding failure. For example,

for the first failure at 79 h (as shown in Column 1 of Table 4.6), its hazard value is 6.25% and there is no cumulative hazard value of the preceding failure because this is the first failure. Thus, in this case, the cumulative hazard simply is 6.25%. However, for the failure at 2037 h, the cumulative hazard is 11.11 (hazard value) + 45.17 (cumulative hazard value of the preceding failure) = 56.28. Similarly, Column 4 of Table 4.6 presents cumulative hazard values for other failures.

It is to be noted that the cumulative hazard values may exceed 100% and have no physical interpretation.

5. Choose a statistical distribution and prepare times to failure and corresponding cumulative hazard data for use in the selected distribution to construct a hazard plot.

   For our case, we select the Weibull distribution; thus, we take natural logarithms of the *times to failure* and of corresponding cumulative hazard values given in Table 4.6. The processed data for $\ln t$ and $\ln z_c$ are presented in Table 4.7.

6. Plot each time to failure against its corresponding cumulative hazard value on a graph paper. Even though the running times are not plotted, they do determine the plotting points of the times to failure through the reverse ranks.

   In our case, we plotted $\ln t$ against $\ln z_c$ as shown in Figure 4.2 using Table 4.7 values.

TABLE 4.7
Processed Data for the Weibull Distribution

lnt (using values for t from Col. 1 Table 4.6)	$\ln z_c$ (using values for z_c from Col. 4 Table 4.6)
4.37	1.83
4.90	2.56
5.22	3.00
5.89	3.32
6.21	3.59
6.44	3.81
—	—
7.62	4.03
8.00	4.23
8.10	4.42
—	—
8.29	4.64
9.21	4.85
—	—
—	—
—	—

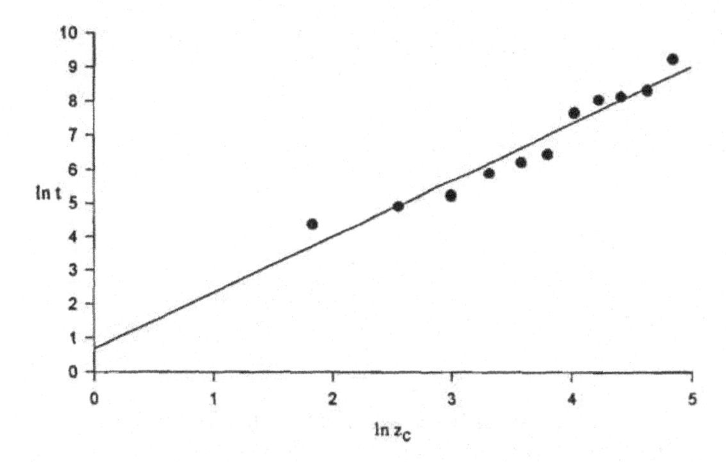

FIGURE 4.2 Failure data fit to a Weibull distribution.

7. Determine if the plotted data points roughly follow a straight line. If they do, it is reasonable to conclude that the selected distribution adequately fits the data and draw a best fit straight line. If they do not, try plotting the given data for another statistical distribution.

   In our case, the given data roughly follow the straight line; thus, it is reasonable to conclude that the data are Weibull distributed.
8. Estimate the values of the distribution parameters using the hazard plot. In our case, using Figure 4.2, we have $\theta = 0.58$, and $\beta = 1.82$ h.

Finally, it is emphasized that the results obtained using this method are valid only when the times to failure of unfailed units are statistically independent of their running or censoring times, if such units were used until their failure.

4.8 UNDERLYING DISTRIBUTION DETERMINATION TESTS

Over the years, there have been many tests developed to determine the underlying statistical distribution of times to failure associated with engineering items when a set of data are given [5, 28–37]. Once the statistical distribution associated with the given failure data of an item is identified, the technique such as maximum likelihood estimation (MLE) can be used to estimate the values of distribution parameters. This section describes three underlying distribution determination tests: (1) Bartlett, (2) general exponential, and (3) Kolmogorov-Smirnov.

4.8.1 Bartlett Test

Often in reliability analysis, it is assumed that an item's times to failure are exponentially distributed. The Bartlett test is a useful method to verify this assumption from a given sample of failure data. The Bartlett test statistic is defined [28] as

$$S_{bm} = 12\ m^2 \left(\ln F - \frac{W}{m} \right) \Big/ (6\ m + m + 1) \quad (4.20)$$

$$W \equiv \sum_{i=1}^{m} \ln t_i \quad (4.21)$$

$$F = \frac{1}{m} \sum_{i=1}^{m} t_i \quad (4.22)$$

where

m is the total number of times to failures in the sample.
t_i is the ith time to failure.

In order for this test to discriminate effectively, a sample of at least 20 times to failure is required. If the times to failures follow the exponential distribution, then S_{bm} is distributed as chi-square with (m-1) degrees of freedom. Thus, a two-tailed chi-square method (criterion) is used [30].

Example 4.2

After testing a sample of identical electronic devices, 25 times to failure as shown in Table 4.8 were observed. Determine if these data belong to an exponential distribution by applying the Bartlett test.

By inserting the given data into Equation (4.22) we get

$$F = \frac{3440}{25}$$

$$= 137.6\ h$$

Similarly, using Equation (4.21) and the given data we have

$$W = 110.6655$$

TABLE 4.8
Times to Failure (in hours) of Electronic Devices

5	34	82	143	270
6	47	87	187	280
21	44	110	183	282
18	61	113	260	350
33	65	140	268	351

Inserting the above values and the given data into Equation (4.20) yields

$$S_{b10} = 12(25)^2 \left(\ln 137.6 - \frac{110.6655}{25} \right) \Big/ \left(6(25) + 25 + 1\right)$$

$$= 21.21$$

From Table 11.1 for a two-tailed test with 98% confidence level, the critical values of

$$\chi^2 \left[\frac{\theta}{2}, (m-1) \right] = \chi^2 \left[\frac{0.02}{2}, (25-1) \right] = 42.98$$

where

$\theta = 1 - (\text{confidence level}) = 1 - 0.98 = 0.2$

and

$$\chi^2 \left[\left(1 - \frac{\theta}{2}\right), (m-1) \right] = \chi^2 \left[\left(1 - \frac{0.02}{2}\right), (25-1) \right] = 10.85$$

The above results indicate that there is no reason to contradict the assumption of exponential distribution.

4.8.2 General Exponential Test

This test is also used to determine whether a given set of data belong to an exponential distribution. The test requires calculating the value of α, the χ^2 variate with 2m degrees of freedom. The α is defined by [28, 29, 32]:

$$\alpha = -2 \sum_{i=1}^{m} \ln \left[\frac{T(\tau_i)}{T(\tau)} \right] \tag{4.23}$$

where

$T(\tau)$ is the total operating time at test termination.
$T(\tau_i)$ is the total operating time at the occurrence of ith failure.
m is the total number of failures in a given sample.

If the value of α lies within

$$\chi^2 \left[\frac{\theta}{2}, 2m \right] < \alpha < \chi^2 \left[1 - \frac{\theta}{2}, 2m \right] \tag{4.24}$$

TABLE 4.9
Times to Failure (in hours) of Electrical Devices

Failure No.	Failure time (h)
1	70
2	85
3	25
4	10
5	20
6	40
7	50
8	45
9	30
10	60

then the assumption of times to failure following an exponential distribution is valid. The symbol θ denotes the risk of rejecting a true assumption and is expressed by

$$\theta = 1 - (\text{confidence level}) \tag{4.25}$$

Example 4.3

Assume that 40 identical electrical devices were tested for 200 h and 10 failed. The failed devices were not replaced. The failure times of the electrical devices are presented in Table 4.9. Determine if the times to failure can be represented by the exponential distribution.

Using the given data, we get

$$T(\tau) = (40-10)\,(200) + 70 + 85 + 25 + 10 + 20 + 40 + 50 + 45 + 30 + 60$$

$$T(\tau) = 6435 \text{ h}$$

Thus, using the above result and the given data in Equation (4.23), we get

$$\alpha = -2\left[\ln\left\{\frac{(10)(40)}{6435}\right\} + \ln\left\{\frac{10+(20)(39)}{6435}\right\} + \ln\left\{\frac{30+(25)(38)}{6435}\right\}\right.$$

$$+ \ln\left\{\frac{55+(30)(37)}{6435}\right\} + \ln\left\{\frac{85+(40)(36)}{6435}\right\}$$

$$+ \ln\left\{\frac{125+(45)(35)}{6435}\right\} + \ln\left\{\frac{170+(50)(34)}{6435}\right\}$$

$$+ \ln \left\{ \frac{220 + (60)(33)}{6435} \right\} + \ln \left\{ \frac{280 + (70)(32)}{6435} \right\}$$

$$+ \ln \left\{ \frac{350 + (85)(31)}{6435} \right\} \Bigg]$$

$$= 30.50$$

For a 98% confidence level from Equation (4.25) we have

$$\theta = 1 - 0.98 = 0.02$$

Using the above value and other given data in relationship (4.24) we get

$$\chi^2 \left[\frac{0.02}{2}, 2(10) \right], \chi^2 \left[1 - \frac{0.02}{2}, 2(10) \right]$$

Thus from Table 11.1, we get

$$\chi^2 \left[0.01, 20 \right] = 37.56$$

$$\chi^2 \left[0.99, 20 \right] = 8.26$$

The above results indicate that the calculated value of α falls in between, i.e.,

$$8.26 < 30.50 < 37.56$$

It means the times to failure data can be represented by the exponential distribution.

4.8.3 Kolmogorov–Smirnov Test

This is another test used to determine the underlying statistical distribution of a given sample of times to failure. The Kolomogorov–Smirnov test can be applied regardless of the assumed failure distribution of a given data set. Also, this test is quite effective regardless of the sample size. However, the test is restricted to continuous statistical distributions only. The Kolmogorov–Smirnov test is composed of the following eight steps [5, 31]:

1. Assume a statistical distribution for the failure data set under consideration.
2. Use the given failure data to estimate the values of the assumed distribution parameters.
3. Determine the level of significance, θ. In more detail, the level of significance may be described as the risk of not accepting the assumed statistical distribution for the given data if it happened to be the real representative distribution.

4. Compute probability, $F(t_i) = P(T \leq t_i)$, using the assumed statistical distribution function and the values of its associated parameters.
5. Compute $\hat{F}(t_i)$ for the ith ordered sample observation by using the given failure data and the following relationship:

$$\hat{F}(t_i) = i/n, \text{ for } i = 1, 2, 3, \ldots, n \tag{4.24}$$

where

n is the number of data points or observations in the given failure data sample.

$\hat{F}(t_i)$ denotes the proportion of the sample data observations that are less than or equal to t_i.

Also, shift i to (i – 1) and calculate

$$\hat{F}(t_{i-1}) = \frac{i-1}{n}, \text{ for } i = 1, 2, 3, \ldots, n \tag{4.25}$$

6. Calculate the statistic:

$$d = \max_i \left\{ \left| F(t_i) - \hat{F}(t_i) \right|, \left| F(t_i) - \hat{F}(t_{i-1}) \right| \right\} \tag{4.26}$$

7. Use Table 4.10 [13, 31, 33, 38] to obtain the value of d_θ for a specified value of θ and the sample size.
8. Make comparison of the value of d with the value of d_θ. If d is greater than d_θ, reject the assumed distribution.

Example 4.4

A medical device manufacturer developed a certain medical device and life tested 12 units of the same device. These units failed after 20, 30, 40, 55, 60, 70, 75, 80, 90, 95, 100, and 110 h of operation. After reviewing the past failure pattern of similar devices, it was assumed that the times to failure follow the normal distribution. Use the Kolomogorov–Smirnov method to test this assumption for 0.05 level of significance.

For the specified data, the mean of the normal distribution is given by

$$\mu = \sum_{i=1}^{m} t_i / m$$

$$= (20 + 30 + 40 + \ldots + 100 + 100)/12$$

$$= 68.75$$

TABLE 4.10
Kolmogorov–Smirnov Critical Values for d_θ

Sample size, n	Level of significance, θ			
	θ = .01	θ = 0.05	θ = 0.10	θ = 0.15
5	0.67	0.57	0.51	0.47
6	0.62	0.52	0.47	0.44
7	0.58	0.49	0.44	0.41
8	0.54	0.46	0.41	0.38
9	0.51	0.43	0.39	0.36
10	0.49	0.41	0.37	0.34
11	0.47	0.39	0.35	0.33
12	0.45	0.38	0.34	0.31
13	0.43	0.36	0.33	0.30
14	0.42	0.35	0.31	0.29
15	0.40	0.34	0.30	0.28
16	0.39	0.33	0.30	0.27
17	0.38	0.32	0.29	0.27
18	0.37	0.31	0.28	0.26
19	0.36	0.30	0.27	0.25
20	0.36	0.29	0.26	0.25
>50	$1.63\ n^{-1/2}$	$1.36\ n^{-1/2}$	$1.22\ n^{-1/2}$	$1.14\ n^{-1/2}$

where

μ is the mean.
m is the number of medical devices life tested.
t_i is the failure time of device i; for i = 1, 2, 3, …, 12.

Similarly, for the given data, the standard deviation, σ, for the normal distribution is

$$\sigma = \left[\sum_{i=1}^{m} (t_i - \mu)^2 / (m-1)\right]^{1/2}$$

$$= \left[\frac{(20-68.75)^2 + (30-68.75)^2 + (40-68.75)^2 + \ldots + (110-68.75)^2}{(12-1)}\right]^{1/2}$$

$$= 28.53$$

Using the above values for μ and σ and cumulative normal distribution function table, the values for $F(t_i)$ as shown in Table 4.11 were obtained.

The values for $\hat{F}(t_i)$ and $\hat{F}(t_{i-1})$ presented in Table 4.12 were obtained using Equations (4.24) and (4.25), respectively.

TABLE 4.11
Values for F (t_i)

i	t_i	$(t_i - \mu)/\sigma$	F (t_i)
1	20	−1.70	0.04
2	30	−1.35	0.08
3	40	−1.00	0.15
4	55	−0.48	0.31
5	60	−0.30	0.38
6	70	0.04	0.51
7	75	0.21	0.58
8	80	0.39	0.65
9	90	0.74	0.77
10	95	0.92	0.82
11	100	1.09	0.86
12	110	1.44	0.92

TABLE 4.12
Computed Values for $\hat{F}$ (t_i) and $\hat{F}$ (t_{i-1})

i	t_i	$\hat{F}$ (t_i)	$\hat{F}$ (t_{i-1})
1	20	1/12 = 0.08	(1 − 1)/12 = 0
2	30	2/12 = 0.16	(2 − 1)/12 = 0.08
3	40	3/12 = 0.25	(3 − 1)/12 = 0.16
4	55	= 0.33	= 0.25
5	60	= 0.41	= 0.33
6	70	= 0.5	= 0.41
7	75	= 0.58	= 0.5
8	80	= 0.66	= 0.58
9	90	= 0.75	= 0.66
10	95	= 0.83	= 0.75
11	100	= 0.91	= 0.83
12	110	12/12 = 1	(12 − 1)/12 = 0.91

Using the values presented in Tables 4.11 and 4.12 for F(t_i), $\hat{F}$ (t_i), and $\hat{F}$ (t_{i-1}), the absolute values of $|F(t_i) - \hat{F}(t_i)|$ and $|F(t_i) - \hat{F}(t_{i-1})|$ listed in Table 4.13 were obtained. By examining Table 4.13 we obtained the maximum absolute difference of 0.11 (i.e., d = 0.11 for Equation (4.26)).

For given data (i.e., $\theta = 0.05$ and n = 12) and using Table 4.10, we get $d_{0.05}$ = 0.38. Thus, we conclude that as $d_{0.05} > d$ (i.e., 0.38 > 0.11), there is no reason to reject the normal distribution assumption.

TABLE 4.13
Absolute Calculated Values for $|F(t_i) - \hat{F}(t_i)|$ and $|F(t_i) - \hat{F}(t_{i-1})|$

i	t_i	$\lvert F(t_i) - \hat{F}(t_i)\rvert$	$\lvert F(t_i) - \hat{F}(t_{i-1})\rvert$
1	20	0.04	0.04
2	30	0.08	0
3	40	0.1	0.01
4	55	0.02	0.06
5	60	0.03	0.05
6	70	0.01	0.1
7	75	0	0.08
8	80	0.01	0.07
9	90	0.02	0.11 max.
10	95	0.01	0.07
11	100	0.05	0.03
12	110	0.08	0.01

4.9 MAXIMUM LIKELIHOOD ESTIMATION METHOD

Once the techniques such as those described in the preceding section identify the failure distribution of a given set of data or the failure distribution of the data under consideration is known through other means, the maximum likelihood estimation method can be used to estimate the values of the distribution parameters. The maximum likelihood estimation method is the most flexible and powerful of modern estimation approaches and is described in detail in References 20 and 39 through 41.

The basis for this approach is to assume a random sample of n failure times, say $t_1, t_2, t_3, \ldots, t_n$ taken from a population with a probability density function, $f(t; \theta)$; where t is time and θ is the distribution parameter to be estimated. The joint density function of these n variables is called the likelihood function, L, and is defined by

$$L = f(t_1;\theta)\, f(t_2;\theta)\, f(t_3;\theta) \ldots f(t_n;\theta) \tag{4.27}$$

The value of θ that maximizes the ln L or the L is known as the maximum likelihood estimator (MLE) of θ. Usually, the θ is estimated by solving the following expression:

$$\frac{\partial \ln L}{\partial \theta} = 0 \tag{4.28}$$

One of the important properties of the MLE is as follows:

For the large size n, the variance is simply given by

$$\text{var}\,\theta = -\left[\frac{\partial^2 \ln L}{\partial \hat{\theta}^2}\right]^{-1} \tag{4.29}$$

where

$\hat{\theta}$ denotes the estimated value of the θ.

The application of this method to selected statistical distributions is presented below.

4.9.1 Exponential Distribution

The probability density function is defined by

$$f(t;\theta) = \theta e^{-\theta t} \qquad t \geq 0 \tag{4.30}$$

where

t is time.
θ is the distribution parameter.

Substituting Equation (4.30) into Equation (4.27) yields

$$L = \theta^n \, e^{-\sum_{i=1}^{n} \theta t_i} \tag{4.31}$$

By taking natural logarithms of Equation (4.31), we get

$$\ln L = n \ln \theta - \sum_{i=1}^{n} \theta \, t_i \tag{4.32}$$

Substituting Equation (4.32) into Equation (4.28) yields

$$\frac{\partial \ln L}{\partial \theta} = \frac{n}{\theta} - \sum_{i=1}^{n} t_i = 0 \tag{4.33}$$

Rearranging Equation (4.33) leads to

$$\hat{\theta} = n \Big/ \sum_{i=1}^{n} t_i \tag{4.34}$$

where

$\hat{\theta}$ is the estimated value of θ.

By differentiating Equation (4.32) with respect to θ twice, we get

$$\frac{\partial^2 \ln L}{\partial \theta^2} = -n/\theta^2 \tag{4.35}$$

For a large sample size n, using Equation (4.35) in Equation (4.29) yields [40]

$$\begin{aligned} \text{Var } \hat{\theta} &= \theta^2/n \\ &\simeq \hat{\theta}^2/n \end{aligned} \tag{4.36}$$

4.9.2 Normal Distribution

The probability density is defined by

$$f(t;\mu,\sigma) = \frac{1}{\sigma\sqrt{2\pi}} e^{-\left[(t-\mu)^2/2\sigma^2\right]} \quad \text{for } -\infty < t < \infty \tag{4.37}$$

where

μ and σ are the distribution parameters.
t is time.

By inserting Equation (4.37) into relationship (4.27) and then taking logarithms, we get

$$\log L = -\frac{n}{2}\ln\sqrt{2\pi} - \frac{n}{2}\ln\sigma^2 - \frac{1}{2\sigma^2}\sum_{i=1}^{n}(t_i-\mu)^2 \tag{4.38}$$

Differentiating Equation (4.28) with respect to μ and equating the resulting expression to zero yields

$$-\sum_{i=1}^{n}\frac{2(t_i-\mu)(-1)}{2\sigma^2} = \sum_{i=1}^{n}\frac{t_i-\mu}{\sigma} = \frac{1}{\sigma}\left(\sum_{i=1}^{n}t_i - n\mu\right) = 0 \tag{4.39}$$

Rearranging Equation (4.39) yields

$$\hat{\mu} = \left(\sum_{i=1}^{n} t_i\right) \Big/ n \tag{4.40}$$

where

$\hat{\mu}$ is the estimated value of μ.

Similarly, differentiating Equation (4.38) with respect to σ^2 and setting the resulting expression equal to zero, we have

$$\frac{1}{\sigma^2} \sum_{i=1}^{n} (t_i - \mu)^2 - n = 0 \tag{4.41}$$

By rearranging Equation (4.41), we get

$$\hat{\sigma}^2 = \frac{1}{n} \sum_{i=1}^{n} (t_i - \hat{\mu})^2 \tag{4.42}$$

where

$\hat{\sigma}^2$ is a biased estimator for σ^2.

4.9.3 Weibull Distribution

The probability density function is defined by

$$f(t; \lambda, b) = \lambda b t^{b-1} e^{-\lambda t^b} \quad \text{for } t \geq 0 \tag{4.43}$$

where

λ is the scale parameter.
b is the shape parameter.
t is time.

By inserting Equation (4.43) into Equation (4.27) and then taking natural logarithms, we get

$$\ln L = n \ln \lambda + n \ln b - \lambda \sum_{i=1}^{n} t_i^b + (b-1) \sum_{i=1}^{n} \ln t_i \tag{4.44}$$

Now we set $\partial \ln L/\partial \lambda = 0$, $\partial \ln L/\partial b = 0$ to get

$$\frac{n}{\lambda} - \sum_{i=1}^{n} t_i^b = 0 \tag{4.45}$$

and

$$\sum_{i=1}^{n} \ln t_i - \lambda \sum_{i=1}^{n} t_i^b \ln t_i + \frac{n}{b} = 0 \tag{4.46}$$

By rearranging Equations (4.45) and (4.46), we get

$$\hat{\lambda} = n \Big/ \sum_{i=1}^{n} t_i^{\hat{b}} \tag{4.47}$$

and

$$\hat{b} = \frac{n}{\hat{\lambda} \sum_{i=1}^{n} t_i^{\hat{b}} \ln t_i - \sum_{i=1}^{n} \ln t_i} \tag{4.48}$$

where

$\hat{\lambda}$ is the estimated value of λ.
$\hat{b}$ is the estimated value of b.

Equations (4.47) and (4.48) can be solved by using an iterative process [39].

4.10 PROBLEMS

1. List important uses of failure data.
2. Discuss the sources of collecting failure related data during an equipment's life cycle.
3. Discuss the typical information to be included on a failure data collection form.
4. What are the important guidelines to be followed during the design of a failure reporting and documentation system?
5. Discuss the following U.S. bodies with respect to obtaining failure data:
    - Reliability Analysis Center (RAC)
    - National Technical Information Service (NTIS)
    - Government Industry Data Exchange Program (GIDEP)

TABLE 4.14
Failure and Censoring Times for 20 Identical Electronic Devices

Device No.	Failure and censoring times in hours
1	200
2	600
3	1000[a]
4	1500
5	2000
6	400
7	600[a]
8	2500
9	3000
10	2600
11	1300[a]
12	1400
13	900
14	800[a]
15	500
16	600
17	1800[a]
18	1000
19	700
20	1600

[a] Censoring time.

6. Write an essay on MIL-HDBK-217.
7. Assume that 20 identical electronic devices were put on test at time $t = 0$ and Table 4.14 presents failure and censoring times of these devices. Determine the statistical distribution fit to the given data and estimate values for its parameters by using the hazard plotting method.
8. A sample of 20 identical electrical parts were tested and their times to failure, in hours, are given in Table 4.15. Determine if these times to failure follow exponential distribution by employing the Bartlett test.

TABLE 4.15
Failure Times (hours) of Electrical Parts

10	25	70	115
12	37	75	140
15	33	95	145
13	55	98	180
20	60	110	200

TABLE 4.16
Parts' Failure Times

Failure No.	Failure time (h)
1	40
2	100
3	25
4	30
5	110
6	140
7	80
8	150

9. A set of 35 identical parts were tested for 300 h and 8 failed. The failed parts were not replaced and their times to failure are given in Table 4.16. Determine if these failure times can be represented by the exponential distribution.
10. Assume that a sample of 10 identical electrical parts were life tested and they failed after 50, 60, 70, 90, 100, 115, 120, 125, 130, and 140 h. After reviewing the failure pattern of similar parts, it was assumed that the times to failure are normally distributed. Use the Kolmogorov–Smirnov method to test this assumption for 0.1 level of significance.
11. The probability density function of the gamma distribution is defined by

$$f(t; \lambda, \theta) = \frac{\lambda^{\theta}\, t^{\theta-1}\, e^{-\lambda t}}{\Gamma(\theta)}, \quad \text{for } t > 0 \tag{4.49}$$

where

t is time.
λ is the scale parameter.
θ is the shape parameter.
$\Gamma(\theta)$ is the gamma function. For positive integer θ, $\Gamma(\theta) = (\theta - 1)!$.

Develop expressions for estimating distribution parameter values by employing the maximum likelihood estimation method.

4.11 REFERENCES

1. Dhillon, B.S. and Viswanath, H.C., Bibliography of literature on failure data, *Microelectronics and Reliability,* 30, 723-750, 1990.
2. Grant Ireson, W., Coombs, C.F., and Moss, R.Y., *Handbook of Reliability Engineering and Management,* McGraw-Hill, New York, 1996.
3. Kletz, T., The uses, availability and pitfalls of data on reliability, *Process Technology,* 18, 111-113, 1973.

4. Mitchell, R.L. and Rutter, R.R., A study of automotive reliability and associated cost of maintenance in the U.S.A., *Soc. Automotive Eng.* (SAE) Paper No. 780277, Jan. 1978.
5. Dhillon, B.S., *Mechanical Reliability: Theory, Models, and Applications,* American Institute of Aeronautics and Astronautics, Washington, D.C., 1988.
6. Hahn, R.F., Data collection techniques, *Proc. Annu. Reliability Maintainability Symp.*, 38-43, 1972.
7. Parascos, E.T., A new approach to the establishment and maintenance of equipment failure rate data bases, in *Failure Prevention and Reliability,* Bennett, S.B., Ross, A.L., and Zemanick, P.Z., Eds., American Society of Mechanical Engineers, New York, 1977, 263-268.
8. A Reliability Guide to Failure Reporting, Analysis, and Corrective Action Systems, Committee on Reliability Reporting, American Society for Quality Control (ASQC), Milwaukee, 1977.
9. Dhillon, B.S., *Advanced Design Concepts for Engineers,* Technomic Publishing Company, Lancaster, PA, 1998.
10. MIL-STD-1556, Government/Industry Data Exchange Program (GIDEP), Department of Defense, Washington, D.C.
11. MIL-HDBK-217F (Notice 2), Reliability Prediction of Electronic Equipment, Department of Defense, Washington, D.C., 1995.
12. TD-84-3, Reliability and Maintainability Data for Industrial Plants, A.P. Harris and Associates, Ottawa, Canada, 1984.
13. Schafer, R.E., Angus, J.E., Finkelstein, J.M., Yerasi, M., and Fulton, D.W., RADC Nonelectronic Reliability Notebook, Reliability Analysis Center, Rome Air Development Center, Griffiss Air Force Base, Rome, New York, 1985. Report No. RADC-TR-85-194.
14. Rossi, M.J., Nonelectronic Parts Reliability Data, Reliability Analysis Center, Rome Air Development Center, Griffiss Air Force Base, Rome, New York, 1985. Report No. NPRD-3.
15. Green, A.E. and Bourne, A.J., *Reliability Technology,* John Wiley & Sons, Chichester, England, 1972.
16. Dhillon, B.S., *Human Reliability: With Human Factors,* Pergamon Press, New York, 1986.
17. Joos, D.W., Sabri, Z.A., and Husseiny, A.A., Analysis of gross error rates in operation of commercial nuclear power stations, *Nuclear Eng. Design,* 52, 265-300, 1979.
18. Peters, G.A., Human error: analysis and control, *Am. Soc. Safety Eng. J.*, XI, 9-15, 1966.
19. Recht, J.L., Systems safety analysis: error rates and costs, *National Safety News,* February 1966, pp. 20-23.
20. Dhillon, B.S., *Reliability Engineering in Systems Design and Operation,* Van Nostrand Reinhold Company, New York, 1983.
21. Nelson, W., Hazard plot analysis of incomplete failure data, *Proc. Annu. Symp. Reliability,* 391-403, 1969.
22. Nelson, W., Theory and applications of hazard plotting for censored failure data, *Technometrics,* 14, 945-966, 1972.
23. Nelson, W., Hazard plotting for incomplete failure data, *J. Quality Technol.*, 1, 27-52, 1969.
24. Nelson, W., Life data analysis by hazard plotting, *Evaluation Eng.*, 9, 37-40, 1970.
25. Nelson, W., *Applied Life Data Analysis,* John Wiley & Sons, New York, 1982.
26. Dhillon, B.S., A hazard rate model, *IEEE Trans. Reliability,* 28, 150, 1979.

27. Dhillon, B.S., Life distributions, *IEEE Trans. Reliability,* 30, 457-460, 1981.
28. Epstein, B., Tests for the validity of the assumption that the underlying distribution of life is exponential, *Technometrics,* 2, 83-101, 1960.
29. Epstein, B., Tests for the validity of the assumption that the underlying distribution of life is exponential, *Technometrics,* 2, 327-335, 1960.
30. Lamberson, L.R., An evaluation and comparison of some tests for the validity of the assumption that the underlying distribution of life is exponential, *AIIE Trans.,* 12, 327-335, 1974.
31. Massey, F., The Kolmogorov-Smirnov test for goodness of fit, *J. Am. Stat. Assoc.,* 46, 70, 1951.
32. Dhillon, B.S., *Quality Control, Reliability, and Engineering Design,* Marcel Dekker, New York, 1985.
33. AMC Pamphlet 706-198, Development Guide for Reliability, Part IV, U.S. Army Materiel Command, Department of Defense, Washington, D.C., 1976.
34. Reliability and Maintainability Handbook for the U.S. Weather Bureau, Publication No. 530- 01-1-762, ARINC Research Corporation, Annapolis, MD, April 1967.
35. Ehrenfeld, S. and Mauer, S.B.L., *Introduction to Statistical Method,* McGraw-Hill, New York, 1964.
36. Klerer, M. and Korn, G.A., *Digital Computer User's Handbook,* McGraw-Hill, New York, 1967.
37. Conover, W.J., *Practical Nonparametric Statistics,* John Wiley & Sons, New York, 1971.
38. AMCP 702-3, Quality Assurance Handbook, U.S. Army Material Command, Washington, D.C., 1968.
39. Lloyd, M. and Lipow, M., *Reliability: Management, Methods, and Mathematics,* Prentice-Hall, Englewood Cliffs, NJ, 1962.
40. Shooman, M.L., *Probablistic Reliability: An Engineering Approach,* McGraw-Hill, New York, 1968.
41. Mann, N., Schafer, R.E., and Singpurwalla, N.D., *Methods for Statistical Analysis of Reliability and Life Data,* John Wiley & Sons, New York, 1974.

5 Basic Reliability Evaluation and Allocation Techniques

5.1 INTRODUCTION

As engineering systems become more complex and sophisticated, the importance of reliability evaluation and allocation techniques during their design is increasing. Usually, in the design of engineering systems, reliability requirements are specified. These requirements could be in the form of system reliability, failure rate, mean time between failures (MTBF), and availability. Normally, in order to determine the fulfillment of such requirements, various reliability evaluation and allocation techniques are employed.

Over the years, many reliability evaluation techniques have been developed but their effectiveness and advantages may vary quite considerably. Some of these methods are known as block diagram, decomposition, delta-star, and Markov modeling [1, 2].

Reliability allocation may be described as the top-down process used to subdivide a system reliability requirement or goal into subsystem and component requirements or goals. Its basic objectives are to translate the system reliability requirement into more manageable, lower level (subsystem and component) requirements and to establish an individual reliability requirement for each subsystem/hardware designer or supplier. There are many methods and techniques available to perform reliability allocation [3].

This chapter not only describes basic reliability evaluation and allocation techniques, but also associated areas because a clear understanding of these areas is considered essential prior to learning the evaluation and allocation methods.

5.2 BATHTUB HAZARD RATE CURVE

The bathtub hazard rate curve shown in Figure 5.1 is used to describe failure rate for many engineering components. This curve is the result of three types of failures: (1) quality, (2) stress-related, and (3) wear out. The quality failures decrease with time, the wear out increase with time, and the stress-related remain constant with respect to time [4]. Nonetheless, the bathtub hazard rate curve may be divided into three distinct parts: (1) burn-in period, (2) useful life period, and (3) wear out period. During the burn-in period, the hazard rate decreases and some of the reasons for the occurrence of failures during this region are poor quality control, inadequate manufacturing methods, poor processes, human error, substandard materials and workmanship, and inadequate debugging. Other terms used for this decreasing hazard rate region are "infant mortality region", "break-in region", and "debugging region".

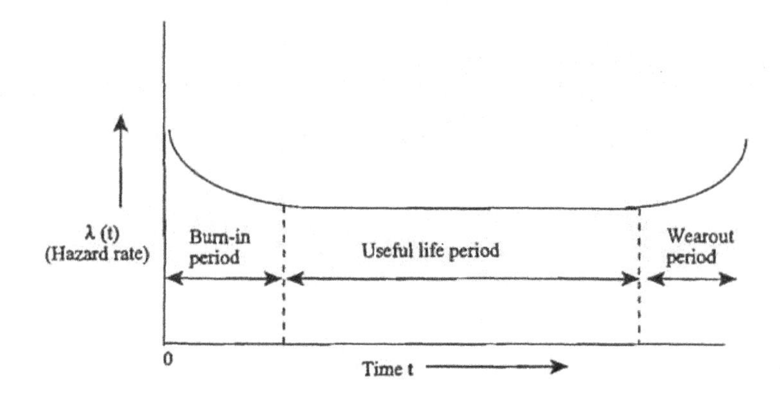

FIGURE 5.1 Bathtub hazard rate curve.

During the useful life period, the hazard rate remains constant and there are various reasons for the occurrence of failures in this region: undetectable defects, low safety factors, higher random stress than expected, abuse, human errors, natural failures, explainable causes, etc.

During the wear out period, the hazard rate increases and the causes for the "wear out region" failures include wear due to aging, corrosion and creep, short designed-in life of the item under consideration, poor maintenance, wear due to friction, and incorrect overhaul practices.

The following hazard rate function can be used to represent the entire bathtub hazard rate curve [5]:

$$\lambda(t) = \gamma \lambda b t^{b-1} + (1-\gamma) c t^{c-1} \theta e^{\theta t^c} \tag{5.1}$$

for

$$c, b, \lambda, \theta > 0$$

$$0 \leq \gamma \leq 1$$

$$b = 0.5$$

$$c = 1$$

$$t \geq 0$$

where

t is time.
$\lambda(t)$ is the hazard rate.
b and c are the shape parameters.
λ and θ are the scale parameters.

5.3 GENERAL RELIABILITY ANALYSIS RELATED FORMULAS

There are a number of formulas often used in conducting reliability analysis. This section presents four of these formulas based on the reliability function.

5.3.1 Failure Density Function

This is defined by

$$\frac{dR(t)}{dt} = -f(t) \tag{5.2}$$

where

R(t) is the item reliability at time t.
f(t) is the failure (or probability) density function.

5.3.2 Hazard Rate Function

This is expressed by

$$\lambda(t) = f(t)/R(t) \tag{5.3}$$

where

λ (t) is the item hazard rate or time dependent failure rate.

Substituting Equation (5.2) into Equation (5.3) yields

$$\lambda(t) = -\frac{1}{R(t)} \cdot \frac{dR(t)}{dt} \tag{5.4}$$

5.3.3 General Reliability Function

This can be obtained by using Equation (5.4). Thus, we have

$$\frac{1}{R(t)} \cdot dR(t) = -\lambda(t)dt \tag{5.5}$$

Integrating both sides of Equation (5.5) over the time interval [o, t], we get

$$\int_1^{R(t)} \frac{1}{R(t)} dR(t) = -\int_o^t \lambda(t)dt \tag{5.6}$$

since at t = 0, R (t) = 1.

Evaluating the left-hand side of Equation (5.6) yields

$$\ln R(t) = -\int_0^t \lambda(t)\,dt \tag{5.7}$$

Thus, from Equation (5.7) we get

$$R(t) = e^{-\int_0^t \lambda(t)\,dt} \tag{5.8}$$

The above equation is the general expression for the reliability function. Thus, it can be used to obtain reliability of an item when its times to failure follow any known statistical distribution, for example, exponential, Rayleigh, Weibull, and gamma.

5.3.4 Mean Time to Failure

This can be obtained by using any of the following three formulas [6]:

$$\text{MTTF} = E(t) = \int_0^\infty t f(t)\,dt \tag{5.9}$$

or

$$\text{MTTF} = \int_0^\infty R(t)\,dt \tag{5.10}$$

or

$$\text{MTTF} = \lim_{s \to 0} R(s) \tag{5.11}$$

where

MTTF is the item mean time to failure.
$E(t)$ is the expected value.
s is the Laplace transform variable.
$R(s)$ is the Laplace transform for the reliability function, $R(t)$.

Example 5.1

Assume that the failure rate of a microprocessor, λ, is constant. Obtain expressions for the microprocessor reliability, mean time to failure, and using the reliability function prove that the microprocessor failure rate is constant.

Thus, substituting λ for $\lambda(t)$ into Equation (5.8) yields

$$R(t) = e^{-\int_0^t \lambda dt} = e^{-\lambda t} \quad (5.12)$$

By inserting Equation (5.12) into Equation (5.10), we get

$$\text{MTTF} = \int_0^\infty e^{-\lambda t} dt = \frac{1}{\lambda} \quad (5.13)$$

Using Equation (5.12) in Equation (5.4) leads to

$$\lambda(t) = -\frac{1}{e^{-\lambda t}} \cdot (-\lambda) e^{-\lambda t} = \lambda \quad (5.14)$$

Thus, Equations (5.12) and (5.13) represent expressions for microprocessor reliability and mean time to failure, respectively, and Equation (5.14) proves that the microprocessor failure rate is constant.

Example 5.2

Using Equation (5.12), prove that the result obtained by utilizing Equation (5.11) is the same as the one given by Equation (5.13).

Thus, the Laplace transform of Equation (5.12) is

$$R(s) = 1/(s + \lambda) \quad (5.15)$$

Using Equation (5.15) in Equation (5.11) yields

$$\text{MTTF} = \lim_{s \to 0} 1/(s + \lambda) = \frac{1}{\lambda} \quad (5.16)$$

Equations (5.13) and (5.16) results are identical; thus, it proves that Equations (5.10) and (5.11) give the same result.

Example 5.3

Assume that the failure rate of an automobile is 0.0004 failures/h. Calculate the automobile reliability for a 15-h mission and mean time to failure.

Using the given data in Equation (5.8) yields

$$R(15) = e^{-(0.0004)(15)}$$
$$= 0.994$$

Similarly, inserting the specified data for the automobile failure rate into Equation (5.10), we get

$$MTTF = \int_0^{\infty} e^{-(0.0004)t}\, dt$$
$$= \frac{1}{(0.0004)}$$
$$= 2{,}500 \text{ h}$$

Thus, the reliability and mean time to failure of the automobile are 0.994 and 2,500 h, respectively.

5.4 RELIABILITY NETWORKS

An engineering system can form various different configurations in conducting reliability analysis. This section presents such commonly occurring configurations.

5.4.1 Series Network

This is the simplest reliability network and its block diagram is shown in Figure 5.2. Each block in the figure represents a unit/component. If any one of the unit fails, the series system fails. In other words, all the units must work normally for the success of the series system.

If we let E_i denote the event that the ith unit is successful, then the reliability of the series system is expressed by

$$R_S = P(E_1\, E_2\, E_3 --- E_n) \tag{5.17}$$

where

R_S is the series system reliability.
$P(E_1\, E_2\, E_3 --- E_n)$ is the occurrence probability of events E_1, E_2, E_3, - - -, and E_n.

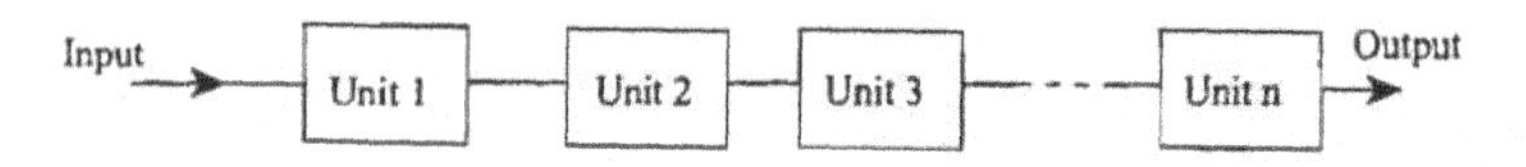

FIGURE 5.2 A series system with n units.

For independent units, Equation (5.17) becomes

$$R_S = P(E_1)(E_2)(E_3)---P(E_n) \tag{5.18}$$

where

$P(E_i)$ is the probability of occurrence of event E_i; for i = 1, 2, 3, ..., n.

If we let $R_i = P(E_i)$ for i = 1, 2, 3, ..., n in Equation (5.18) becomes

$$R_S = R_1 R_2 R_3 \ldots R_n = \prod_{i=1}^{n} R_i \tag{5.19}$$

where

R_i is the unit reliability; for i = 1, 2, 3, ..., n.

Since normally the value of R_i is between zero and one, the series system reliability decreases with the increasing value of n.

For constant failure rate, λ_i, of unit i, the reliability of the unit i is given by

$$R_i = e^{-\lambda_i t} \tag{5.20}$$

where

$R_i(t)$ is the reliability of unit i at time t.

Thus, inserting Equation (5.20) into Equation (5.19) yields

$$R_S(t) = e^{-\sum_{i=1}^{n} \lambda_i t} \tag{5.21}$$

where

$R_s(t)$ is the series system reliability at time t.

By substituting the above equation into Equation (5.10), we get the following expression for the series system mean time to failure:

$$MTTF_S = \int_0^{\infty} e^{-\sum_{i=1}^{n} \lambda_i t} dt = 1 \Big/ \sum_{i=1}^{n} \lambda_i \tag{5.22}$$

Using Equation (5.21) in Equation (5.4) yields the following expression for the series system hazard rate:

$$\lambda_S(t) = -\frac{1}{e^{-\sum_{i=1}^{n}\lambda_i t}}\left(-\sum_{i=1}^{n}\lambda_i\right)e^{-\sum_{i=1}^{n}\lambda_i t} = \sum_{i=1}^{n}\lambda_i \tag{5.23}$$

As the right-hand side of Equation (5.23) is independent of time t, its left-hand side is simply λ_s, the failure rate of the series system. It means whenever we add up failure rates of items, we automatically assure they act in series, a worst case design scenario with respect to reliability.

Example 5.4

A car has four identical and independent tires and the failure rate of each tire is 0.0004 failures per hour. Obviously, when any one of the tires is punctured, the car cannot be driven. Calculate the following:

- Car reliability for a 15-h mission with respect to tires.
- Car failure rate with respect to tires.
- Car mean time to failure with respect to tires.

Inserting the given data into Equation (5.21) yields

$$R(15) = e^{-(0.0016)(15)} = 0.9763$$

Using the specified failure data in Equation (5.23), we get

$$\lambda_S = 4(0.0004) = 0.0016 \text{ failure/h}$$

Substituting the given data into Equation (5.22), we get

$$MTTF_S = 1/4\ (0.0004) = 625 \text{ h}$$

Thus, the car reliability, failure rate, and mean time to failure with respect to tires are 0.9763, 0.0016 failures/h, and 625 h, respectively.

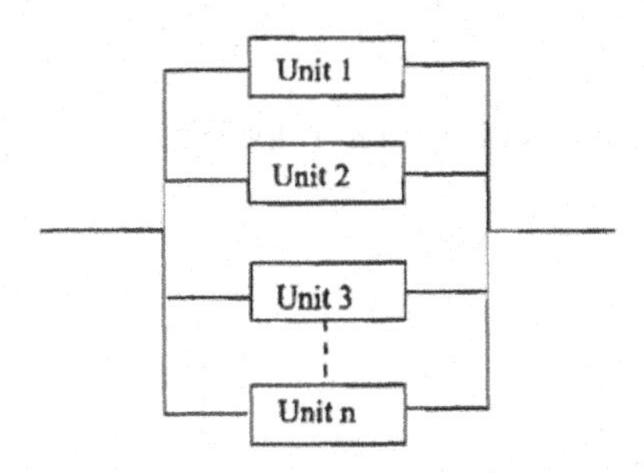

FIGURE 5.3 Block diagram of a parallel system with n units.

5.4.2 Parallel Network

In this case all the n units operate simultaneously and at least one such unit must work normally for the system success. The n-unit parallel system block diagram is shown in Figure 5.3. Each block in the figure denotes a unit/component.

If we let $\bar{E}_i$ denote the event that the ith unit is unsuccessful, then the failure probability of the parallel system is expressed by

$$F_p = P\left(\bar{E}_1 \, \bar{E}_2 \, \bar{E}_3 \ldots \bar{E}_n\right) \tag{5.24}$$

where

F_p is the parallel system failure probability.
$P\,(\bar{E}_1 \, \bar{E}_2 \, \bar{E}_3 \ldots \bar{E}_n)$ is the occurrence probability of failure events $\bar{E}_1$, $\bar{E}_2$, $\bar{E}_3$, …, $\bar{E}_n$.

For independent units, Equation (5.24) is written as

$$F_p = P\left(\bar{E}_1\right) P\left(\bar{E}_2\right) P\left(\bar{E}_3\right) \ldots P\left(\bar{E}_n\right) \tag{5.25}$$

where

$P\,(\bar{E}_1)$ is the probability of occurrence of failure event $\bar{E}_1$; for i = 1, 2, 3, …, n.
F_p is the failure probability of the parallel system.

If we let $F_i = P(\bar{E}_1)$ for i = 1, 2, 3, …, n in Equation (5.25) becomes

$$\begin{aligned} F_p &= F_1 \, F_2 \, F_3 \ldots F_n \\ &= \prod_{i=1}^{n} F_i \end{aligned} \tag{5.26}$$

where

F_i is the failure probability of unit i; for i = 1, 2, 3, ..., n.

Substracting Equation (5.26) from unity we get

$$R_p = 1 - F_p = 1 - \prod_{i=1}^{n} F_i \tag{5.27}$$

where

R_p is the parallel system reliability.

For constant failure rates of units, subtracting Equation (5.20) from unity and then substituting it into Equation (5.27) yields

$$R_p(t) = 1 - \prod_{i=1}^{n} \left(1 - e^{-\lambda_i t}\right) \tag{5.28}$$

where

R_p (t) is the parallel system reliability at time t.

For identical units, integrating Equation (5.28) over the time interval [0, ∞] yields the following formula for the parallel system mean time to failure:

$$\begin{aligned} MTTF_p &= \int_0^{\infty} \left[1 - \left(1 - e^{-\lambda t}\right)^n\right] dt \\ &= \frac{1}{\lambda} \sum_{i=1}^{n} 1/i \end{aligned} \tag{5.29}$$

where

$MTTF_p$ is the parallel system mean time to failure.
λ is the unit failure rate.

Example 5.5

An aircraft has two independent and identical engines. At least one of the engines must operate normally for the aircraft to fly successfully. Calculate the aircraft's reliability with respect to engines, if each engine's probability of success is 0.95.

Subtracting the engine probability of success from unity, we get the following engine failure probability:

$$F_e = 1 - 0.95$$
$$= 0.05$$

Inserting the above value and the other given data into Equation (5.27) yields

$$R_p = 1 - (0.05)^2$$
$$= 0.9975$$

Thus, the aircraft's reliability with respect to engines is 0.9975.

Example 5.6

A system is composed of two independent units in parallel. The failure rates of units A and B are 0.002 failures per hour and 0.004 failures per hour, respectively. Calculate the system reliability for a 50-h mission and mean time to failure.

Let λ_A be the failure rate of unit A and λ_B the failure rate of unit B. Thus, for n = 2, using Equation (5.28) we get

$$R_p(t) = e^{-\lambda_A t} + e^{-\lambda_B t} - e^{-(\lambda_A + \lambda_B)t} \quad (5.30)$$

By substituting the given data into Equation (5.30) we get

$$R_p(50) = e^{-(0.002)(50)} + e^{-(0.004)(50)} - e^{-(0.002+0.004)(50)}$$
$$= 0.9827$$

Integrating Equation (5.30) over the time interval [0, ∞], we get the following result for system mean time to failure:

$$MTTF = \int_0^{\infty} \left[e^{-\lambda_A t} + e^{-\lambda_B t} - e^{-(\lambda_A + \lambda_B)t}\right] dt$$

$$MTTF = \frac{1}{\lambda_A} + \frac{1}{\lambda_B} - \frac{1}{\lambda_A + \lambda_B}$$
$$= \frac{1}{(0.002)} + \frac{1}{(0.004)} - \frac{1}{(0.002 + 0.004)}$$
$$= 583.33 \text{ h}$$

Thus, the system reliability and mean time to failure are 0.9827 and 583.33 h, respectively.

5.4.3 r-out-of-n Network

This is another form of redundancy in which at least r units out of a total of n units must work normally for the system success. Furthermore, all the units in the system are active. The parallel and series networks are special cases of this network for $r = 1$ and $r = n$, respectively.

Using the Binomial distribution, for independent and identical units, the reliability of the r-out-of-n network is given by

$$R_{r/n} = \sum_{i=r}^{n} \binom{n}{i} R^i (1 - R)^{n-i} \tag{5.31}$$

where

$$\binom{n}{i} = \frac{n!}{(n-i)!\,i!} \tag{5.32}$$

$R_{r/n}$ is the r-out-of-n network reliability and R the unit reliability.

For constant failure rates of units, Equation (5.31) becomes

$$R_{r/n}(t) = \sum_{i=r}^{n} \binom{n}{i} e^{-i\lambda t} \left(1 - e^{-\lambda t}\right)^{n-i} \tag{5.33}$$

By inserting Equation (5.33) into Equation (5.10) we get

$$\begin{aligned} MTTF_{r/n} &= \int_0^{\infty} \left[\sum_{i=r}^{n} \binom{n}{r} e^{-i\lambda t} \left(1 - e^{-\lambda t}\right)^{n-i} \right] dt \\ &= \frac{1}{\lambda} \sum_{i=r}^{n} 1/i \end{aligned} \tag{5.34}$$

where

$MTTF_{r/n}$ is the mean time to failure of the r-out-of-n system.

Example 5.7

A computer system has three independent and identical units in parallel. At least two units must work normally for the system success. Calculate the computer system mean time to failure, if the unit failure rate is 0.0004 failures per hour.

Substituting the given data into Equation (5.34) yields

$$MTTF_{2/3} = \frac{1}{(0.0004)} \sum_{i=2}^{3} 1/i$$

$$= \frac{1}{(0.0004)} \left[\frac{1}{2} + \frac{1}{3}\right]$$

$$= 2083.33 \text{ h}$$

Thus, the computer system mean time to failure is 2083.33 h.

5.4.4 Standby Redundancy

This is another type of redundancy used to improve system reliability. In this case, one unit operates and m units are kept in their standby mode. The total system contains (m + 1) units and as soon as the operating unit fails, the switching mechanism detects the failure and turns on one of the standby units. The system fails when all the standbys fail. For independent and identical units, perfect switching and standby units, and unit time dependent failure rate, the standby system reliability is given by

$$R_{sd}(t) = \sum_{i=0}^{m} \left[\left[\int_0^t \lambda(t)\,dt\right]^i e^{-\int_0^t \lambda(t)\,dt} \right] \Big/ i! \tag{5.35}$$

where

$R_{sd}(t)$ is the standby system reliability.
m is the number of standby units.

For constant unit failure rate, Equation (5.35) becomes

$$R_{sd}(t) = \sum_{i=0}^{m} (\lambda t)^i e^{-\lambda t} / i! \tag{5.36}$$

By substituting Equation (5.36) into Equation (5.10) we get

$$\begin{aligned} MTTF_{sd} &= \int_0^{\infty} \left[\sum_{i=0}^{m} (\lambda t)^i e^{-\lambda t} / i! \right] dt \\ &= (m+1)/\lambda \end{aligned} \tag{5.37}$$

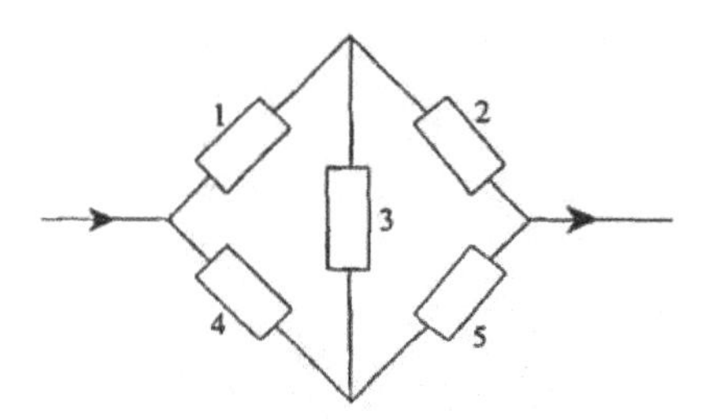

FIGURE 5.4 A five dissimilar unit bridge network.

where

$MTTF_{sd}$ is the standby system mean time to failure.

Example 5.8

Assume that a standby system has two independent and identical units: one operating, another on standby. The unit failure rate is 0.006 failures per hour. Calculate the system reliability for a 200-h mission and mean time to failure, if the switching mechanism never fails and the standby unit remains as good as new in its standby mode.

By substituting the specified data into Equation (5.36) yields

$$R_{sd}(200) = \sum_{i=0}^{1} [(0.006)(200)]^{i} e^{-(0.006)(200)} / i!$$

$$= 0.6626$$

Similarly, using the given data in Equation (5.37) we get

$$MTTF_{sd} = 2/(0.006)$$

$$= 333.33 \text{ h}$$

Thus, the standby system reliability and mean time to failure are 0.6626 and 333.333 h, respectively.

5.4.5 Bridge Network

In some engineering systems, units may be connected in a bridge configuration as shown in Figure 5.4. Each block in the figure denotes a unit.

For independent units, Figure 5.4 reliability is given by [7]

$$R_{br} = 2\prod_{i=1}^{5} R_i + \prod_{i=2}^{4} R_i + R_1 R_3 R_5 + R_1 R_4 + R_2 R_5$$
$$-\prod_{i=2}^{5} R_i - \prod_{i=1}^{4} R_i - R_5 \prod_{i=1}^{3} R_i - R_1 \prod_{i=3}^{5} R_i - R_1 R_2 R_4 R_5 \quad (5.38)$$

where

R_{br} is the bridge network reliability.
R_i is the ith unit reliability; for i = 1, 2, 3, ..., 5.

For identical units, Equation (5.38) simplifies to

$$R_{br} = 2\,R^5 - 5\,R^4 + 2\,R^3 + 2\,R^2 \quad (5.39)$$

For constant failure rates of units, from Equation (5.39) we write

$$R_{br}(t) = 2\,e^{-5\lambda t} - 5\,e^{-4\lambda t} + 2e^{-3\lambda t} + 2\,e^{-2\lambda t} \quad (5.40)$$

where

λ is the unit failure rate.

By integrating Equation (5.40) over the time interval [0, ∞], we get the following formula for the bridge network mean time to failure:

$$MTTF_{br} = 49/60\,\lambda \quad (5.41)$$

where

$MTTF_{br}$ is the bridge network mean time to failure.

Example 5.9

Assume that five independent and identical units form a bridge configuration. The failure rate of each unit is 0.0002 failures per hour. Calculate the configuration reliability for a 500-h mission.

Substituting the given values into Equation (5.40) yields

$$R_{br}(500) = 2\,e^{-5(0.0002)(500)} - 5\,e^{-4(0.0002)(500)} + 2\,e^{-3(0.0002)(500)} + 2\,e^{-2(0.0002)(500)}$$
$$= 0.9806$$

Thus, the bridge configuration reliability is 0.9806.

5.5 RELIABILITY EVALUATION METHODS

Over the years many reliability evaluation methods have been developed [2, 8, 9]. This section presents some of those techniques.

5.5.1 Network Reduction Approach

This is probably the simplest approach to determine the reliability of systems made up of independent series and parallel subsystems. However, the subsystems forming bridge configurations can also be handled by first applying the delta-star method [10]. Nonetheless, the network reduction approach sequentially reduces the series and parallel subsystems to equivalent hypothetical single units until the entire system under consideration itself becomes a single hypothetical unit. The following example demonstrates this method.

Example 5.10

An independent unit network representing an engineering system is shown in Figure 5.5 (i). The reliability R_i of unit i; for i = 1, 2, 3, …, 6 is specified. Calculate the network reliability by using the network reduction approach.

First we have identified subsystems A, B, C, and D of the network as shown in Figure 5.5 (i). The subsystems A and B have their units in series; thus, we reduce them to single hypothetical units as follows:

$$R_A = R_1 R_3 = (0.4)(0.6) = 0.24$$

and

$$R_B = R_2 R_4 = (0.5)(0.8) = 0.40$$

where

R_A is the reliability of subsystem A.
R_B is the reliability of subsystem B.

The reduced network is shown in Figure 5.5 (ii). Now, this network is composed of two parallel subsystems C and D. Thus, we reduce both subsystems to single hypothetical units as follows:

$$R_C = 1-(1-R_A)(1-R_B) = 1-(1-0.24)(1-0.4)$$
$$= 0.5440$$

and

$$R_D = 1-(1-R_5)(1-R_6) = 1-(1-0.7)(1-0.9)$$
$$= 0.97$$

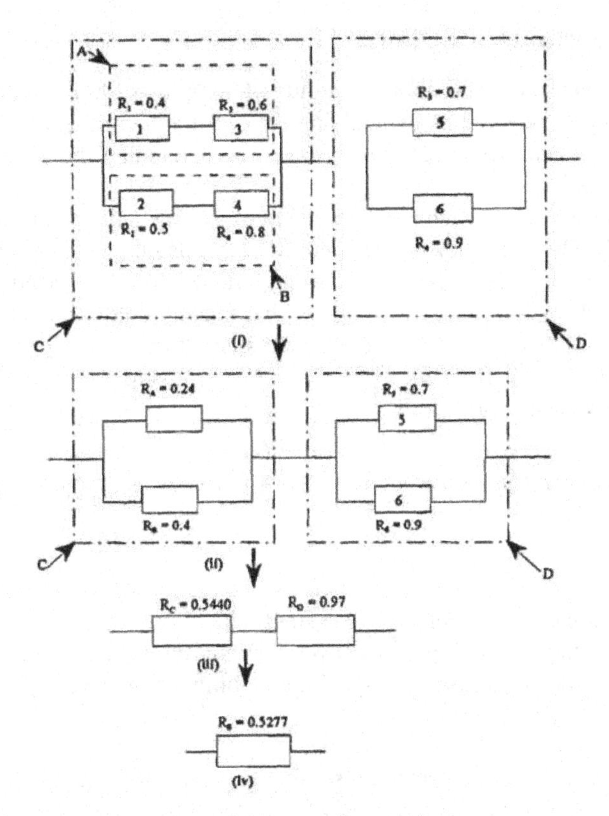

FIGURE 5.5 Diagrammatic steps of the network reduction approach: (i) Original network; (ii) reduced network I; (iii) reduced network II; (iv) single hypothetical unit.

where

R_C is the subsystem C reliability.
R_D is the subsystem D reliability.

Figure 5.5. (iii) depicts the reduced network with the above calculated values. This resulting network is a two unit series system and its reliability is given by

$$R_S = R_C R_D = (0.5440)(0.97)$$
$$= 0.5277$$

The single hypothetical unit shown in Figure 5.5 (iv) represents the reliability of the whole network given in Figure 5.5 (i). More specifically, the entire network is reduced to a single hypothetical unit. Thus, the total network reliability is 0.5277.

5.5.2 Decomposition Approach

This approach is used to determine reliability of complex systems, which it decomposes into simpler subsystems by applying the conditional probability theory. Subsequently, the system reliability is determined by combining the subsystems' reliability measures.

The basis for the approach is the selection of the key unit used to decompose a given network. The efficiency of the approach depends on the selection of this key unit. The past experience usually plays an important role in its selection.

The method starts with the assumption that the key unit, say k, is replaced by another unit that is 100% reliable or never fails and then it assumes the key unit k is completely removed from the network or system. Thus, the overall system or network reliability is expressed by

$$R_n = P(k)\, P(\text{system good/k good}) + P(\bar{k})\, P(\text{system good/k fails}) \tag{5.42}$$

where

R_n is the overall network or system reliability.
$P(k)$ is the probability of success or reliability of the key unit k.
$P(\bar{k})$ is the failure probability of unreliability of the key unit k.
$P(\cdot)$ is the probability.

This approach is demonstrated by solving the following example.

Example 5.11

An independent and identical units bridge network is shown in Figure 5.6. The letter R in the figure denotes unit reliability. Obtain an expression for the bridge network reliability by using the decomposition method.

With the aid of past experience, we choose the unit falling between nodes A and B, shown in Figure 5.6, as our **Key** unit, say k.

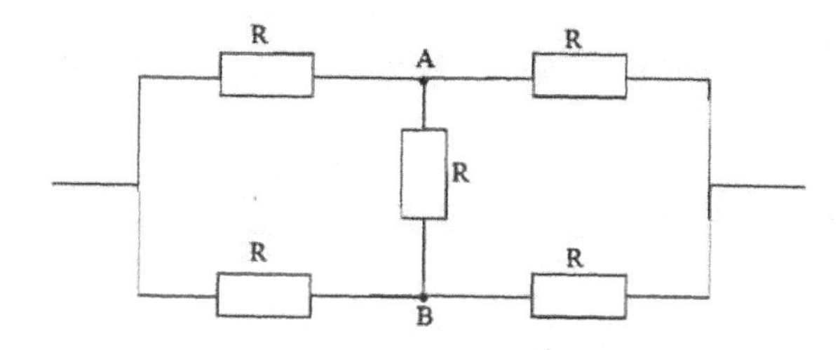

FIGURE 5.6 An identical unit bridge network.

Next, we replace this key unit k with a unit that never fails. Consequently, the Figure 5.6 network becomes a series-parallel system whose reliability is expressed by

$$R_{sp} = \left[1-(1-R)^2\right]^2 = \left[2\,R-R^2\right]^2 \tag{5.43}$$

Similarly, we totally remove the key unit k from Figure 5.6 and the resulting network becomes a parallel-series system. This parallel-series system reliability is given by

$$R_{ps} = 1-\left(1-R^2\right)^2 = 2\,R^2-R^4 \tag{5.44}$$

where

R_{ps} is the parallel-series system reliability.

The reliability and unreliability of the key unit k, respectively, are given by

$$P(k) = R \tag{5.45}$$

and

$$P(\bar{k}) = (1-R) \tag{5.46}$$

Rewriting Equation (5.42) in terms of our example, we get

$$R_n = R\,R_{sp} + (1-R)\,R_{ps} \tag{5.47}$$

Substituting Equations (5.43) and (5.44) into Equation (5.47) yields

$$R_n = R\left(2\,R-R^2\right)^2 + (1-R)\left(2\,R^2-R^4\right) = 2\,R^5-5\,R^4+2\,R^3+2\,R^2 \tag{5.48}$$

The above equation is for the reliability of the bridge network shown in Figure 5.6. Also, it is identical to Equation (5.39).

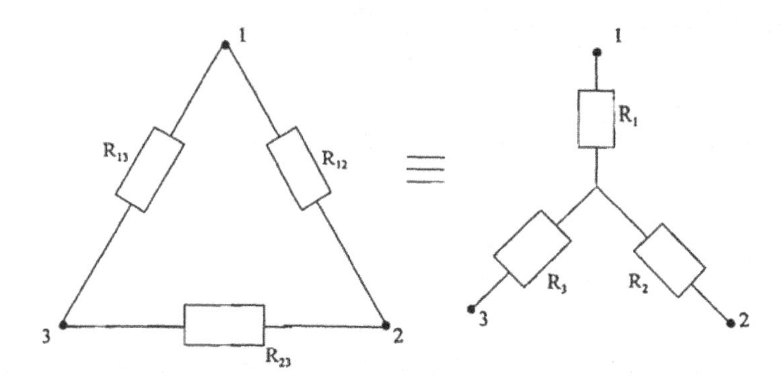

FIGURE 5.7 Delta to star equivalent reliability diagram.

5.5.3 Delta-Star Method

This is the simplest and very practical approach to evaluate reliability of bridge networks. This technique transforms a bridge network to its equivalent series and parallel form. However, the transformation process introduces a small error in the end result, but for practical purposes it should be neglected [2].

Once a bridge network is transformed to its equivalent parallel and series form, the network reduction approach can be applied to obtain network reliability. Nonetheless, the delta-star method can easily handle networks containing more than one bridge configurations. Furthermore, it can be applied to bridge networks composed of devices having two mutually exclusive failure modes [10, 11].

Figure 5.7 shows delta to star equivalent reliability diagram. The numbers 1, 2, and 3 denote nodes, the blocks the units, and $R_{(.)}$ the respective unit reliability.

In Figure 5.7, it is assumed that three units of a system with reliabilities R_{12}, R_{13}, and R_{23} form the delta configuration and its star equivalent configuration units' reliabilities are R_1, R_2, and R_3.

Using Equations (5.19) and (5.27) and Figure 5.7, we write down the following equivalent reliability equations for network reliability between nodes 1, 2; 2, 3; and 1, 3, respectively:

$$R_1 R_2 = 1-(1-R_{12})(1-R_{13} R_{23}) \tag{5.49}$$

$$R_2 R_3 = 1-(1-R_{23})(1-R_{12} R_{13}) \tag{5.50}$$

$$R_1 R_3 = 1-(1-R_{13})(1-R_{12} R_{23}) \tag{5.51}$$

Solving Equations (5.49) through (5.51), we get

$$R_1 = \sqrt{AC/B} \tag{5.52}$$

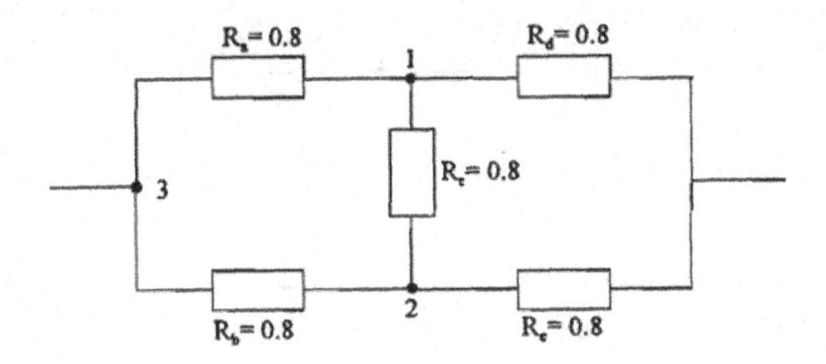

FIGURE 5.8 A five unit bridge network with specified unit reliabilities.

where

$$A = 1 - (1 - R_{12})(1 - R_{13} R_{23}) \tag{5.53}$$

$$B = 1 - (1 - R_{23})(1 - R_{12} R_{13}) \tag{5.54}$$

$$C = 1 - (1 - R_{13})(1 - R_{12} R_{23}) \tag{5.55}$$

$$R_2 = \sqrt{AB/C} \tag{5.56}$$

$$R_3 = \sqrt{BC/A} \tag{5.57}$$

Example 5.12

A five independent unit bridge network with specified unit reliability R_i; for i = a, b, c, d, and e is shown in Figure 5.8. Calculate the network reliability by using the delta-star method and also use the specified data in Equation (5.39) to obtain the bridge network reliability. Compare both results.

In Figure 5.8 nodes labeled 1, 2, and 3 denote delta configurations. Using Equations (5.52) through (5.57) and the given data, we get the following star equivalent reliabilities:

$$R_1 = \sqrt{AC/B}$$
$$= 0.9633$$

where

$$A = B = C = 1 - (1 - 0.8)[1 - (0.8)(0.8)] = 0.9280$$

$$R_2 = 0.9633$$

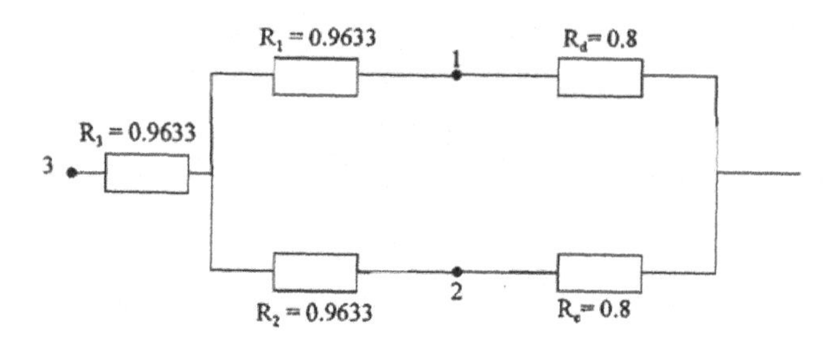

FIGURE 5.9 Equivalent network to bridge configuration of Figure 5.8.

and

$$R_3 = 0.9633$$

Using the above results, the equivalent network to Figure 5.8 bridge network is shown in Figure 5.9.

The reliability of Figure 5.9 network, R_{br}, is

$$R_{br} = R_3 \left[1-\left(1-R_1 R_d\right)\left(1-E_2 R_e\right)\right]$$
$$= 0.9126$$

By substituting the given data into Equation (5.39) we get

$$R_{br} = 2(0.8)^5 - 5(0.8)^4 + 2(0.8)^3 + 2(0.8)^2$$
$$= 0.91114$$

Both the reliability results are basically same, i.e., 0.9126 and 0.9114. All in all, for practical purposes the delta-star approach is quite effective.

5.5.4 Parts Count Method

This is a very practically inclined method used during bid proposal and early design phases to estimate equipment failure rate [12]. The information required to use this method includes generic part types and quantities, part quality levels, and equipment use environment. Under single use environment, the equipment failure rate can be estimated by using the following equation [12]:

$$\lambda_E = \sum_{i=1}^{m} Q_i \left(\lambda_g F_q\right)_i \tag{5.58}$$

where

λ_E is the equipment failure rate, expressed in failures/10^6 h.
m is the number of different generic part/component classifications in the equipment under consideration.
λ_g is the generic failure rate of generic part i expressed in failures/10^6 h.
Q_i is the quantity of generic part i.
F_q is the quality factor of generic part i.

Reference 12 presents tabulated values for λ_g and F_q.

Failure Rate Estimation of an Electronic Part

As the design matures, more information becomes available, the failure rates of equipment components are estimated. Usually, in the case of electronic parts, the MIL-HDBK-217 [12] is used to estimate the failure rate of electronic parts. The failure rates are added to obtain total equipment failure rate. This number provides a better picture of the actual failure rate of the equipment under consideration than the one obtained through using Equation (5.58).

An equation of the following form is used to estimate failure rates of many electronic parts [12]:

$$\lambda_p = \lambda_b \, \theta_e \, \theta_q \ldots \tag{5.59}$$

where

λ_p is the part failure rate.
λ_b is the base failure rate and is normally defined by a model relating the influence of temperature and electrical stresses on the part under consideration.
θ_e is the factor that accounts for the influence of environment.
θ_q is the factor that accounts for part quality level.

For many electronic parts, the base failure rate, λ_b, is calculated by using the following equation:

$$\lambda_b = C \exp\left[-E/kT\right] \tag{5.60}$$

where

C is a constant.
E is the activation energy for the process.
k is the Boltzmann's constant.
T is the absolute temperature.

5.5.5 Markov Method

This is a powerful reliability analysis tool and it can be used in more cases than any other method. The method is quite useful to model systems with dependent failure and repair modes. Markov method is widely used to model repairable systems with constant failure and repair rates. However, with the exception of a few special cases, the technique breaks down for a system having time dependent failure and repair rates. In addition, a problem may occur in solving a set of differential equations for large and complex systems. The following assumptions are associated with the Markov approach [8]:

- All occurrences are independent of each other.
- The probability of transition from one system state to another in the finite time interval Δt is given by $\lambda \Delta t$, where the λ is the transition rate (i.e., failure or repair rate) from one system state to another.
- The probability of more than one transition occurrence in time interval Δt from one state to another is very small or negligible (e.g., $(\lambda \Delta t)(\lambda \Delta t) \rightarrow 0$).

The Markov method is demonstrated by solving the following example.

Example 5.13

Assume that an engineering system can either be in an operating or a failed state. It fails at a constant failure rate, λ, and is repaired at a constant repair rate, μ. The system state space diagram is shown in Figure 5.10. The numerals in box and circle denote the system state. Obtain expressions for system time dependent and steady state availabilities and unavailabilities by using the Markov method.

Using the Markov method, we write down the following equations for state 0 and state 1, respectively, shown in Figure 5.10.

$$P_0(t+\Delta t) = P_0(t)(1-\lambda \Delta t) + P_1(t)\mu \Delta t \tag{5.61}$$

$$P_1(t+\Delta t) = P_1(t)(1-\mu \Delta t) + P_0(t)\lambda \Delta t \tag{5.62}$$

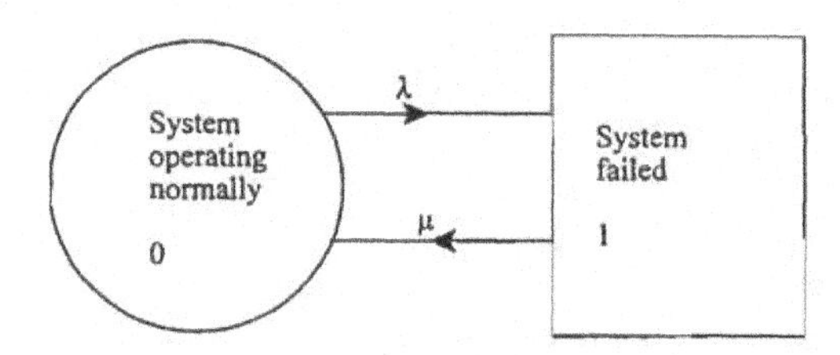

FIGURE 5.10 System transition diagram.

where

t	is time.
$\lambda\Delta t$	is the probability of system failure in finite time interval Δt.
$(1 - \lambda\Delta t)$	is the probability of no failure in finite time interval Δt.
$\mu\Delta t$	is the probability of system repair in finite time interval Δt.
$(1 - \mu\Delta t)$	is the probability of no repair in finite time interval Δt.
$P_0\,(t + \Delta t)$	is the probability of the system being in operating state 0 at time $(t + \Delta t)$.
$P_1\,(t + \Delta t)$	is the probability of the system being in failed state 1 at time $(t + \Delta t)$.
$P_i\,(t)$	is the probability that the system is in state i at time t, for $i = 0, 1$.

In the limiting case, Equations (5.61) and (5.62) become

$$\frac{dP_0(t)}{dt} + \lambda P_0(t) = P_1(t)\mu \tag{5.63}$$

$$\frac{dP_1(t)}{dt} + \mu P_1(t) = P_0(t)\lambda \tag{5.64}$$

At time $t = 0$, $P_0\,(0) = 1$ and $P_1\,(0) = 0$.

Solving Equations (5.63) and (5.64), we get

$$P_0(t) = \frac{\mu}{(\lambda+\mu)} + \frac{\lambda}{(\lambda+\mu)}\,e^{-(\lambda+\mu)t} \tag{5.65}$$

$$P_1(t) = \frac{\lambda}{(\lambda+\mu)} - \frac{\lambda}{(\lambda+\mu)}\,e^{-(\lambda+\mu)t} \tag{5.66}$$

Thus, the system time dependent availability and unavailability, respectively, are

$$A(t) = P_0(t) = \frac{\mu}{(\lambda+\mu)} + \frac{\lambda}{(\lambda+\mu)}\,e^{-(\lambda+\mu)t} \tag{5.67}$$

and

$$UA(t) = P_1(t) = \frac{\lambda}{(\lambda+\mu)} - \frac{\lambda}{(\lambda+\mu)}\,e^{-(\lambda+\mu)t} \tag{5.68}$$

where

A (t) is the system time dependent availability.
UA (t) is the system time dependent unavailability.

The system steady state availability and unavailability can be obtained by using any of the following three approaches:

- **Approach I:** Letting time t go to infinity in Equations (5.67) and (5.68), respectively.
- **Approach II:** Setting the derivatives of Equations (5.63) and (5.64) equal to zero and then discarding any one of the resulting two equations and replacing it with $P_0 + P_1 = 1$. The solutions to the ultimate equations will be system steady state availability (i.e., $A = P_0$) and unavailability (i.e., $UA = P_1$).
- **Approach III:** Taking Laplace transforms of Equations (5.63) and (5.64) and then solving them for $P_0(s)$, the Laplace transform of probability that the system is in operating state at time t, and $P_1(s)$, the Laplace transform of probability that the system is in failed state at time t. Multiplying $P_0(s)$ and $P_1(s)$ with the Laplace transform variable s and then letting s in $sP_0(s)$ and $sP_1(s)$ go to zero result in system steady state availability (i.e., $A = P_0$) and unavailability (i.e., $UA = P_1$), respectively.

Thus, in our case applying Approach I to Equations (5.67) and (5.68), we get

$$A = \lim_{t \to \infty} A(t) = \frac{\mu}{(\lambda + \mu)} \tag{5.69}$$

and

$$UA = \lim_{t \to \infty} UA(t) = \frac{\lambda}{(\lambda + \mu)} \tag{5.70}$$

where

A is the system steady state availability.
UA is the system steady state unavailability.

Since $\lambda = \frac{1}{MTTF}$ and $\mu = \frac{1}{MTTR}$, Equations (5.69) and (5.70) become

$$A = \frac{MTTF}{MTTF + MTTR} = \frac{\text{System uptime}}{\text{System uptime} + \text{System downtime}} \tag{5.71}$$

where

MTTF is the system mean time to failure.
MTTR is the system mean time to repair.

and

$$UA = \frac{MTTR}{MTTF + MTTR} = \frac{\text{System downtime}}{\text{System uptime} + \text{System downtime}} \quad (5.72)$$

Thus, the system time dependent and steady state availabilities are given by Equations (5.67) and (5.69) or (5.71), respectively. Similarly, the system time dependent and steady state unavailabilities are given by Equations (5.68) and (5.70) or (5.72), respectively.

5.6 RELIABILITY ALLOCATION

Reliability allocation may simply be described as the process of assigning reliability requirements to individual parts or components to achieve the specified system reliability. There are many different methods used to perform reliability allocation [1, 2, 13]. Nonetheless, the reliability allocation problem is not that simple and straightforward but quite complex. Some of the associated reasons are as follows:

- Role the component plays for the operation of the system.
- Component complexity.
- The chargeable component reliability with the type of function to be conducted.
- Approaches available for accomplishing the given allocation task.
- Lack of detailed information on many of the above factors in the early design phase.

Nonetheless, there are many benefits of the reliability allocation including it forces individuals involved in design and development to clearly understand and develop the relationships between reliabilities of components, subsystems, and systems, it forces the design engineer to seriously consider reliability equally with other design parameters such as performance, weight, and cost, and it ensures satisfactory design, manufacturing approaches, and test methods.

Two reliability allocation methods are described below.

5.6.1 Hybrid Method

This method is the result of combining two approaches: similar familiar systems and factors of influence. The resulting method incorporates benefits of the other two methods; thus, it becomes more attractive.

The basis for the similar familiar systems reliability allocation approach is the familiarity of the designer with similar systems or sub-systems. In addition, failure data collected on similar systems from various sources can also be used during the allocation process. The main drawback of this approach is to assume that reliability and life cycle cost of previous similar designs were adequate.

The factors of influence method is based upon the following factors that are considered to effect the system reliability:

- **Complexity/Time:** The complexity relates to the number of subsystem parts and the time to the relative operating time of the item during the entire system functional period.
- **Failure criticality.** This factor considers the criticality of the item failure on the system. For example, some auxiliary instrument failure in an aircraft may not be as crucial as the engine failure.
- **Environment.** This factor takes into consideration the susceptibility or exposure of items to environmental conditions such as temperature, humidity, and vibration.
- **State-of-the-Art.** This factor takes into consideration the advancement in the state-of-the-art for a specific item.

In using the above influence factors, each item is rated with respect to each of these influence factors by assigning a number from 1 to 10. The assignment of 1 means the item under consideration is least affected by the factor in question and 10 means the item is most affected by the same influence factor. Ultimately, the reliability is allocated by using the weight of these assigned numbers for all influence factors considered.

Now, it should be obvious to the reader that the hybrid method is better than similar familiar systems and factors of influence methods because it uses data from both of these approaches.

5.6.2 Failure Rate Allocation Method

This method is concerned with allocating failure rates to system components when the system required failure rate is known. The following assumptions are associated with this method:

- System components form a series configuration.
- System components fail independently.
- Time to component failure is exponentially distributed.

Thus, the system failure rate using Equation (5.23) is

$$\lambda_S = \sum_{i=1}^{n} \lambda_i \tag{5.73}$$

where

n is the number of components in the system.
λ_S is the system failure rate.
λ_i is the failure rate of system component i; for i = 1, 2, 3, …, n.

If the system required failure rate is λ_{sr}, then allocate component failure rate such that

$$\sum_{i=1}^{n} \lambda_i^* \leq \lambda_{sr} \tag{5.74}$$

where

λ_i^* is the failure rate allocated to component i; for i = 1, 2, 3, ..., n.

The following steps are associated with this method:

1. Estimate the component failure rates λ_i for i = 1, 2, 3, ..., n, using the past data.
2. Calculate the relative weight, θ_i, of component i using the preceding step failure rate data and the following equation:

$$\theta_i = \frac{\lambda_i}{\sum_{i=1}^{n} \lambda_i}, \quad \text{for } i = 1, 2, \ldots, n \tag{5.75}$$

It is to be noted that θ_i denotes the relative failure vulnerability of the component i and

$$\sum_{i=1}^{n} \theta_i = 1 \tag{5.76}$$

3. Allocate failure rate to component i using the following relationship:

$$\lambda_i^* = \theta_i \lambda_{sr}, \quad \text{for } i = 1, 2, \ldots, n \tag{5.77}$$

It must be remembered that Equation (5.77) is subject to the condition that the equality holds in Equation (5.74).

Example 5.14

Assume that a military system is composed of five independent subsystems in series and its specified failure rate is 0.0006 failures/h. The estimated failure rates from past experience for subsystems 1, 2, 3, 4, and 5 are $\lambda_1 = 0.0001$ failures/h, $\lambda_2 = 0.0002$ failures/h, $\lambda_3 = 0.0003$ failures/h, $\lambda_4 = 0.0004$ failures/h, and $\lambda_5 = 0.0005$ failures/h, respectively. Allocate the specified system failure rate to five subsystems.

Using Equation (5.73) and the given data, we get the following estimated military system failure rate:

$$\lambda_s = \sum_{i=1}^{5} \lambda_i$$

$$= (0.0001) + (0.0002) + (0.0003) + (0.0004) + (0.0005)$$

$$= 0.0015 \text{ failures/h}$$

Thus, utilizing Equation (5.75) and calculated and given values, we get the following relative weights for subsystems 1, 2, 3, 4, and 5, respectively:

$$\theta_1 = (0.0001/0.0015) = 0.0667$$

$$\theta_2 = (0.0002/0.0015) = 0.1333$$

$$\theta_3 = (0.0003/0.0015) = 0.2$$

$$\theta_4 = (0.0004/0.0015) = 0.2667$$

$$\theta_5 = (0.0005/0.0015) = 0.3333$$

Using Equation (5.77) and calculated and given values, the subsystems 1, 2, 3, 4, and 5 allocated failure rates, respectively, are as follows:

$$\lambda_1^* = \theta_1 \, \lambda_{sr} = (0.0667)\,(0.0006)$$

$$= 0.00004 \text{ failures/h}$$

$$\lambda_2^* = \theta_2 \, \lambda_{sr} = (0.1333)\,(0.0006)$$

$$= 0.00007 \text{ failures/h}$$

$$\lambda_3^* = \theta_3 \, \lambda_{sr} = (0.2)\,(0.0006)$$

$$= 0.00012 \text{ failures/h}$$

$$\lambda_4^* = \theta_4 \, \lambda_{sr} = (0.2667)\,(0.0006)$$

$$= 0.00016 \text{ failures/h}$$

$$\lambda_5^* = \theta_5 \, \lambda_{sr} = (0.333)\,(0.0006)$$

$$= 0.00019 \text{ failures/h}$$

5.7 PROBLEMS

1. Describe the bathtub hazard rate curve and the reasons for its useful life region failures.
2. Prove that the item mean time to failure (MTTF) is given by

$$MTTF = \lim_{s \to 0} R(s) \tag{5.78}$$

where s is the Laplace transform variable.
R(s) is the Laplace transform of the item reliability.

3. Prove that the mean time to failure of a system is given by

$$MTTF = \frac{1}{\lambda_1 + \lambda_2 + \ldots + \lambda_n} \tag{5.79}$$

where

n is the number of units.
λ_i is the unit failure rate; for i = 1, 2, 3, …, n.

Write down your assumptions.

4. A parallel system has n independent and identical units. Obtain an expression for the system hazard rate, if each unit's times to failure are exponentially distributed. Compare the end result with the one for the series system under the same condition.
5. Prove that when an item's times to failure are exponentially distributed, its failure rate is constant.
6. A system has four independent and identical units in parallel. Each unit's failure rate is 0.0005 failures/h. Calculate the system reliability, if at least two units must operate normally for the system success during a 100-h mission.
7. Compare the mean time to failure of k independent and identical unit series and standby systems when unit times to failure are exponentially distributed.
8. Calculate the reliability of the Figure 5.11 network using the delta-star approach. Assume that each block in the figure denotes a unit with reliability 0.8 and all units fail independently.

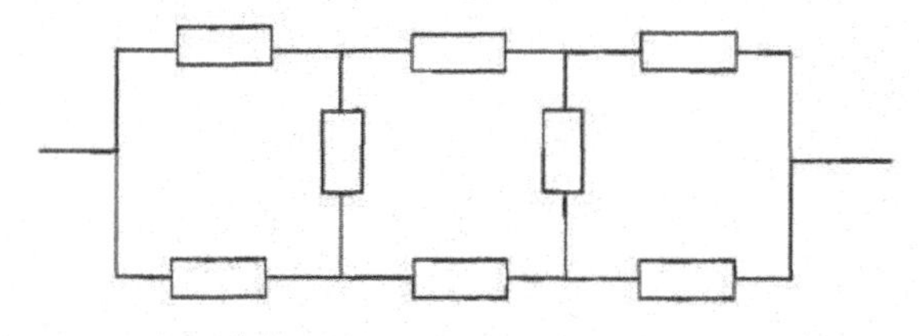

FIGURE 5.11 An eight unit network.

9. For the system whose transition diagram is shown in Figure 5.10, obtain steady state probability equations by applying the final value theorem to Laplace transforms (i.e., Approach III of Examples 5.13).
10. An aerospace system is made up of seven independent subsystems in series and it specified failure rate 0.009 failures/h. Subsystems 1, 2, 3, 4, 5, 6, and 7 estimated failure rates from previous experience are 0.001 failures/h, 0.002 failures/h, 0.003 failures/h, 0.004 failures/h, 0.005 failures/h, 0.006 failures/h, and 0.007 failures/h, respectively. Allocate the specified system failure rate to seven subsystems.

5.8 REFERENCES

1. Grant Ireson, W., Coombs, C.F., and Moss, R.Y., Eds., *Handbook of Reliability Engineering and Management,* McGraw-Hill, New York, 1996.
2. Dhillon, B.S., *Reliability in Systems Design and Operation,* Van Nostrand Reinhold Company, New York, 1983.
3. AMCP 706-196, Engineering Design Handbook: Development Guide for Reliability, Part II: Design for Reliability, U.S. Army Material Command (AMC), Washington, D.C., 1976.
4. Reliability Design Handbook, RDG-376, Reliability Analysis Center, Rom Air Development Center, Griffis Air Force Base, Rome, New York, 1976.
5. Dhillon, B.S., A hazard rate model, *IEEE Trans. Reliability,* 28, 150, 1979.
6. Dhillon, B.S., *Mechanical Reliability: Theory, Models, and Applications,* American Institute of Aeronautics and Astronautics, Washington, D.C., 1988.
7. Lipp, J.P., Topology of switching elements vs. reliability, *Trans. IRE Reliability Quality Control,* 7, 21-34, 1957.
8. Shooman, M.L., *Probabilistic Reliability: An Engineering Approach,* McGraw-Hill, New York, 1968.
9. Dhillon, B.S. and Singh, C., *Engineering Reliability: New Techniques and Applications,* John Wiley & Sons, New York, 1981.
10. Dhillon, B.S., The Analysis of the Reliability of Multistate Device Networks, Ph.D., Dissertation, 1975. Available from the National Library of Canada, Ottawa.
11. Dhillon, B.S. and Proctor, C.L. Reliability analysis of multistate device networks, *Proc. Annu. Reliability Maintainability Symp.,* 31-35, 1976.
12. MIL-HDBK-217, Reliability Prediction of Electronic Equipment, Department of Defense, Washington, D.C.
13. Dhillon, B.S., *Systems Reliability, Maintainability and Management,* Petrocelli Books, New York, 1983.

6 Failure Modes and Effect Analysis

6.1 INTRODUCTION

Failure modes and effect analysis (FMEA) is a powerful design tool to analyze engineering systems and it may simply be described as an approach to perform analysis of each potential failure mode in the system to examine the results or effects of such failure modes on the system [1]. When FMEA is extended to classify each potential failure effect according to its severity (this incorporates documenting catastrophic and critical failures), the method is known as failure mode effects and criticality analysis (FMECA).

The history of FMEA goes back to the early 1950s with the development of flight control systems when the U.S. Navy's Bureau of Aeronautics, in order to establish a mechanism for reliability control over the detail design effort, developed a requirement called "Failure Analysis" [2]. Subsequently, Coutinho [3] coined the term "Failure Effect Analysis" and the Bureau of Naval Weapons (i.e., successor to the Bureau of Aeronautics) introduced it into its new specification on flight controls. However, FMECA was developed by the National Aeronautics and Astronautics Administration (NASA) to assure the desired reliability of space systems [4].

In the 1970s, the U.S. Department of Defense directed its effort to develop a military standard entitled "Procedures for Performing a Failure Mode, Effects, and Criticality Analysis" [5]. Today FMEA/FMECA methods are widely used in the industry to conduct analysis of systems, particularly for use in aerospace, defense, and nuclear power generation. Reference 6 presents a comprehensive list of publications on FMEA/FMECA. This chapter discusses different aspects of FMEA/FMECA.

6.2 TERMS AND DEFINITIONS

There are many terms used in performing FMEA/FMECA and some of them are as follows [5]:

- **Failure cause.** The factors such as design defects, quality defects, physical or chemical processes, or part misapplication are the primary reason for failure or they start the physical process through which deterioration progresses to failure.
- **Failure mode.** The notion or manner through which a failure is perceived.
- **Failure effect.** The consequence or consequences a failure mode has on an item's function, operation, or status.

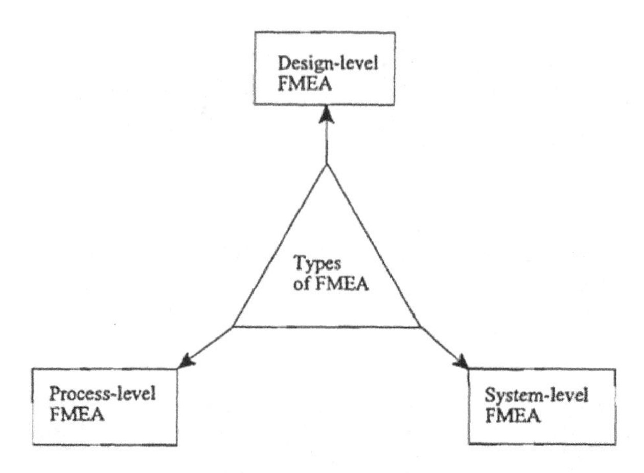

FIGURE 6.1 Types of FMEA.

- **Single failure point.** An item's malfunction that would lead to system failure and is not compensated through redundancy or through other operational mechanisms.
- **Criticality.** A relative measure of a failure mode's consequences and its occurrence frequency.
- **Severity.** A failure mode's consequences, taking into consideration the worst case scenario of a failure, determined by factors such as damage to property, the degree of injury, or ultimate system damage.
- **Corrective action.** A change such as design, process, procedure, or materials implemented and validated to rectify failure cause/design deficiency.
- **Criticality analysis.** An approach through which each possible failure mode is ranked with respect to the combined influence of occurrence probability and severity.
- **Undetectable failure.** A postulated failure mode in the FMEA for which no failure detection approach is available through which the concerned operator can be alerted of the failure.
- **Local effect.** The consequence or conseqences a failure mode has on the function, operation, or status of the item currently being analyzed.

6.3 TYPES OF FMEA AND THEIR ASSOCIATED BENEFITS

FMEA may be grouped under three distinct classifications as shown in Figure 6.1 [7]. These are design-level FMEA, system-level FMEA, and process-level FMEA. Each of these categories is described below.

6.3.1 Design-Level FMEA

The purpose of performing design-level FMEA is to help identify and stop product failures related to design. This type of FMEA can be carried out upon component-level/subsystem-level/system-level design proposal and its intention is to validate the design parameters chosen for a specified functional performance requirement. The advantages of performing design-level FMEA include identification of potential design-related failure modes at system/subsystem/component level, identification of important characteristics of a given design, documentation of the rationale for design changes to guide the development of future product designs, help in the design requirement objective evaluation and evaluation of design alternatives, systematic approach to reduce criticality and risk, accumulated data serve as a useful historical record of the thought processes and the actions taken during the product development effort, and a useful tool to establish priority for design improvement actions.

6.3.2 System-Level FMEA

This is the highest-level FMEA that can be performed and its purpose is to identify and prevent failures related to system/subsystems during the early conceptual design. Furthermore, this type of FMEA is carried out to validate that the system design specifications reduce the risk of functional failure to the lowest level during the operational period.

Some benefits of the system-level FMEA are identification of potential systemic failure modes due to system interaction with other systems and/or by subsystem interactions, selection of the optimum system design alternative, identification of potential system design parameters that may incorporate deficiencies prior to releasing hardware/software to production, a systematic approach to identify all potential effects of subsystem/assembly part failure modes for incorporation into design-level FMEA, and a useful data bank of historical records of the thought processes as well as of actions taken during product development efforts.

6.3.3 Process-Level FMEA

This identifies and prevents failures related to the manufacturing/assembly process for a certain item. The benefits of the process-level FMEA include identification of potential process-level failure modes at system/subsystem/operation level, identification of important characteristics associated with the process, identification of potential process shortcomings early in the process planning cycle, development of priorities for process improvement actions, and documentation of rationale for process changes to help guide the establishment of potential manufacturing processes.

6.4 STEPS FOR PERFORMING FMEA

FMEA can be performed in six steps as shown in Figure 6.2 [4, 8]. These steps are as follows:

- define system and its associated requirements,
- establish ground rules,
- describe the system and its associated functional blocks,
- identify failure modes and their associated effects,

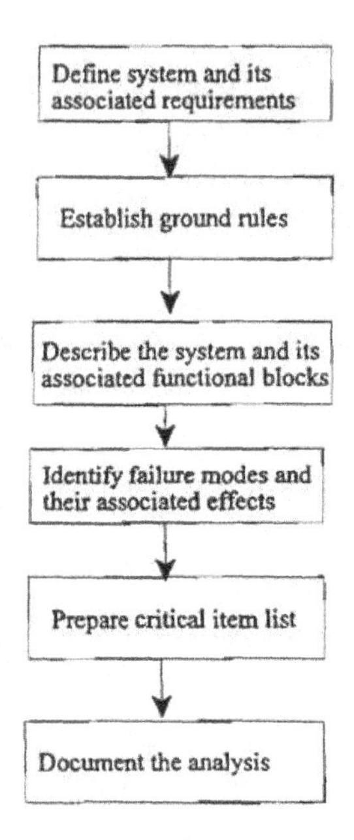

FIGURE 6.2 Steps for performing FMEA.

- prepare critical items list,
- document the analysis.

Each of these steps is described below.

6.4.1 Define System and Its Associated Requirements

This is concerned with defining the system under consideration and the definition normally incorporates a breakdown of the system into blocks, block functions, and the interface between them. Usually this early in the program a good system definition does not exist and the analyst must develop his/her own system definition using documents such as trade study reports, drawings, and development plans and specifications.

6.4.2 Establish Ground Rules

These are established as to which the FMEA is conducted. Usually, developing the ground rules is a quite straightforward process when the system definition and mission requirements are reasonably complete. Nonetheless, examples of the ground rules include primary and secondary mission objectives statement, limits of environmental and operational stresses, statement of analysis level, delineation of mission phases, definition of what constitutes failure of system hardware parts, and the detail of the coding system used.

6.4.3 Describe the System and Its Associated Functional Blocks

This is concerned with the preparation of the description of the system under consideration. Such description may be grouped into two parts:

- **Narrative functional statement.** This is prepared for each subsystem and component as well as for the total system. It provides narrative description of each item's operation for each operational mode/mission phase. The degree of the description detail depends on factors such as an item's application and the uniqueness of the functions performed.
- **System block diagram.** The purpose of this block diagram is to determine the success/failure relationships among all the system components; thus, it graphically shows total system components to be analyzed as well as the series and redundant relationships among the components. In addition, the block diagram shows the entire system's inputs and outputs and each system element's inputs and outputs.

6.4.4 Identify Failure Modes and Their Associated Effects

This is concerned with performing analysis of the failure modes and their effects. A form such as that shown in Figure 6.3 is used as a worksheet to assure systematic and thorough coverage of all failure modes. Even though all the terms used in that form are considered self-explanatory, the terms "Compensating provisions" and "Criticality classification" are described below.

- **Compensating provisions.** These provisions, i.e., design provisions or operator actions, concerning circumventing or mitigating the failure effect should be identified and evaluated [5].
- **Criticality classification.** This is concerned with the categorization of potential effect of failure. For example [4],
  - People may lose their lives due to a failure.
  - Failure may cause mission loss.
  - Failure may cause delay in activation.
  - Failure has no effect.

6.4.5 Prepare Critical Items List

The critical items list is prepared to facilitate communication of important analysis results to management. A typical critical items list worksheet is shown in Figure 6.4.

6.4.6 Document the Analysis

This is the final step of the FMEA performing process and is concerned with the documentation of analysis. This step is at least as important as the other previous five steps because poor documentation can lead to ineffectiveness of the FMEA

System: Subsystem: Drawing:			FAILURE MODE AND EFFECTS ANALYSIS							Analyst: Date: Page: of Approved by:	
Item identification	Item function	Failure modes and causes	Mission phase/operational mode	Failure Effects on				Method for failure detection	Compensating provisions	Criticality classification	Remarks
				System	Sub-system	Mission	Personnel				

FIGURE 6.3 A typical FMEA worksheet.

CRITICAL ITEMS LIST					
Item identification	Failure mode concise statement	Degree of loss effect	Criticality classification	Retention rationale	FMEA page no.

FIGURE 6.4 A critical items list worksheet.

process. Nonetheless, the FMEA report incorporates items such as system definition, FMEA associated ground rules, failure modes and their effects, description of the system (i.e., including functional descriptions and block diagrams), and critical items list.

6.5 CRITICALITY ASSESSMENT

The objective of the criticality assessment is to prioritize the failure modes discovered during the system analysis on the basis of their effects and occurrence likelihood [9]. Thus, for making an assessment of the severity of an item failure, two commonly used methods are Risk Priority Number (RPN) Technique and Military Standard Technique. The former is widely used in the automotive industrial sector and the latter in defense, nuclear, and aerospace industries. Both approaches are described below [9-11].

6.5.1 RPN Technique

This method calculates the risk priority number for a part failure mode using three factors: (1) failure effect severity, (2) failure mode occurrence probability, and

TABLE 6.1
Failure Detection Ranking

Item No.	Likelihood of detection	Rank meaning	Rank
1	Very high	Potential design weakness will almost certainly be detected	1, 2
2	High	There is a good chance of detecting potential design weakness	3, 4
3	Moderate	There is a possibility of detecting potential design weakness	5, 6
4	Low	Potential design weakness is unlikely to be detected	7, 8
5	Very low	Potential design weakness probably will not be detected	9
6	Detectability absolutely uncertain	Potential design weakness cannot be detected	10

(3) failure detection probability. More specifically, the risk priority number is computed by multiplying the rankings (i.e., 1–10) assigned to each of these three factors. Thus, mathematically the risk priority number is expressed by

$$RPN = (DR)(OR)(SR) \tag{6.1}$$

where

RPN is the risk priority number.
DR is the detection ranking.
OR is the occurrence ranking.
SR is the severity ranking.

Since the above three factors are assigned rankings from 1 to 10, the value of the RPN will vary from 1 to 1000. Failure modes with a high RPN are considered to be more critical; thus, they are given a higher priority in comparison to the ones with lower RPN. Nonetheless, rankings and their interpretation may vary from one organization to another. Tables 6.1 through 6.3 present rankings for failure detection, failure mode occurrence probability, and failure effect severity used in one organization [12], respectively.

6.5.2 Military Standard Technique

This technique is often used in defense, aerospace, and nuclear industries to prioritize the failure modes of the item under consideration so that appropriate corrective measures can be undertaken [5]. The technique requires the categorization of the failure-mode effect severity and then the development of a "critical ranking". Table 6.4 presents classifications of failure mode effect severity.

In order to assess the likelihood of a failure-mode occurrence, either a qualitative or a quantitative approach can be used. This is determined by the availability of specific parts configuration data and failure rate data. The qualitative method is used

TABLE 6.2
Failure Mode Occurrence Probability

Item No.	Ranking term	Ranking term meaning	Occurrence probability	Rank
1	Remote	Occurrence of failure is quite unlikely	<1 in 10^6	1
2	Low	Relatively few failures are expected	1 in 20,000	2
			1 in 4,000	3
3	Moderate	Occasional failures are expected	1 in 1,000	4
			1 in 400	5
			1 in 80	6
4	High	Repeated failures will occur	1 in 40	7
			1 in 20	8
5	Very high	Occurrence of failure is almost inevitable	1 in 8	9
			1 in 2	10

TABLE 6.3
Severity of the Failure-Mode Effect

Item No.	Failure effect severity category	Severity category description	Rank
1	Minor	No real effect on system performance and the customer may not even notice the failure	1
2	Low	The occurrence of failure will only cause a slight customer annoyance	2, 3
3	Moderate	Some customer dissatisfaction will be caused by failure	4, 5, 6
4	High	High degree of customer dissatisfaction will be caused by failure but the failure itself does not involve safety or noncompliance with government rules and regulations	7, 8
5	Very high	The failure affects safe item operation, involves noncompliance with government rules and regulations	9, 10

when there are no specific failure rate data. Thus, in this approach the individual occurrence probabilities are grouped into distinct logically defined levels, which establish the qualitative failure probabilities. Table 6.5 presents occurrence probability levels [5, 9]. For the purpose of identifying and comparing each failure mode to all other failure modes with respect to severity, a critical matrix is developed as shown in Figure 6.3.

The criticality matrix is developed by inserting item/failure mode identification number values in matrix locations denoting the severity category classification and either the criticality number (K_i) for the failure modes of an item or the occurrence level probability. The distribution of criticality of item failure modes is depicted by the resulting matrix and the matrix serves as a useful tool for assigning corrective measure priorities. The direction of the arrow, originating from the origin, shown in

TABLE 6.4
Failure Mode Effect Severity Classifications

Item No.	Classification term or name	Classification description	Classification No.
1	Catastrophic	The occurrence of failure may result in death or equipment loss	A
2	Critical	The occurrence of failure may result in major property damage/severe injury/major system damage ultimately leading to mission loss	B
3	Marginal	The occurrence of failure may result in minor injury/minor property damage/minor system damage, etc.	C
4	Minor	The failure is not serious enough to lead to injury/system damage/property damage, but it will result in repair or unscheduled maintenance	D

TABLE 6.5
Qualitative Failure Probability Levels

Item No.	Probability level	Term for the level	Term description
1	I	Frequent	High probability of occurrence during the item operational period
2	II	Reasonably probable	Moderate probability of occurrence during the item operational period
3	III	Occasional	Occasional probability of occurrence during the item operational period
4	IV	Remote	Unlikely probability of occurrence during the item operational period
5	V	Extremely unlikely	Essentially zero chance of occurrence during the item operational period

Figure 6.5, indicates the increasing criticality of the item failure and the darkened region in the figure shows the approximate desirable design region. For severity classifications A and B, the desirable design region has low occurrence probability or criticality number. On the other hand, for severity classifications C and D failures, higher probabilities of occurrence can be tolerated. Nonetheless, failure modes belonging to classifications A and B should be eliminated altogether or at least their probabilities of occurrence be reduced to an acceptable level through design changes.

The quantitative approach is used when failure mode and probability of occurrence data are available. Thus, the failure mode critical number is calculated using the following equation:

$$K_{fm} = F\,\theta\,\lambda\,T \tag{6.2}$$

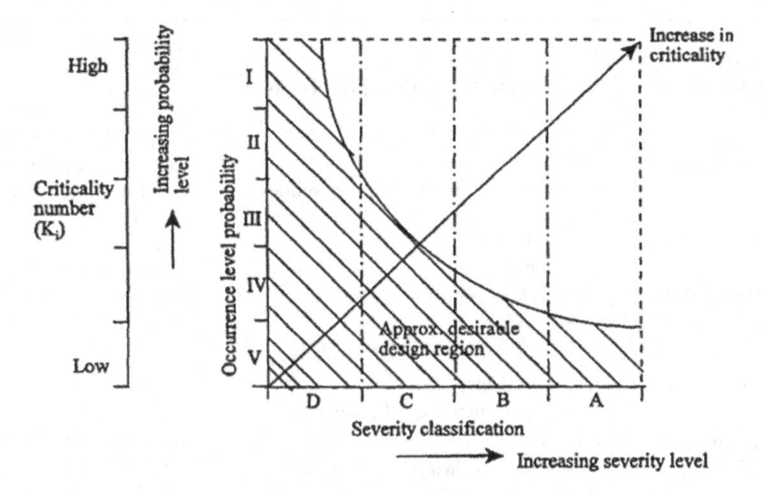

FIGURE 6.5 Criticality matrix.

where

- K_{fm} is the failure-mode criticality number.
- θ is the failure-mode ratio or the probability that the component/part fails in the particular failure mode of interest. More specifically, it is the fraction of the part failure rate that can be allocated to the failure mode under consideration and when all failure modes of an item under consideration are specified, the sum of the allocations equals unity. Table 6.6 presents failure mode apportionments for certain parts [10, 11, 13].
- F is the conditional probability that the failure effect results in the indicated severity classification or category, given that the failure mode occurs. The values of F are based on an analyst's judgment and these values are quantified according to Table 6.7 guidelines.
- T is the item operational time expressed in hours or cycles.
- λ is the item/part failure rate.

The item criticality number K_i is calculated for each severity class, separately. Thus, the total of the criticality numbers of all failure modes of an item resulting in failures in the severity class of interest is given by

$$K_i = \sum_{j=1}^{n} (k_{fm})_j = \sum_{j=1}^{n} (F\theta\lambda T)_j \tag{6.3}$$

where

n is the item failure modes that fall under the severity classification under consideration.

TABLE 6.6
Failure Mode Apportionments for Selective Items

Item no.	Item description	Item important failure modes	Approximate probability of occurrence (or probability value for θ)
1	Relay	• Contact failure	0.75
		• Open coils	0.05
2	Relief valve	• Premature open	0.77
		• Leaking	0.23
3	Incandescent lamp	• Catastrophic (filament breakage, glass breakage)	0.10
		• Degradation (loss of filament emission)	0.90
4	Fixed resistor	• Open	0.84
		• Parameter change	0.11
		• Short	0.05
5	Transformer	• Shorted turns	0.80
		• Open circuit	0.05
6	Diode	• Short circuit	0.75
		• Intermittent circuit	0.18
		• Open circuit	0.06
		• Other	0.01
7	Hydraulic valve	• Stuck closed	0.12
		• Stuck open	0.11
		• Leaking	0.77

TABLE 6.7
Failure Effect Probability Guideline Values

Item no.	Failure effect description	Probability value of F
1	No effect	0
2	Actual loss	1.00
3	Probable loss	$0.10 < F < 1.00$
4	Possible loss	$0 < F \leq 0.10$

It is to be noted that when item/part failure mode results in multiple severity-class effects each with its own occurrence probability, then in the calculation of K_i only the most important is used [14]. This can lead to erroneously low K_i values for the less critical severity categories. In order to rectify this error, it is recommended to compute F values for all severity categories associated with a failure mode and ultimately include only contributions of K_i for category B, C, and D failures [9, 14].

6.6 FMECA INFORMATION NEEDS, DATA SOURCES, AND USERS

In order to perform effective FMECA, various types of information are needed. Design related information required for the FMECA includes system schematics, functional block diagrams, equipment/part drawings, design descriptions, relevant company/customer/commercial/military specifications, reliability data, field service data, effects of environment on item under consideration, configurations management data, operating specifications and limits, and interface specifications.

Some specific FMECA related factors and their corresponding data retrieval sources in parentheses, are as follows [9]:

- Failure modes, causes, and rates (factory database, field experience database).
- Failure effects (design engineer, reliability engineer, safety engineer).
- Item identification numbers (parts list).
- Failure detection method (design engineer, maintainability engineer).
- Function (customer requirements, design engineer).
- Failure probability/severity classification (safety engineer).
- Item nomenclature/functional specifications (parts list, design engineer).
- Mission phase/operational mode (design engineer).

Usually, FMECA is expected to satisfy the needs of many groups during the design process including design, quality assurance, reliability and maintainability, customer representatives, internal company regulatory agency, system engineering, testing, logistics support, system safety, and manufacturing.

6.7 FMEA IMPLEMENTATION RELATED FACTORS AND GENERAL GUIDELINES

The implementation of FMEA is not as easy as it may appear to be. Nonetheless, prior to its implementation, the following areas must be explored [15, 16]:

- Obtaining engineer's support and approval.
- Examination of every conceivable failure mode by the involved professionals.
- Making decisions based on the RPN.
- Measuring FMEA cost/benefits.
- Use of FMEA to select the concept.

There are a number of guidelines/facts associated with FMEA including: avoid evaluating every conceivable failure mode, FMEA has limitations, FMEA is not designed to supersede the engineer's work, RPN could be misleading, avoid developing majority of the FMEA in a meeting, a very small RPN may justify corrective measure, FMEA is not the tool for choosing the optimum design concept, tailor severity, occurrence, and detection scales to reflect the company's product and processes, and applying the Pareto principle to the RPN is a misapplication of the Pareto principle [15].

6.8 ADVANTAGES OF FMEA

There are many benefits of performing FMEA, including a systematic approach to classify hardware failures, reduces development time and cost, reduces engineering changes, easy to understand, serves as a useful tool for more efficient test planning, highlights safety concerns to be focused on, improves customer satisfaction, an effective tool to analyze small, large, and complex systems, useful in the development of cost-effective preventive maintenance systems, provides safeguard against repeating the same mistakes in the future, useful to compare designs, a visibility tool for manager, a useful approach that starts from the detailed level and works upward, and useful to improve communication among design interface personnel [8, 15].

6.9 PROBLEMS

1. Write a short essay on the history of FMEA.
2. Define the following terms:
   - Failure mode
   - Single point failure
   - Severity
3. Discuss design-level FMEA and its associated benefits.
4. Describe the FMEA process.
5. Discuss the elements of a typical critical items list worksheet.
6. Describe the RPN method.
7. Identify sources for obtaining data for the following factors:
   - Item identification numbers
   - Failure effects
   - Failure probability/severity classification
8. List at least 10 advantages of performing FMEA.
9. What is the difference between FMEA and FMECA?
10. Who are the users of FMEA/FMECA?

6.10 REFERENCES

1. Omdahl, T.P., Ed., *Reliability, Availability and Maintainability (RAM) Dictionary*, American Society for Quality Control (ASQC) Press, Milwaukee, WI, 1988.
2. MIL-F-18372 (Aer), General Specification for Design, Installation, and Test of Aircraft Flight Control Systems, Bureau of Naval Weapons, Department of the Navy, Washington, D.C., Para. 3.5.2.3.
3. Coutinho, J.S., Failure effect analysis, *Trans. NY Acad. Sci.*, 26, Series II, pp. 564-584, 1963-1964.
4. Jordan, W.E., Failure modes, effects and criticality analyses, *Proc. Annu. Reliability Maintainability Symp.*, 30-37, 1972.

5. MIL-STD-1629, Procedures for Performing a Failure Mode, Effects, and Criticality Analysis, Department of Defense, Washington, D.C., 1980.
6. Dhillon, B.S., Failure mode and effects analysis — bibliography, *Microelectronics and Reliability,* 32, 719-731, 1992.
7. Grant Ireson, W., Coombs, C.F., and Moss, R.Y., *Handbook of Reliability Engineering and Management,* McGraw-Hill, New York, 1996.
8. Dhillon, B.S., *Systems Reliability, Maintainability, and Management,* Petrocelli Books, New York, 1983.
9. Bowles, J.B. and Bonnell, R.D., Failure mode, effects, and criticality analysis, in *Tutorial Notes, Annual Reliability and Maintainability Symposium,* 1994, pp. 1-34. Available from the Annual Reliability and Maintainability Symposium, C/O Evans Associates, 804 Vickers Avenue, Durham, NC.
10. AMCP 706-196, *Engineering Design Handbook: Development Guide for Reliability, Part II, Design for Reliability,* Army Material Command, Department of the Army, Washington, D.C., 1976.
11. AE-9, *Automotive Electronics Reliability Handbook,* Society of Automotive Engineers, Warrendale, PA, 1987.
12. Potential Failure Mode and Effects Analysis in Design (Design FMEA) and for Manufacturing and Assembly Process (Process FMEA) Instruction Manual, Ford Motor Company, Detroit, MI, September 1988.
13. Failure Mode/Mechanism Distributions, Report No. FMD-91, Reliability Analysis Center, Rome Air Development Center, Griffiss Air Force Base, Rome, NY, 1991.
14. Agarwala, A.S., Shortcomings in MIL-STD-1629A guidelines for criticality analysis, *Proc. Annu. Reliability Maintainability Symp.,* 494-496, 1990.
15. Palady, P., *Failure Modes and Effects Analysis,* PT Publications, West Palm Beach, FL, 1995.
16. Mcdermott, R.E., Mikulak, R.J., and Beauregard, M.R., *The Basics of FMEA,* Quality Resources, New York, 1996.

7 Fault Tree Analysis

7.1 INTRODUCTION

Fault tree analysis (FTA) is widely performed in the industry to evaluate engineering systems during their design and development, in particular the ones used in nuclear power generation. A fault tree may simply be described as a logical representation of the relationship of primary events that lead to a specified undesirable event called the "top event" and is depicted using a tree structure with OR, AND, etc. logic gates. The fault tree method was developed in the early 1960s by H.A. Watson of Bell Telephone Laboratories to perform analysis of the Minuteman Launch Control System. A study team at the Bell Telephone Laboratories further refined it and Haasl [1] of the Boeing company played a pivotal role in its subsequent development.

In 1965 several papers related to the technique were presented at the System Safety Symposium held at the University of Washington, Seattle [1]. In 1974, a conference on "Reliability and Fault Tree Analysis" was held at the University of California, Berkeley [2]. A paper appeared in 1978 providing a comprehensive list of publications on Fault Trees [3]. The three books that described FTA in considerable depth appeared in 1981 [4–6]. Needless to say, since the inception of the fault tree technique, many people have contributed to its additional developments.

This chapter describes different aspects of FTA in considerable depth.

7.2 FTA PURPOSES AND PREREQUISITES

There are many purposes in performing FTA, including identifying critical areas and cost-effective improvements, understanding the functional relationship of system failures, meeting jurisdictional requirements, providing input to test, maintenance, and operational policies and procedures, understanding the level of protection that the design concept provides against failures, evaluating performance of systems/equipment for bid-evaluation purposes, providing an integrated picture of some aspects of system operation, confirming the ability of the system to meet its imposed safety requirements, and providing input to cost-benefit trade-offs.

There are many prerequisites associated with FTA including thorough understanding of design, operation, and maintenance aspects of system/item under consideration; clear definition of what constitutes system failure: the undesirable event, clearly defined analysis scope and objectives; a comprehensive review of system/item operational experience; well-defined level of analysis resolution; clear identification of associated assumptions; and clearly defined system/item physical bounds and system interfaces.

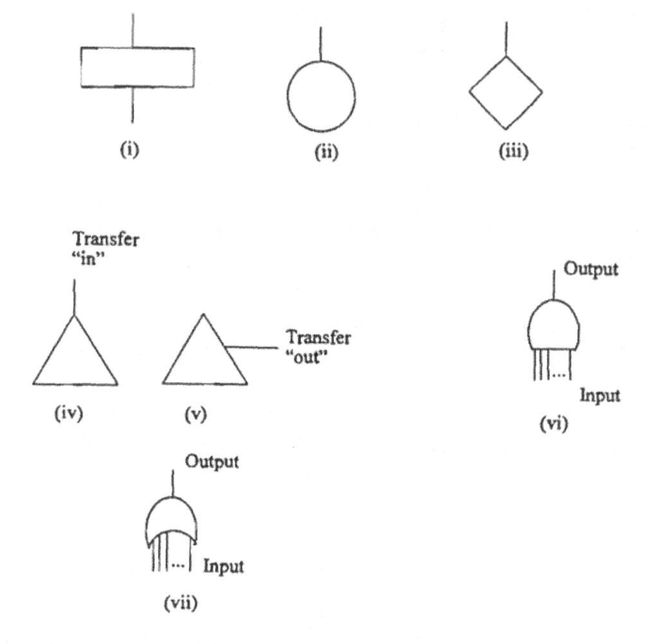

FIGURE 7.1 Commonly used fault tree symbols: (i) rectangle, (ii) circle, (iii) diamond, (iv) triangle A, (v) triangle B, (vi) AND gate, (vii) OR gate.

7.3 FAULT TREE SYMBOLS

The fault tree method makes use of many different symbols to develop fault trees of engineering systems. Seven such symbols are shown in Figure 7.1. Each one of these symbols is described below.

- **Rectangle**. Represents a fault event that results from the logical combination of fault events through the input of the logic gate.
- **Circle**. Denotes a basic fault event or the failure of an elementary part. The fault event's probability of occurrence, failure, and repair rates are usually obtained from empirical data.
- **Diamond.** Denotes a fault event whose causes have not been fully developed either due to lack of interest or due to lack of information.
- **Triangle A.** Denotes "transfer in" and is used to avoid repeating segments of the fault tree.
- **Triangle B.** Denotes "transfer out" and is used to avoid repeating segments of the fault tree.
- **AND gate.** Denotes that an output fault event occurs only if all of the input fault events occur.
- **OR gate.** Denotes that an output fault event occurs if one or more of the input fault events occur.

From time to time, there are many other symbols used to perform FTA. Most of these symbols are discussed in References 6 through 8.

7.4 FUNDAMENTAL APPROACH TO FTA

The development or construction of a fault tree is top-down, in that the undesirable or top event is the tree root and the logical combination of sub-events are employed to map out the tree until reaching the basic initiating fault events. Nonetheless, steps such as those listed below are involved in performing FTA [9]:

- Define system, analysis associated assumptions, what constitutes a failure, etc.
- If the simplification of the scope of the analysis is necessary, develop a simple system block diagram showing relevant inputs, outputs, and interfaces.
- Identify undesirable or top fault events to be analyzed and if necessary develop fault trees for all top-level events.
- Identify all the causes that can make the top event occur using fault tree symbols and the logic tree format. More specifically, using deductive reasoning highlight event that can lead to the occurrence of the top event.
- Assuming the causes of the previous step as intermediate effects, continue developing the logic tree by identifying the causes of these intermediate events.
- Develop the fault tree to the lowest level of detail as required.
- Perform analysis of the completed fault tree with respect to understanding the logic and the interrelationships among various fault paths, gaining insight into the unique modes of product faults, etc.
- Determine appropriate corrective measures.
- Prepare documentation of the analysis process and follow up on identified corrective measures.

Example 7.1

Assume that a windowless room has a switch and four light bulbs. Develop a fault tree for the top or undesired fault event "dark room" (i.e., room without light). Assume that the room can only be dark if all the light bulbs burn out, there is no electricity, or the switch fails to close.

A fault tree for this example is shown in Figure 7.2. Each fault event in the figure is labeled as B_1, B_2, B_3, ..., B_{10}.

7.5 BOOLEAN ALGEBRA RULES

A fault tree may be considered as a pictorial representation of those Boolean relationships among fault events that lead to the occurrence of the top event. Furthermore, a fault tree may be translated into an entirely equivalent set of Boolean expressions. Usually, prior to the probability evaluation of the top event, the simplification of the Boolean expressions is required. The Boolean algebra rules play an important role in this aspect. Selective rules of Boolean algebra are presented below [4, 5, 8].

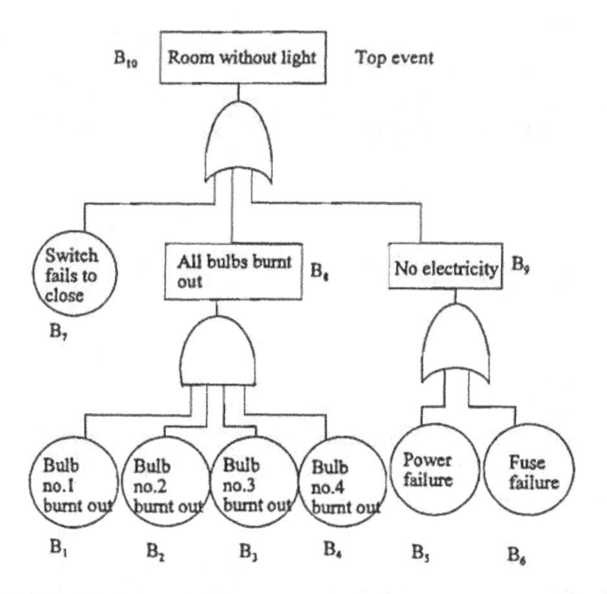

FIGURE 7.2 Fault tree for the top event: dark room or room without light.

- Idempotent law

$$A \cdot A = A \tag{7.1}$$

$$A + A = A \tag{7.2}$$

- Distributive law

$$X(Y + Z) = XY + XZ \tag{7.3}$$

$$X + YZ = (X + Y)(X + Z) \tag{7.4}$$

- Commutative law

$$AB = BA \tag{7.5}$$

$$A + B = B + A \tag{7.6}$$

- Absorption law

$$X + XY = X \tag{7.7}$$

$$X(X + Y) = X \tag{7.8}$$

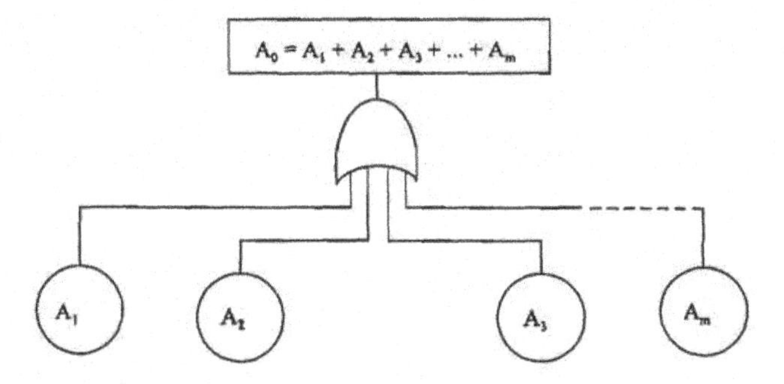

FIGURE 7.3 An m input OR gate with Boolean expression.

7.6 ANALYTICAL DEVELOPMENTS OF BASIC GATES, REPEATED FAULT EVENTS, AND MINIMAL CUT SETS

A fault tree is developed using logic gates such as OR and AND that relate logically various basic fault events to the undesirable or the top event. Boolean algebra is an invaluable tool to represent a fault tree diagram in a mathematical form. Boolean expressions for OR and AND gates are presented below.

OR Gate

An m input fault events $A_1, A_2, A_3, \ldots, A_m$ OR gate along with its output fault event A_0 in a Boolean expression is shown in Figure 7.3. Thus, mathematically, the output fault event A_0 of the m input fault event OR gate is expressed by

$$A_0 = A_1 + A_2 + A_3 + \ldots + A_m \tag{7.9}$$

where

A_i is the ith input fault event; for i = 1, 2, 3, …, m.

AND Gate

A k input fault event $X_1, X_2, X_3, \ldots, X_k$ AND gate along with its output fault event X_0 in a Boolean expression is shown in Figure 7.4. Thus, mathematically, the output fault event X_0 of the k input fault event AND gate is expressed by

$$X_0 = X_1 X_2 X_3 \ldots X_k \tag{7.10}$$

where

X_i is the ith input fault event; for i = 1, 2, 3, …, k.

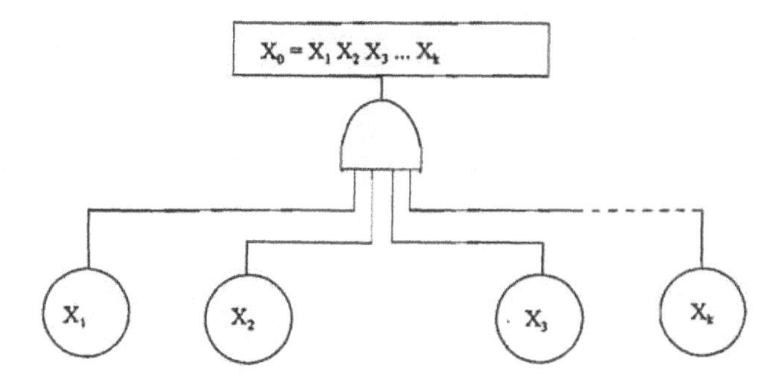

FIGURE 7.4 A k input AND gate with Boolean expression.

Example 7.2

For the fault tree shown in Figure 7.2, develop a Boolean expression for the top event: Room without light (i.e., fault event B_{10}) using fault event identification symbols B_1, B_2, B_3, ..., B_9.

Using relationship (7.10), the Boolean expression for the intermediate fault event B_8 is

$$B_8 = B_1 B_2 B_3 B_4 \tag{7.11}$$

Similarly, utilizing relationship (7.9), the Boolean equation for the intermediate fault event B_9 is given by

$$B_9 = B_5 + B_6 \tag{7.12}$$

Again using relationship (7.9), the Boolean expression for the top fault event B_{10} is expressed by

$$B_{10} = B_7 + B_8 + B_9 \tag{7.13}$$

7.6.1 Repeated Fault Events

A hypothetical fault tree with repeated fault events is shown in Figure 7.5 (i.e., the basic fault event A is repeated). Thus, in this case, the repetition of A must be eliminated prior to obtaining the quantitative reliability parameter results for the fault tree. Otherwise, the quantitative values will be incorrect. The elimination of repeated events can either be achieved by applying the Boolean algebra properties such as presented in Section 7.5 or algorithms especially developed for this purpose [6, 10-12].

One such algorithm is represented in Section 7.6.2.

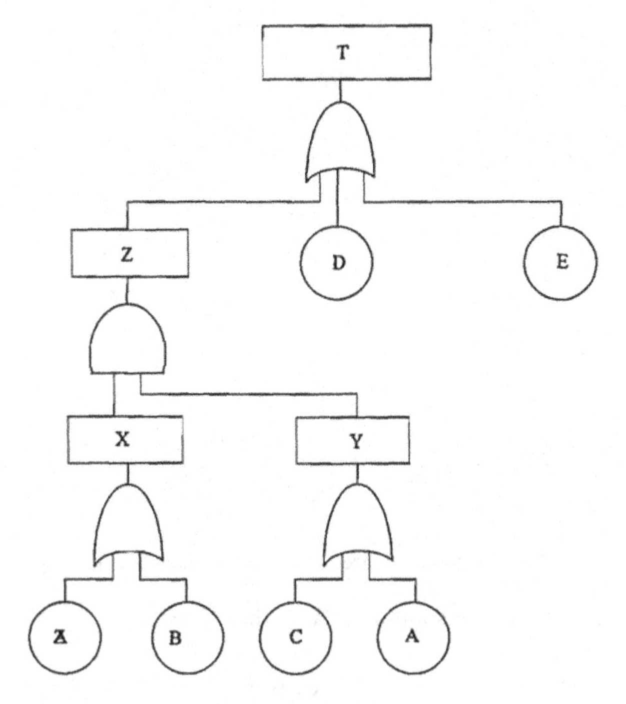

FIGURE 7.5 A hypothetical fault tree with repeated events.

Example 7.3

Use Boolean algebra properties to eliminate the repetition of the occurrence of event A shown in Figure 7.5. Construct the repeated event free fault tree using the simplified Boolean expression for the top event.

Using Figure 7.5, we write down the following Boolean expressions:

$$X = A + B \tag{7.14}$$

$$Y = A + C \tag{7.15}$$

$$Z = XY \tag{7.16}$$

$$T = Z + D + E \tag{7.17}$$

Inserting relationships (7.14) and (7.15) into expression (7.16), we get

$$Z = (A + B)(A + C) \tag{7.18}$$

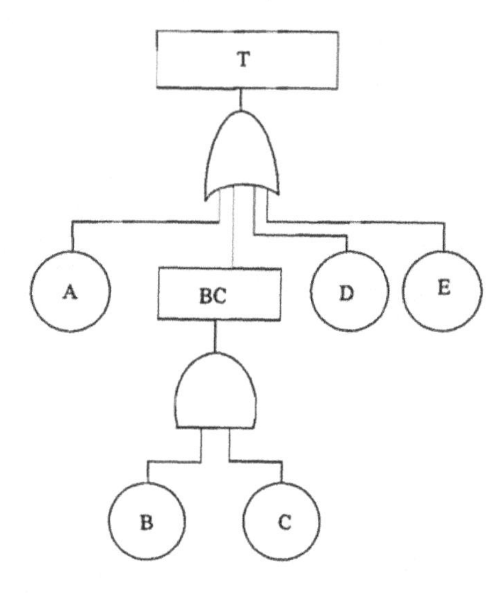

FIGURE 7.6 The repeated event free fault tree developed using Equation (7.20).

Utilizing Equation (7.4), we simplify relationship (7.18) to

$$Z = A + BC \tag{7.19}$$

Substituting Equation (7.19) into Equation (7.17) yields

$$T = A + BC + D + E \tag{7.20}$$

The above equation represents the simplified Boolean expression for the top event of the fault tree shown in Figure 7.5. The repeated event free fault tree constructed using Equation (7.20) is shown in Figure 7.6.

7.6.2 Algorithm for Obtaining Minimal Cut Sets

One of the difficult problems facing the fault tree method is to obtain minimal cut sets or eliminate repeated events of a fault tree. This section presents one algorithm to obtain minimal cut sets of a fault tree [6, 10–12].

A *cut set* may be described as a collection of basic events that will cause the top fault tree event to occur. Furthermore, a cut set is said to be minimal if it cannot be further minimized or reduced but it can still ensure the occurrence of the top fault tree event.

The algorithm under consideration can either be used manually for simple fault trees or computerized for complex fault trees with hundreds of gates and basic fault events. In this algorithm, the AND gate increases the size of a cut set and the OR gate increases the number of cut sets. The algorithm is demonstrated by solving the following example:

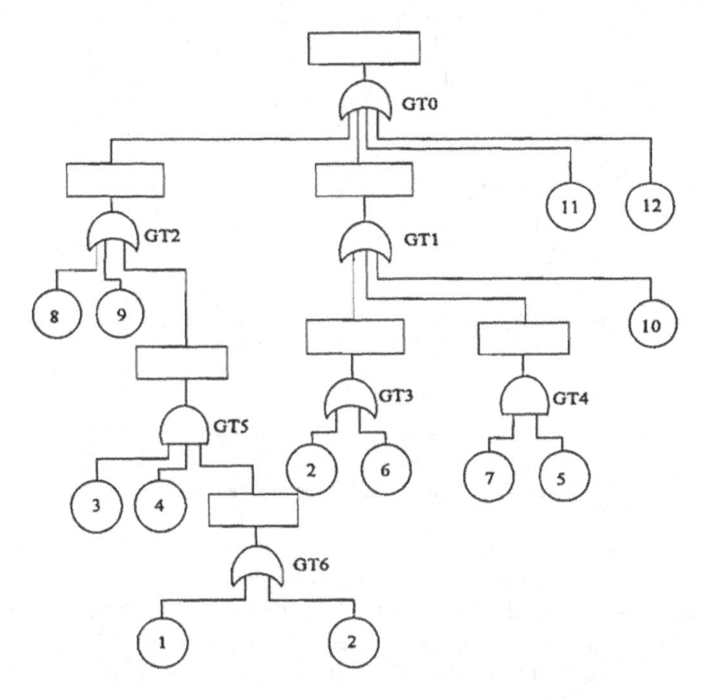

FIGURE 7.7 An event tree with a repeated event.

Example 7.4

Obtain a repeated event free fault tree of the fault tree shown in Figure 7.7. In this figure, the basic fault events are identified by the numerals and the logic gates are labeled as GT0, GT1, GT2, ..., GT6.

The algorithm starts from the gate, GT0, just below the top event of the fault tree shown in Figure 7.7. Usually, in a fault tree this gate is either OR or AND. If this gate is an OR gate, then each of its inputs represents an entry for each row of the list matrix. On the other hand, if this gate is an AND, then each of its inputs denotes an entry for each column of the list matrix.

In our case, GT0 is an OR gate; thus, we start the formulation of the list matrix by listing the gate inputs: 11, 12, GT1, and GT2 (output events) in a single column but in separate rows as shown in Figure 7.8 (i). As any one input of the GT0 could cause the occurrence of the top event, these inputs are the members of distinct cut sets.

One simple rule associated with this algorithm is to replace each gate by its inputs until all the gates in a given fault tree are replaced with the basic event entries. The inputs of a gate could be the basic events or the outputs of other gates. Consequently, in our case, in order to obtain a fully developed list matrix, we proceed to replace the OR gate GT1 of list matrix of Figure 7.8 (i) by its input events as separate rows, as indicated by the dotted line in Figure 7.8 (ii).

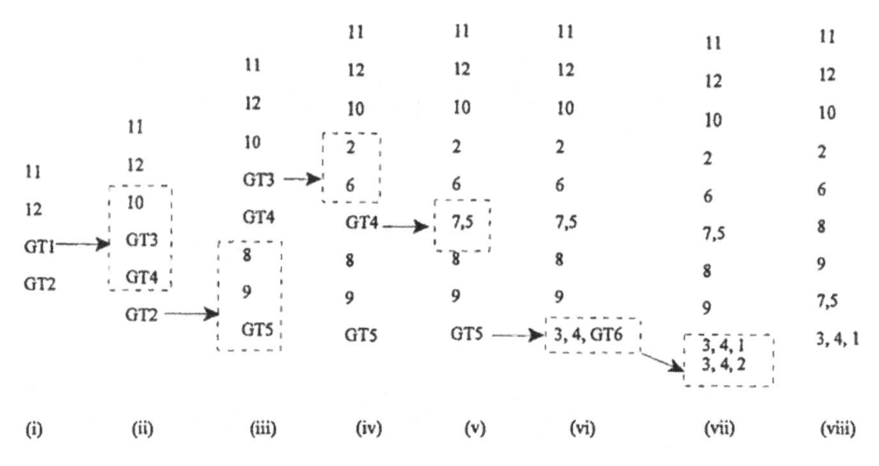

FIGURE 7.8 List matrix under different conditions.

Next, we replace the OR gate GT2 of the list matrix in Figure 7.8 (ii) by its input events as indicated by the dotted line in Figure 7.8 (iii).

Similarly, we replace the OR gate GT3 of the list matrix in Figure 7.8 (iii) by its input events 5 and 6 as identified by the dotted line in Figure 7.8 (iv).

The next gate GT4 to be replaced with its input events is an AND gate. The dotted line in Figure 7.8 (v) indicates its input events. Again, the next gate, GT5, to be replaced by its input events is an AND gate.

Its input events are indicated by the dotted line in Figure 7.8 (vi).

The last gate to be replaced by its input events is an OR gate GT6. Its inputs are indicated by the dotted line in the list matrix of Figure 7.8 (vii). As shown in the list matrix of Figure 7.8 (vii), the cut set 2 is a single event cut set. As only its occurrence will result in the occurrence of the top event, we eliminate cut set {3, 4, 2} from the list matrix because the occurrence of this cut set requires all events 3, 4, and 2 to occur. Consequently, the list matrix shown in Figure 7.8 (viii) represents the minimal cut sets of the fault tree given in Figure 7.7.

The fault tree of the Figure 7.8 (viii) list matrix is shown in Figure 7.9. Now this fault tree can be used to obtain the quantitative measures of the top or undesirable event.

7.7 PROBABILITY EVALUATION OF FAULT TREES WITH REPAIRABLE AND NONREPAIRABLE COMPONENTS

After obtaining minimal cut sets or the redundancy free events of a given fault tree, one can proceed to determine the probability of occurrence of the top event. This probability can be obtained by first estimating the probability of occurrence of the output events of lower and intermediate logic gates such as OR and AND. The output event occurrence probability of these gates can be obtained as follows:

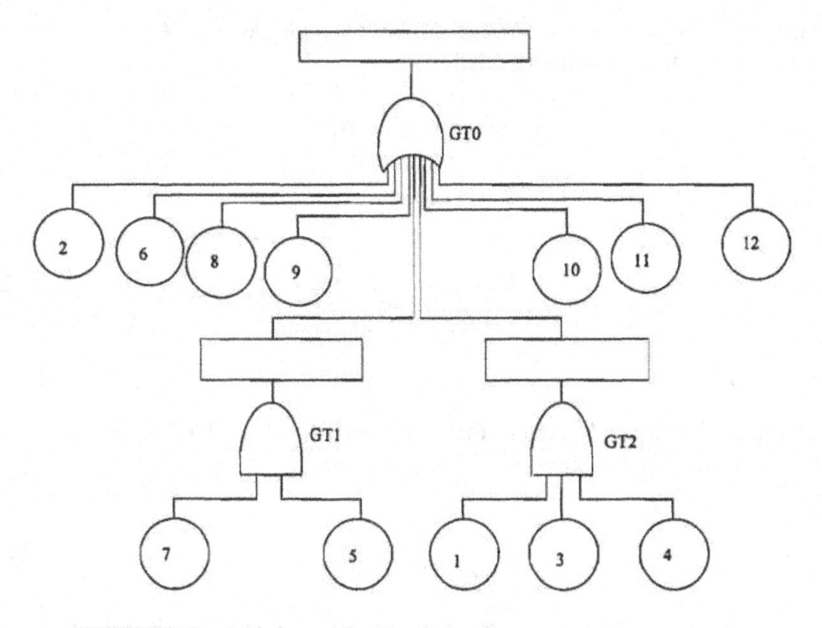

FIGURE 7.9 A fault tree for the minimal cut sets of Figure 7.8 (viii).

OR GATE

Using Figure 7.3 for independent input events, the probability of occurrence of the output fault event A_0 is given by

$$P(A_0) = 1 - \prod_{i=1}^{m} \{1 - P(A_i)\} \tag{7.21}$$

where

> $P(A_i)$ is the probability of occurrence of input fault event A_i; for i = 1, 2, 3, …, m.

For m = 2, Equation (7.21) reduces to

$$\begin{aligned} P(A_0) &= 1 - \prod_{i=1}^{2} \{1 - P(A_i)\} \\ &= P(A_1) + P(A_2) - P(A_1)\,P(A_2) \end{aligned} \tag{7.22}$$

For the probabilities of occurrence of fault events A_1 and A_2 less than 0.1, Equation (7.22) may be approximated by

$$P(A_0) \simeq P(A_1) + P(A_2) \tag{7.23}$$

For m fault events, Equation (7.23) becomes

$$P(A_0) \simeq \sum_{i=1}^{m} P(A_i) \tag{7.24}$$

AND GATE

Using Figure 7.4, for independent input fault events, the probability of occurrence of the output fault event X_0 is

$$P(X_0) = \prod_{i=1}^{k} P(X_i) \tag{7.25}$$

where

$P(X_i)$ is the probability of occurrence of input fault event X_i; for i = 1, 2, 3, ..., k.

7.7.1 PROBABILITY EVALUATION OF FAULT TREES WITH NON-REPAIRABLE COMPONENTS

In this case, the basic fault events, say, representing component failures of a system, are not repaired but their probabilities of occurrence are known. Under such conditions, the probability evaluation of fault trees is demonstrated through Example 7.4.

Example 7.4

Assume that in Figure 7.2, the probability of occurrence of basic events B_1, B_2, B_3, B_4, B_5, B_6, and B_7 are 0.15, 0.15, 0.15, 0.15, 0.06, 0.08, and 0.04, respectively. Calculate the probability of occurrence of the top event: room without light.

By inserting the given data into Equation (7.25), the probability of occurrence of event B_8 is

$$\begin{aligned} P(B_8) &= P(B_1)\,P(B_2)\,P(B_3)\,P(B_4) \\ &= (0.15)\,(0.15)\,(0.15)\,(0.15) \\ &\simeq 0.0005 \end{aligned}$$

Using the given data with Equation (7.22), we get the following probability of occurrence of event B_9:

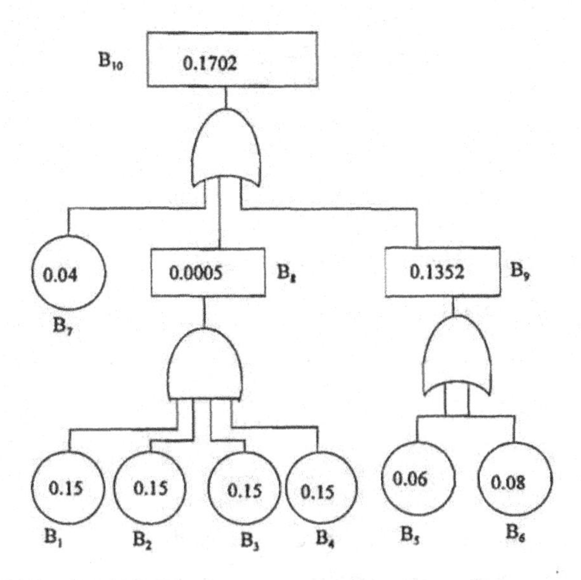

FIGURE 7.10 Calculated fault event probability values of Figure 7.2 fault tree.

$$P(B_9) = P(B_5) + P(B_6) - P(B_5)\,P(B_6)$$
$$= 0.06 + 0.08 - (0.06)\,(0.08)$$
$$= 0.1352$$

Substituting the above calculated values and the given data into Equation (7.21), we get the following value for the probability of occurrence of event B_{10} (i.e., top event):

$$P(B_{10}) = 1 - [1 - P(B_7)][1 - P(B_8)][1 - P(B_9)]$$
$$= 1 - [1 - 0.04]\,[1 - 0.0005]\,[1 - 0.1352]$$
$$= 0.1702$$

Thus, the probability of occurrence of the top event, room without light, is 0.1702. The Figure 7.2 fault tree is shown in Figure 7.10 with given and calculated fault event probability values.

7.7.2 Probability Evaluation of Fault Trees with Repairable Components

In this case, the basic fault events, say, representing component failures of a system, are repaired and the failure and repair rates of the components are known. Thus, using the Markov method, the unavailability of a component is given by

$$UA(t) = \frac{\lambda}{\lambda+\mu}\left[1-e^{-(\lambda+\mu)t}\right] \tag{7.26}$$

where

UA (t) is the component unavailability at time t.
λ is the component constant failure rate.
μ is the component constant repair rate.

For large t, Equation (7.26) reduces to

$$UA = \frac{\lambda}{\lambda+\mu} \tag{7.27}$$

where

UA is the component steady state unavailability.

The above equation yields steady state probability of a component (that it will be unavailable for service) when its failure and repair rates are known. Thus, substituting Equation (7.27) into Equation (7.21), we get the steady state probability of occurrence of the output fault event A_0 of the OR gate as follows:

$$\begin{aligned} P_{OS} &= 1-\prod_{i=1}^{m}(1-UA_i) \\ &= 1-\prod_{i=1}^{m}\left(1-\frac{\lambda_i}{\lambda_i+\mu_i}\right) \end{aligned} \tag{7.28}$$

where

P_{OS} is the steady state probability of occurrence of the OR gate output fault event.
UA_i is the steady state unavailability of component i; for i = 1, 2, 3, ..., m.
λ_i is the constant failure rate of component i; for i = 1, 2, 3, ..., m.
μ_i is the constant failure rate of component i; for i = 1, 2, 3, ..., m.

Similarly, inserting Equation (7.27) into Equation (7.25) yields the following result:

$$\begin{aligned} P_{aS} &= \prod_{i=1}^{k} UA_i \\ &= \prod_{i=1}^{k}\frac{\lambda_i}{\lambda_i+\mu_i} \end{aligned} \tag{7.29}$$

where

P_{as} is the steady state probability of occurrence of the AND gate output fault event.

The probability evaluation of fault trees with repairable components using the above equations is demonstrated by solving the following example [13]:

Example 7.5

Assume that in Figure 7.2 failure and repair (in parentheses) rates associated with basic events B_1, B_2, B_3, B_4, B_5, B_6, and B_7 are 0.0002 failures/h (4 repairs/h), 0.0002 failures/h (4 repairs/h), 0.0002 failures/h (4 repairs/h), 0.0002 failures/h (4 repairs/h), 0.0001 failures/h (0.004 repairs/h), 0.00015 failures/h (4 repairs/h), and 0.0001 failures/h (0.004 repairs/h), respectively. Calculate the steady state probability of occurrence of the top event: room without light.

Substituting the given data into Equation (7.27), we get the following steady state probability of occurrence of events B_1, B_2, B_3, and B_4:

$$UA = \frac{0.0002}{0.0002+4}$$

$$\simeq 4.99 \times 10^{-5}$$

Similarly, inserting the specified data into Equation (7.27), the following steady state probability of occurrence of events B_5 and B_7 is obtained:

$$UA = \frac{0.0001}{0.0001+0.004}$$

$$= 0.0244$$

Finally, substituting the given data into Equation (7.27), we get the following steady state probability of occurrence of event B_6:

$$UA = \frac{0.00015}{0.00015+4}$$

$$= 3.75 \times 10^{-5}$$

Using the UA calculated value for events B_1, B_2, B_3, and B_4 in Equation (7.29), we get the following steady state probability of occurrence of event B_8:

$$P(B_8) = \left(4.99 \times 10^{-5}\right)^4$$

$$= 6.25 \times 10^{-18}$$

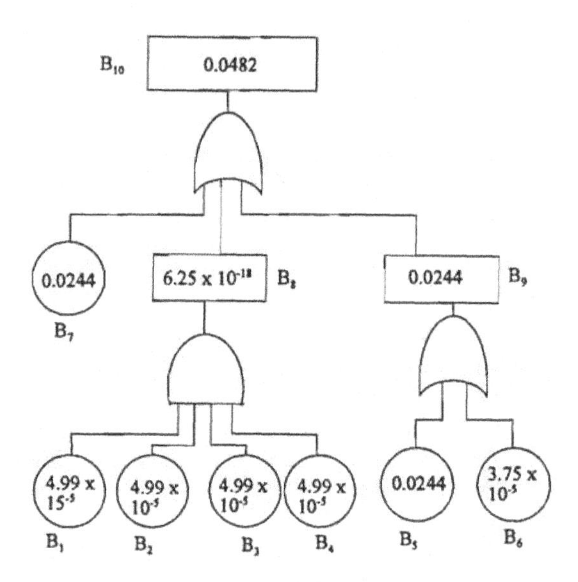

FIGURE 7.11 Calculated steady state probability values of basic repairable events and of intermediate and top events for the fault tree of Figure 7.2.

Substituting the UA calculated values for events B_5 and B_6 into Equation (7.28), we obtain the following steady state probability of occurrence of event B_9:

$$P(B_9) = 1 - (1 - 0.0244)\left(1 - 3.75 \times 10^{-5}\right)$$
$$\simeq 0.0244$$

Inserting the UA calculated values for events B_7, B_8, and B_9 into Equation (7.28), the following steady state probability of occurrence of event B_{10} is obtained:

$$P(B_{10}) = 1 - \left[1 - P(B_7)\right]\left[1 - P(B_8)\right]\left[1 - P(B_9)\right]$$
$$= 1 - [1 - 0.0244]\left[1 - 6.25 \times 10^{-18}\right][1 - 0.0244]$$
$$\simeq 0.0482$$

Thus, the steady state probability of occurrence of the top event, room without light, is 0.0482. The above calculated probability values are shown in the Figure 7.11 fault tree.

7.8 FAULT TREE DUALITY AND FTA APPROACH BENEFITS AND DRAWBACKS

Once a fault tree is developed, it is possible to obtain its dual without much additional effort. The dual of a "fault tree" is the "success tree". Thus, in order to obtain a "success tree" from a "fault tree", replace all OR gates with AND gates in the original fault tree and vice versa, as well as replace all fault events with success events. For example, the top fault event, "room without light", becomes "room lit".

Just like any other reliability analysis method, the FTA approach also has its benefits and drawbacks. Some of the benefits of the FTA approach are as follows:

- It identifies failures deductively.
- It serves as a graphic aid for system management.
- It provides insight into the system behavior.
- It can handle complex systems more easily.
- It allows concentration on one particular failure at a time and requires the analyst to understand thoroughly the system under consideration prior to the starting of FTA.
- It provides options for management and others to conduct either qualitative or quantitative reliability analysis.

On the other hand, some of the drawbacks of the FTA approach include time consuming, costly, end results difficult to check, and considers components in either working or failed state (more specifically, the partial failure states of components are difficult to handle).

7.9 PROBLEMS

1. What are the purposes of performing FTA?
2. Write an essay on the history of the fault tree method.
3. Define the following named symbols used in constructing fault trees:
   - AND gate
   - Diamond
   - Circle
   - Rectangle
4. Discuss the basic process followed in developing fault trees.
5. What are the advantages and disadvantages of the FTA method?
6. Compare the FTA approach with the block diagram method (i.e., network reduction technique) used in reliability analysis.
7. Assume that a windowless room has one switch and one light bulb. Develop a fault tree for the top or undesired event "dark room". Assume the switch only fails to close.
8. Obtain a "success tree" for the fault tree shown in Figure 7.2.
9. Prove the following Boolean expression:

$$X + YZ = (X + Y)(X + Z) \tag{7.30}$$

10. Assume that in Figure 7.2 the basic independent fault events' failure and repair rates are 0.009 failures/h and 0.04 repairs/h, respectively. Calculate the steady state probability of occurrence of the top event, room without light.

7.10 REFERENCES

1. Haasl, D.F., Advanced Concepts in Fault Tree Analysis, System Safety Symposium, 1965. Available from the University of Washington Library, Seattle.
2. Barlow, R.E., Fussell, J.B., and Singpurwalla, N.D., Eds., *Reliability and Fault Tree Analysis,* Society for Industrial and Applied Mathematics (SIAM), Philadelphia, 1975. (Conference proceedings publication.)
3. Dhillon, B.S. and Singh, C., Bibliography of literature on fault trees, *Microelectronics and Reliability,* 17, 501-503, 1978.
4. *Fault Tree Handbook,* Report No. NUREG-0492, U.S. Nuclear Regulatory Commission, Washington, D.C., 1981.
5. Henley, E.J. and Kumamoto, H., *Reliability Engineering and Risk Assessment,* Prentice-Hall, Englewood Cliffs, NJ, 1981.
6. Dhillon, B.S. and Singh, C., *Engineering Reliability: New Techniques and Applications,* John Wiley & Sons, New York, 1981.
7. Schroder, R.J., Fault tree for reliability analysis, *Proc. Annu. Symp. Reliability,* 170-174, 1970.
8. Risk Analysis Using the Fault Tree Technique, Flow Research Report, Flow Research, Inc., Washington, D.C., 1973.
9. Grant Ireson, W., Coombs, C.F., and Moss, R.Y., Eds., *Handbook of Reliability Engineering and Management,* McGraw-Hill, New York, 1996.
10. Barlow, R.E. and Proschan, F., *Statistical Theory of Reliability and Life Testing,* Holt, Rinehart and Winston, New York, 1975.
11. Fussell, J.B. and Vesely, W.E., A new methodology for obtaining cut sets for fault trees, *Trans. Am. Nucl. Soc.,* 15, 262-263, 1972.
12. Semanderes, S.N., Elraft, a computer program for efficient logic reduction analysis of fault trees, *IEEE Trans. Nuclear Sci.,* 18, 310-315, 1971.
13. Dhillon, B.S., Proctor, C.L., and Kothari, A., On repairable component fault trees, *Proc. Annu. Reliability Maintainability Symp.,* 190-193, 1979.

8 Common Cause Failures and Three State Devices

8.1 INTRODUCTION

Over the years, common cause failures have been receiving increasing attention because of the realization that the assumption of independent failures may be violated in the real life environment. For example, according to Reference 1, in the U.S. power reactor industry "of 379 components failures or groups of failures arising from independent causes, 78 involved common cause failures of two or more components". A common cause failure may simply be described as any instance where multiple items fail due to a single cause. Thus, the occurrence of common cause failures leads to lower system reliability.

The late 1960s and early 1970s may be regarded as the beginning of the serious realization of common cause failure problems when a number of publications on the subject appeared [2-6]. In 1975, a number of methods were proposed to evaluate reliablity of systems with common cause failures [7, 8]. In 1978, an article presented most of the publications on common cause failures [9]. The first book that gave an extensive covereage to common cause failures appeared in 1981 [10]. In 1994, an article entitled "Common Cause Failures in Engineering Systems: A Review" presented a comprehensive review of the subject [11]. Nonetheless, ever since the late 1960s many people have contributed to the subject of common cause failures and over 350 contributions from these individuals can be found in Reference 11.

A device is said to have three states if it operates satisfactorily in its normal mode but fails in either of two mutually exclusive modes. Two examples of a three state device are an electronic diode and a fluid flow valve. The two failure modes pertaining to both these devices are open, short and open, close, respectively. In systems having these devices, the redundancy may increase or decrease the system reliability. This depends on three factors: type of system configuration, number of redundant components, and the dominant mode of component failure.

The history of reliaiblity studies of three state devices may be traced back to the works of Moore and Shannon [12] and Creveling [13] in 1956. In 1975, the subject of three state device systems was studied in depth and a delta-star method to evaluate reliability of complex systems made up of three state devices was developed [14]. In 1977, an article entitled "Literature Survey on Three State Device Reliability Systems" provided a list of publications on the subject [15]. A comprehensive review on the subject appeared in 1992 [16]. Needless to say, many other individuals have also contributed to the topic and most of their contributions are listed in Reference 16.

This chapter presents the topics of common cause failures and three state devices, separately.

8.2 COMMON CAUSE FAILURES

There have been numerous instances of the occurrence of common cause failures in industry, particularly in the nuclear power generation. Three examples of the occurrence of the common cause failures in the nuclear power generation are as follows [17]:

1. Two motorized valves were placed in a failed state due to a maintenance error of wrongly disengaging the clutches.
2. Some spring loaded relays forming a parallel configuration failed simultaneously due to a common cause.
3. A steam line rupture caused mutliple circuit board failures. In this case, the common cause was the steam line rupture.

In passing, it is to be noted that, in some cases, instead of triggering a total failure of redundant systems (i.e., simultaneous failure), a common cause may lead to less severe but common degradation of redundant units. Obviously, this will increase the joint probability of the system units' failure and the redundant unit may fail sometime later than the first unit failure. Thus, the failure of the second unit is dependent and coupled to the first unit failure because of the common morose environment.

There are many causes for the occurrence of common cause failures. Some of them are as follows [10]:

- **Common external environment.** This includes temperature, humidity, vibrations, dust, and moisture.
- **Operations and maintenance errors.** These are due to factors such as carelessness, incorrect maintenance, and wrong adjustment or calibration by humans.
- **Deficiency in equipment design.** This is the result of oversights during the design phase. For example, failure to consider interdependence between electrical and mechanical parts.
- **External catastrophic events.** These include natural phenomena such as earthquake, fire, tornado, and flood. The occurrence of natural events such as these may cause the simultaneous failure of redundant units.
- **Functional deficiency.** This may result due to poorly designed protective action or inappropriate instrumentation.
- **Common external power source.** Redundant units fed from the same power source (directly or indirectly) may fail simultaneously in the event of the power source failure.
- **Common manufacturer.** The redundant units purchased from the same manufacturer may have the same fabrication, design, or other defects. Some examples of the fabrication defects are wiring a circuit board backward, use of incorrect material, and poor soldering. Thus, the manufacturing defects such as these may cause the simultaneous failure of redundant units.

Three different methods to handle the occurrence of common cause failures in reliability analysis of engineering systems are presented below.

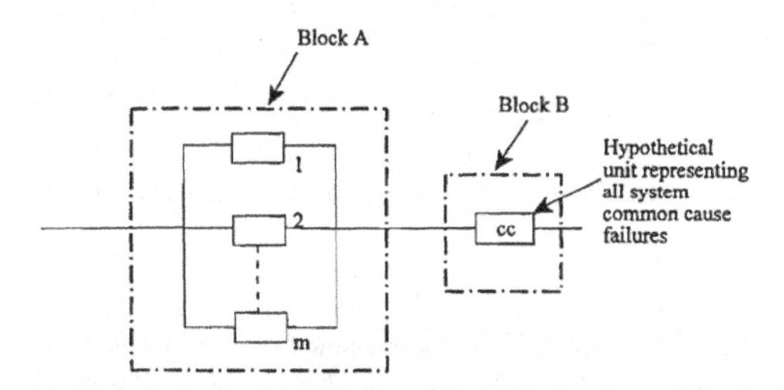

FIGURE 8.1 Block diagram of a parallel system with common cause failures. m is the number of units and cc is the hypothetical unit representing common cause failures.

8.2.1 Block Diagram Method

This is a simple and straightforward approach to analyze redundant systems with common cause failures. The basis for this approach is to represent all common cause failures associated with a system as a single hypothetical unit and then connect it in series with the block diagram of a redundant system representing independent units or failures. From there on the reliability of the system is determined by using this overall block diagram and applying the usual network reduction approach.

The method is demonstrated by applying it to parallel and k-out-of-m networks as discussed below.

Parallel Network

Figure 8.1 shows block diagram of a parallel system with common cause failures. This figure is composed of two blocks (i.e., blocks A and B) in series. Block A represents the normal parallel system with independent failures and block B contains a single hypothetical unit representing all the common cause failures associated with the parallel system.

Assume that a unit failure rate, λ, is expressed by

$$\lambda = \lambda_{in} + \lambda_{cc} \tag{8.1}$$

where

λ_{in} is the unit independent mode failure rate.
λ_{cc} is the unit or system common cause failure rate.

Thus, the fraction of common cause type unit failures is given by

$$\theta = \frac{\lambda_{cc}}{\lambda} \tag{8.2}$$

Therefore,

$$\lambda_{cc} = \theta \, \lambda \tag{8.3}$$

By inserting Equation (8.3) into Equation (8.1) and rearranging, we get

$$\lambda_{in} = (1 - \theta) \, \lambda \tag{8.4}$$

In Figure 8.1 block A represents independent mode failures and block B represents the common cause failures. The complete system can fail only if the hypothetical unit representing all system common cause failures fails or all units of the parallel system fail. The reliability of the independent modified parallel system shown in Figure 8.1 is given by

$$\begin{aligned} R_{mp} &= R_A \cdot R_B \\ &= \left[1 - \left(1 - R_{in}\right)^m\right] R_{cc} \end{aligned} \tag{8.5}$$

where

R_{mp} is the reliability of the modified parallel system (i.e., the parallel system with common cause failures).
R_A is the reliability of block A.
R_B is the reliability of block B.
R_{in} is the unit's independent failure mode relialibility.
R_{cc} is the system common cause failure mode reliaiblity or the reliability of the hypothetical unit representing all system common cause failures.
m is the number of identical units in parallel.

For constant failure rates, λ_{in} and λ_{cc}, we get the following reliability equations, respectively:

$$R_{in}(t) = e^{-(1-\theta)\lambda t} \tag{8.6}$$

$$R_{cc}(t) = e^{-\theta \lambda t} \tag{8.7}$$

where t is time.

Substituting Equations (8.6) and (8.7) into Equation (8.5) yields

$$R_{mp}(t) = \left[1 - \left\{1 - e^{-(1-\theta)\lambda t}\right\}^m\right] e^{-\theta \lambda t} \tag{8.8}$$

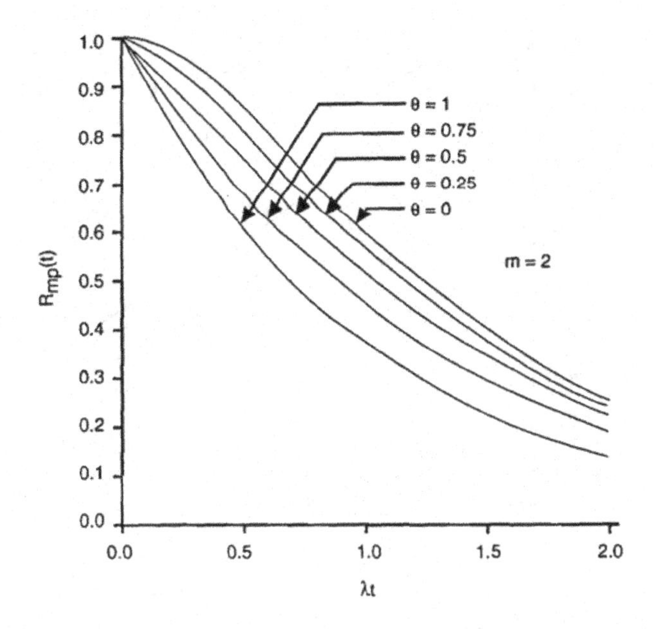

FIGURE 8.2 Reliability plots of a two-unit parallel system with common cause failures.

It can be seen from Equations (8.2) and (8.8) that the parameter θ takes values from 0 to 1. Thus, at $\theta = 0$ Equation (8.8) becomes simply a reliability expression for a parallel system without common cause failures. More specifically, there are no common cause failures associated with the parallel system. At $\theta = 1$, it means all the parallel system failures are of common cause type and the system simply acts as a single unit with failure rate, λ.

Figures 8.2 and 8.3 show plots of Equation (8.8) for various specified values of θ and for m = 2 and 3, respectively. In both the cases, the system reliability decreases with the increasing values of θ.

The system mean time to failure is given by [18]:

$$\begin{aligned} MTTF_{mp} &= \int_0^\infty R_{mp}(t)\,dt \\ &= \sum_{i=1}^{m} (-1)^{i+1} \binom{m}{i} \Big/ \lambda\{i-(i-1)\theta\} \end{aligned} \tag{8.9}$$

where

$MTTF_{mp}$ is the mean time to failure of the parallel system with common cause failures.

$$\binom{m}{i} = m!/i!(m-i)! \tag{8.10}$$

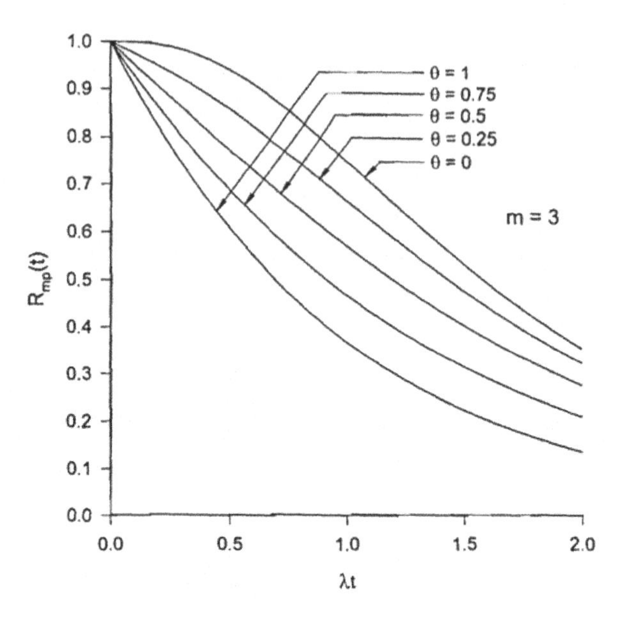

FIGURE 8.3 Reliability plots of a three-unit parallel system with common cause failures.

The system hazard rate is given by [18]:

$$\lambda_{mp}(t) = -\frac{1}{R_{mp}(t)}\frac{dR_{mp}(t)}{dt}$$
$$= \theta\lambda + m\lambda(1-\theta)\left[(\beta-1)/(\beta^m-1)\right] \quad (8.11)$$

where

$\lambda_{mp}(t)$ is the hazard rate of the parallel system with common cause failures.

$$\beta = 1/\left\{1-e^{-(1-\theta)\lambda t}\right\} \quad (8.12)$$

k-out-of-m Network

Just like in the case of the parallel network, a hypothetical unit respresenting all system common cause failures is connected in series with the independent and identical unit k-out-of-m system. More specifically, this modified k-out-of m network is same as the modified parallel network except that at least k units must function normally instead of at least one unit for the parallel case. Thus, the k-out-of-m system with common cause failures reliability is [10]

$$R_{k/m} = \left[\sum_{i=k}^{m}\binom{m}{i} R_{in}^{i}\left(1-R_{in}\right)^{m-i}\right] R_{cc} \tag{8.13}$$

where

k is the required number of units for the k-out-of-m system success.
$R_{k/m}$ is the reliability of the k-out-of-m system with common cause failures.

Inserting Equations (8.6) and (8.7) into Equation (8.13) we get

$$R_{k/m}(t) = \sum_{i=k}^{m}\binom{m}{i} e^{-i(1-\theta)\lambda t}\left[1-e^{-i(1-\theta)\lambda t}\right]^{m-i} e^{-\theta\lambda t} \tag{8.14}$$

In a manner similar to the parallel network case, the expressions for system mean time to failure and hazard rate may be obtained. Also, the block diagram method can be applied to other redundant configurations such as parallel-series and bridge [18].

Example 8.1

Assume that a system is composed of two independent, active and identical units. At least one unit must function normally for the sytem success. The value of the θ is 0.5. Calculate the system reliability for a 100-h mission, if the unit failure rate is 0.004 failures/h.

Substituting the given data into Equation (8.8) yields

$$R_{mp}(100) = \left[1-\left\{1-e^{-(1-0.5)(0.004)(100)}\right\}^{2}\right] e^{-(0.5)(0.004)(100)}$$

$$= 0.7918$$

Thus, the reliability of the two-unit parallel system with common cause failures is 0.7918.

Example 8.2

If the value of θ is zero in Example 8.1, calculate the system reliability and compare it with the system reliability result obtained in that example.

Using the given data in Equation (8.8) we get

$$R_{mp}(100) = 1-\left\{1-e^{-(0.004)(100)}\right\}^{2}$$

$$= 0.8913$$

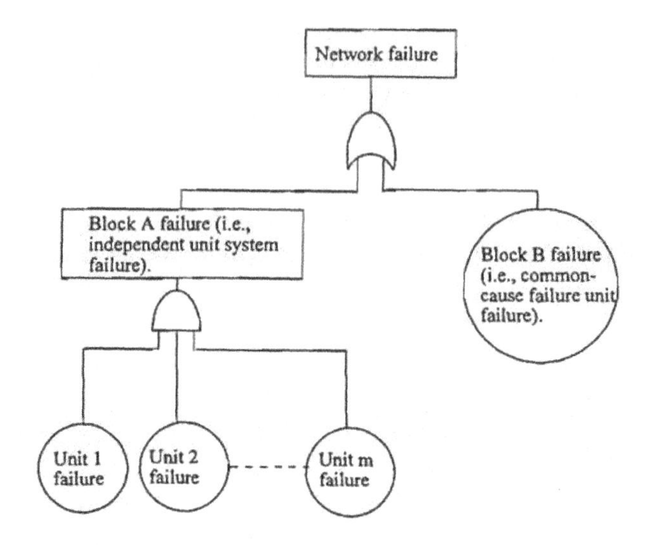

FIGURE 8.4 Fault tree for Figure 8.1 network.

Thus, in this case the system reliability without common cause failures is 0.8913. In comparison to the system reliability result obtained in Example 8.1, i.e., 0.7918, the occurrence of common cause failures leads to lower system reliability.

8.2.2 Fault Tree Method

This is another method that can be used to perform reliability analysis of systems with common cause failures. The application of this method is demonstrated by applying it to the network shown in Figure 8.1. In this case, the top event is *Network Failure*. The fault tree for the Figure 8.1 network is shown in Figure 8.4. The probability of the top event occurrence, i.e., network failure, is given by

$$F_N = F_A + F_B - F_A F_B \tag{8.15}$$

where

F_N is the network failure probability, i.e., failure propability of the parallel system with common cause failures.
F_A is the block A failure probability, i.e., failure probability of the parallel system with independent failures.
F_B is the block B failure probability, i.e., the failure probability of the hypothetical unit representing all system common cause failures.

In turn,

$$F_A = \prod_{i=1}^{m} F_i \tag{8.16}$$

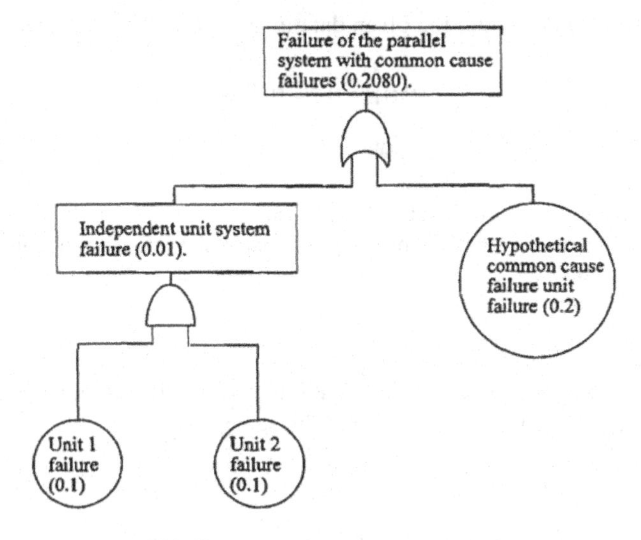

FIGURE 8.5 Fault tree for Example 8.3.

where

F_i is the failure probability of independent unit i; for i = 1, 2, ---, m.

The network reliability (i.e., the reliability of the parallel system with common cause failures) is given by

$$R_{mp} = 1 - F_N \tag{8.17}$$

Example 8.3

Assume that a parallel system with common cause failures is composed of two units and the system common cause failure occurrence probability is 0.2. In addition, each unit's independent failure probability is 0.1. Calculate the failure probability of the parallel system with common cause failures using the fault tree method.

For m = 2, the fault tree shown in Figure 8.4 will represent this example and the specific fault tree is given in Figure 8.5. In this case, the top event is the failure of the parallel system with common cause failures. The occurrence probability of the top event can be calculated by using Equations (8.15) and (8.16).

Substituting the given data into Equation (8.16) yields

$$\begin{aligned} F_A &= F_1 F_2 \\ &= (0.1)(0.1) \\ &= 0.01 \end{aligned}$$

Using the calculated and the given data in Equation (8.15), we get

$$F_N = (0.01) + 0.2 - (0.01)(0.2)$$
$$= 0.2080$$

The given and the calculated probability values are shown in the Figure 8.5 fault tree. The failure probability of the two-unit parallel system with common cause failures is 0.2080.

Example 8.4

By using the given data of Example 8.3 and the block diagram method, prove that the end result obtained in Example 8.3 is the same.

Thus, we have

$$R_{CC} = 1 - 0.2$$
$$= 0.8$$

$$R_{in} = 1 - 0.1$$
$$= 0.9$$

$$m = 2$$

By inserting the above values into Equation (8.5) we get

$$R_{mp} = \left[1 - (1 - 0.9)^2\right](0.8)$$
$$= 0.7920$$

Using the above result in Equation (8.17) yields

$$F_N = 0.2080$$

It proves that Example 8.3 solved through two different methods, fault tree and block diagram, gives the same end result.

8.2.3 Markov Method

This is another method that can be used to perform reliability analysis of systems with common cause failures. The systems can be either repairable or nonrepairable. The following three models demonstrate the applicability of the method to redundant systems with common cause failures:

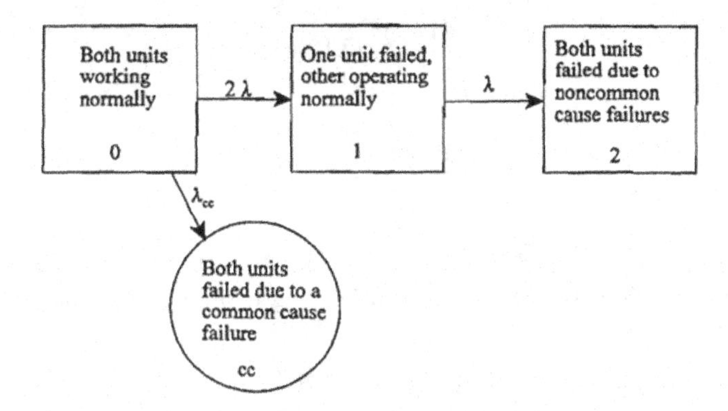

FIGURE 8.6 State space diagram of a two-unit parallel system with common cause failures. The numbers and letters denote system state.

Model I

This model represents a two independent unit parallel system with common cause failures [19]. The system can fail due to a common cause failure when both the units are active. For the system success, at least one unit must operate normally. The system transition diagram is shown in Figure 8.6.

The following assumptions are associated with this model:

- Both units are identical and active.
- Common cause and other failures are statistically independent.
- Common cause and other failure rates are constant.
- No failed unit is repaired.

The following notation is associated with this model:

i is the ith system state: i = 0 (means both units working normally), i = 1 (means one unit failed, other operating normally), i = 2 (means both units failed due to noncommon cause failures), i = cc (means both units failed due to a common cause failure).

$P_i(t)$ is the probability that the system is in state i at time t; for i = 0, 1, 2, cc.

λ_{cc} is the system common cause failure rate.

λ is the unit failure rate.

The following equations are associated with Figure 8.6:

$$\frac{dP_0(t)}{dt}+(2\lambda+\lambda_{cc})P_0(t)=0 \tag{8.18}$$

$$\frac{dP_1(t)}{dt}+\lambda P_1(t)=2\lambda P_0(t) \tag{8.19}$$

$$\frac{dP_2(t)}{dt} = \lambda P_1(t) \tag{8.20}$$

$$\frac{dP_{cc}(t)}{dt} = \lambda_{cc} P_0(t) \tag{8.21}$$

At time $t = 0$, $P_0(0) = 1$ and $P_1(0) = P_2(0) = P_{cc}(0) = 0$.

Solving Equations (8.18) through (8.21) using Laplace transforms [10], we get

$$P_0(t) = e^{-(2\lambda+\lambda_{cc})t} \tag{8.22}$$

$$P_1(t) = X\left[e^{-\lambda t} - e^{-(2\lambda+\lambda_{cc})t}\right] \tag{8.23}$$

where

$$X = 2\lambda/(2\lambda + \lambda_{cc} - \lambda) \tag{8.24}$$

$$P_2(t) = Y + X\left[Ze^{-(2\lambda+\lambda_{cc})t} - e^{-\lambda t}\right] \tag{8.25}$$

where

$$Y = 2\lambda/(2\lambda + \lambda_{cc}) \tag{8.26}$$

$$Z = \lambda/(2\lambda + \lambda_{cc}) \tag{8.27}$$

$$P_{cc}(t) = \frac{\lambda_{cc}}{2\lambda + \lambda_{cc}}\left[1 - e^{-(2\lambda+\lambda_{cc})t}\right]. \tag{8.28}$$

The reliability of the parallel system with common cause failures is

$$\begin{aligned} R_{ps}(t) &= P_0(t) + P_1(t) \\ &= e^{-(2\lambda+\lambda_{cc})t} + X\left[e^{-\lambda t} - e^{-(2\lambda+\lambda_{cc})t}\right] \end{aligned} \tag{8.29}$$

The mean time to failure of the parallel system with common cause failures is given by

$$MTTF_{ps} = \int_0^{\infty} R_{ps}(t)\,dt = 3/(2\lambda + \lambda_{cc}) \tag{8.30}$$

where

$MTTF_{ps}$ is the mean time to failure of the two unit parallel system with common cause failures.

The probability of the system failure due to a common cause failure is given by

$$P_{cc}(t) = \frac{\lambda_{cc}}{2\lambda + \lambda_{cc}}\left[1 - e^{-(2\lambda+\lambda_{cc})t}\right] \tag{8.31}$$

Example 8.5

Assume that a system has two identical units in parallel and it can fail due to the occurrence of a common cause failure. The common cause failure occurrence rate is 0.0005 failures/h. Calculate the following:

if the unit failure rate is 0.004 failures/h:

- System mean time to failure
- System mean time to failure, if the occurrence of common cause failures is equal to zero.

Compare the above two results.

By substituting the given data into Equation (8.30), we get

$$MTTF = 3/[2(0.004) + 0.0005] = 353 \text{ h}$$

For zero common cause failures and the other given data, Equation (8.30) yields

$$MTTF = 3/2(0.004) = 375 \text{ h}$$

It means that the occurrence of common cause failures leads to the reduction in the system mean time to failure (i.e., from 375 h to 353 h).

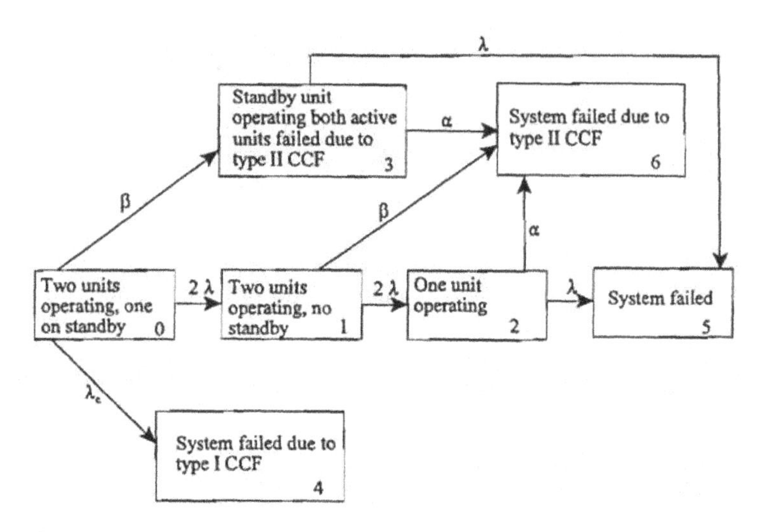

FIGURE 8.7 The state transition diagram of the system without repair. CCF means common cause failure.

Model II

This model represents a system composed of two active parallel units and one standby subject to two types of common cause failures [20]. The type I common cause failure leads to total system failure (active plus the standby units) whereas, type II failures only cause the failure of the active two units. The system units can also malfunction due to a normal failure.

The system transition diagram is shown in Figure 8.7. The system starts operating (two parallel units and one unit on standby) at time $t = 0$. As soon as one of the operating units fails due to a normal failure, it is immediately replaced with the standby unit. After this, the system may fail due to a common cause failure or one of the operating units fails because of the normal failure. The single operating unit may fail either due to a normal failure or a common cause failure.

In the state when two units operate and one is on standby, the system units may also fail due to type I and type II common cause failures. In the case of the occurrence of type II common cause failure, the standby unit is immediatly activated. For the system success, at least one of these units must operate normally.

The following assumptions are associated with the analysis of the Figure 8.7 system transition diagram.

i. All system units are identical and fail independently.
ii. All failure rates associated with the system are constant.
iii. The switching mechanism for the standby is instantaneous and perfect.
iv. The system starts at time $t = 0$ with three good units (i.e., two in parallel and one on standby).
v. The failed system is never repaired.
vi. The standby unit remains as good as new in its standby mode.

The following symbols are associated with this model:

- i The state of the system shown in Figure 8.7: $i = 0$ (two units operating in parallel and one on standby), $i = 1$ (one of the operating units fails and the standby is activated), $i = 2$ (one unit operating, two units failed), $i = 3$ (two active parallel units failed due to a type II common cause failure and the standby unit is activated), $i = 4$ (system failed, all three units failed due to a type I common cause failure), $i = 5$ (system failed due to normal failures), $i = 6$ (system failed due to a common cause failure when operating with two or less good units).
- $P_i(t)$ The probability that the system is in state i at time t; for $i = 0, 1, 2, 3, 4, 5, 6$.
- λ Failure rate of a unit.
- β Type II system common cause failure rate.
- λ_c Type I system common cause failure rate.
- α Common cause failure rate of the system with one good unit.
- $R(t)$ System reliability at time t.
- MTTF System mean time to failure.

The system of equations associated with Figure 8.7 is

$$\frac{dP_0(t)}{dt} + (2\lambda + \lambda_c + \beta) P_0(t) = 0 \tag{8.32}$$

$$\frac{dP_1(t)}{dt} + (2\lambda + \beta) P_1(t) = 2\lambda P_0(t) \tag{8.33}$$

$$\frac{dP_2(t)}{dt} + (\lambda + \alpha) P_2(t) = 2\lambda P_1(t) \tag{8.34}$$

$$\frac{dP_3(t)}{dt} + (\alpha + \lambda) P_3(t) = \beta P_0(t) \tag{8.35}$$

$$\frac{dP_4(t)}{dt} = P_0(t)\lambda_c \tag{8.36}$$

$$\frac{dP_5(t)}{dt} = P_2(t)\lambda + P_3(t)\lambda \tag{8.37}$$

$$\frac{dP_6(t)}{dt} = P_1(t)\beta + P_2(t)\alpha + P_3(t)\alpha \tag{8.38}$$

At time $t = 0$, $P_0(0) = 1$ and $P_1(0) = P_2(0) = P_3(0) = P_4(0) = P_5(0) = P_6(0) = 0$.

Solving Equations (8.32) through (8.38), we get the following state probability expressions:

$$P_0(t) = e^{-At} \tag{8.39}$$

$$P_1(t) = \frac{2\lambda}{\lambda_c}\left(e^{-Bt} - e^{-At}\right) \tag{8.40}$$

$$P_2(t) = 4\lambda^2\left[\frac{e^{-Bt}}{(B-C)(A-C)} + \frac{e^{-BT}}{(C-B)(A-B)} + \frac{e^{-AT}}{(C-A)(B-A)}\right] \tag{8.41}$$

$$P_3(t) = \frac{\beta}{D}\left(e^{-Ct} - e^{-At}\right) \tag{8.42}$$

where

$$A = 2\lambda + \lambda_c + \beta$$

$$B = 2\lambda + \beta$$

$$C = \alpha + \lambda$$

$$D = \lambda + \lambda_c + \beta - \alpha$$

$$P_4(t) = \frac{\lambda_c}{A}\left(1 - e^{-At}\right) \tag{8.43}$$

$$P_5(t) = \frac{\beta\lambda}{CA}\left[1 + \frac{Ce^{-At} - Ae^{-Ct}}{D}\right] + 4\lambda^3\left[\frac{1}{ABC} - \frac{e^{-At}}{AD\lambda_c} - \frac{e^{-Ct}}{CDE} + \frac{e^{-Bt}}{BE\lambda_c}\right] \tag{8.44}$$

where

$$E = \lambda + \beta - \alpha$$

$$P_6(t) = \frac{2\beta\lambda}{AB}\left[1 + \frac{Be^{-At} - Ae^{-Bt}}{\lambda_c}\right] \tag{8.45}$$

$$+\frac{\alpha\beta}{AC}\left[1+\frac{Ce^{-At}-Ae^{-Ct}}{D}\right]$$

$$+4\lambda^2\alpha\left[\frac{1}{ABC}+\frac{e^{-At}}{AD\lambda_c}-\frac{e^{-Ct}}{CDE}+\frac{e^{-Bt}}{BE\lambda_c}\right]$$

The system reliability is given by

$$R(t)=\sum_{i=0}^{3}P_i(t)$$

$$=e^{-At}+\frac{2\lambda}{\lambda_c}\left(e^{-Bt}-e^{-At}\right)$$

$$+4\lambda^2\left[\frac{e^{-Ct}}{(B-C)(A-C)}+\frac{e^{-Bt}}{(C-B)(A-B)}+\frac{e^{-At}}{(C-A)(B-A)}\right] \tag{8.46}$$

$$+\frac{B}{D}\left(e^{-Ct}-e^{-At}\right)$$

The system mean time to failure is

$$MTTF=\int_0^\infty R(t)dt$$

$$=\frac{1}{A}\left[1+\frac{2\lambda}{B}+\frac{4\lambda^2}{BC}+\frac{\beta}{C}\right] \tag{8.47}$$

Model III

This model is the same as Model II except that the repair process is initiated as soon as one of the two active units fails and α is set equal to zero. Thus, the failed unit/units in state 1, 2, and 3 of Figure 8.7 are repaired at a constant repair rate μ.

The additional symbols used in this model are as follows:

- s is the Laplace transform variable.
- $R(s)$ is the Laplace transform of the system reliability, $R(t)$.
- $P_i(s)$ is the Laplace transform of $P_i(t)$, for $i = 0, 1, 2, \ldots, 6$.

Thus, for this model, the Laplace transforms of the state probabilities are as follows:

$$P_0(s) = \frac{s^2 + sG + F}{s^3 + s^2 J(H + GA) + I} \tag{8.48}$$

where

$$\begin{aligned}
F &= 2\lambda^2 + \beta\lambda + \beta\mu + \mu\lambda + \mu^2 \\
G &= 3\lambda + 2\mu + \beta \\
H &= 2\lambda^2 + \beta\lambda + \beta\mu + \mu^2 - \mu\lambda \\
I &= \lambda_c F + 4\lambda^3 + 4\lambda^2\beta + 3\lambda\mu\beta + \beta^2\lambda + \beta^2\mu + \mu^2\beta \\
J &= 5\lambda + 2\mu + 2\beta + \lambda_c
\end{aligned} \tag{8.49}$$

$$P_1(s) = \frac{P_0(s) 2\lambda(s + \lambda + \mu)}{s^2 + sG + F} \tag{8.50}$$

$$P_2(s) = \frac{P_0(s) 4\lambda^2}{s^2 + sG + F} \tag{8.51}$$

$$P_3(s) = \frac{P_0(s)\beta}{s + \lambda + \mu} \tag{8.52}$$

$$P_4(s) = P_0(s)\lambda_c/s \tag{8.53}$$

$$P_5(s) = \lambda P_2(s) + \lambda P_3(s) \tag{8.54}$$

$$P_6(s) = \frac{2\lambda\beta(s + \lambda + \mu) P_0(s)}{s(s^2 + sG + F)} \tag{8.55}$$

The Laplace transform of system reliability with repair is

$$R_r(s) = \sum_{i=0}^{3} P_i(s) \tag{8.56}$$

The system mean time to failure with repair is

$$\begin{aligned}
MTTF_r &= \lim_{s \to 0} R_r(s) \\
&= \frac{1}{I}\left[F + 2\lambda(\lambda + \mu) + 4\lambda^2 + \frac{F\beta}{\lambda + \mu}\right]
\end{aligned} \tag{8.57}$$

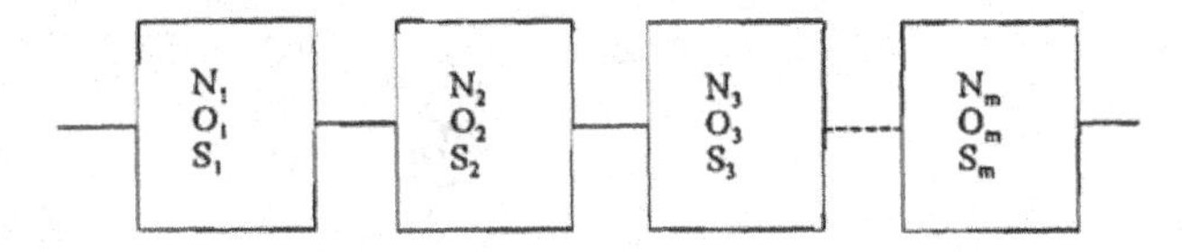

FIGURE 8.8 A three state device series system.

8.3 THREE STATE DEVICES

Three state devices may form configurations such as series, parallel, series-parallel, and bridge. The reliability evaluation mechanism of these configurations is different than the configurations made up of two state devices (a device is said to have two states, if it either operates or fails). The ordinary reliability configurations fall under the two state device category. Nonetheless, the three state device configurations are more complex to analyze and redundancy may lead to a decrease in the overall system reliability. This reduction in system reliability is due to factors such as dominant mode of the device failure, number of redundant units, and the redundant configuration type.

The reliability evaluation and optimization of series and parallel networks composed of three state devices are presented below.

8.3.1 Reliability Evaluation of a Series Network

The block diagram of a series network composed of m three state devices is shown in Figure 8.8. Each block represents a three state device i that can operate normally (N_i) or fail either open (O_i) or short (S_i) circuited. The system operates normally when either all the devices function successfully or until at least one device is working normally and the remaining ones have failed in their short circuit mode. This is based on the assumption that the current can still flow through a short circuited device. On the other hand, the system fails when either any one of the devices fails in open mode or all the system devices are short circuited. A reliability equation for this network can be developed by using the probability tree method [21]. A probability tree for a two component three state device network is given in Figure 8.9.

The dot on the extreme left side of Figure 8.9 signifies the point of origin of paths and the dots on the extreme right side indicate the paths' termination points. The total number of different paths in a three state device probability tree is given by 3^m. The m denotes the number of units (i.e., three-state state devices) in a network. The following symbols are used in the probability tree analysis:

- N_i denotes the normal mode (i.e., success) of the ith unit (i.e., ith path link's success)
- O_i denotes the open mode failure state of the ith unit (i.e., ith path link's failure)
- S_i denotes the short mode failure state of the ith unit (i.e., ith path link's failure)
- P_i probability of being in State N_i

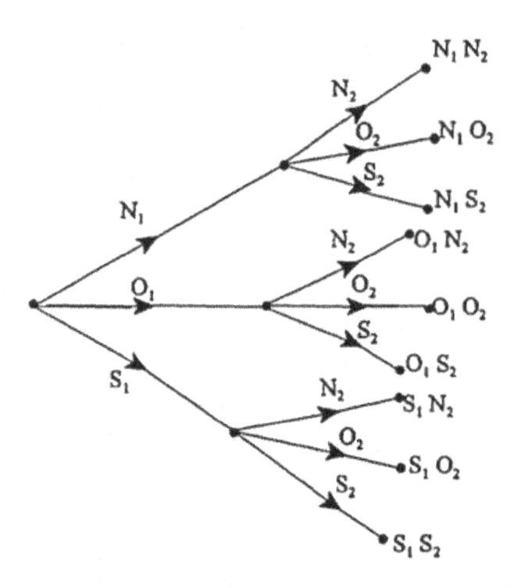

FIGURE 8.9 Probability tree diagram for two non-identical unit three state device network.

q_{Oi} probability of being in State O_i
q_{Si} probability of being in State S_i

From Figure 8.9, the two non-identical unit series system success paths are: N_1N_2, N_1S_2, and S_1N_2. With the aid of these paths and the relationships $P_1 + q_{O1} + q_{S1} = 1$ and $P_2 + q_{O2} + q_{S2} = 1$, the independent series system reliability is

$$\begin{aligned} R_{S2} &= P_1 P_2 + P_1 q_{S2} + P_2 q_{S1} \\ &= (1 - q_{O1})(1 - q_{O2}) - q_{S1} q_{S2} \end{aligned} \tag{8.58}$$

Similarly, the three non-identical unit parallel system success paths are as follows:

$N_1 N_2 N_3$, $N_1 N_2 S_3$, $N_1 S_2 N_3$, $N_1 S_2 S_3$, $S_1 N_2 N_3$, $S_1 N_2 S_3$, $S_1 S_2 N_3$

Thus, the reliability of the three independent non-identical unit series system is given by

$$\begin{aligned} R_{S3} &= P_1 P_2 P_3 + P_1 P_2 q_{S3} + P_1 q_{S2} P_3 + P_1 q_{S2} q_{S3} + q_{S1} P_2 P_3 \\ &\quad + q_{S1} P_2 q_{S3} + q_{S1} q_{S2} P_3 \\ &= (1 - q_{O1})(1 - q_{O2})(1 - q_{O3}) - q_{S1} q_{S2} q_{S3} \end{aligned} \tag{8.59}$$

Thus, for an m unit or device series system, we can generalize Equation (8.59) to get

$$R_{Sm} = \prod_{i=1}^{m} (1 - q_{Oi}) - \prod_{i=1}^{m} q_{Si} \tag{8.60}$$

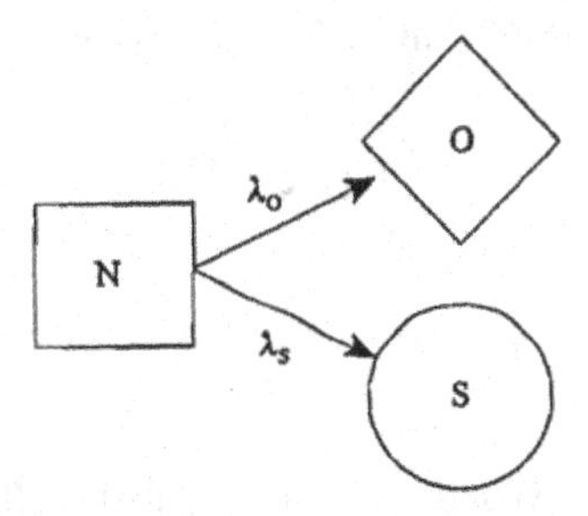

FIGURE 8.10 State space diagram of a three state device. N, O, and S in rectangle, diamond, and circle mean device operating normally, device failed in open mode, and device failed in short mode, respectively.

For identical units or devices, Equation (8.60) simplifies to

$$R_{Sm} = (1 - q_O)^m - q_S^m \tag{8.61}$$

where

q_o is the open mode failure probability of the device.
q_S is the short mode failure probability of the device.

For constant open and short mode failure rates, we use the Markov method to obtain the open and short failure probabilities of a three state device. Figure 8.10 presents the state space diagram of a three state device.

The following symbols are associated with the Figure 8.10 model:

- i is the ith state of the three state device: i = N (means the three state device is operating normally), i = O (means the three state device failed in open mode), i = S (means the three state device failed in short mode).
- $P_i(t)$ is the probability that the three state device is in state i at time t; for i = N, O, S.
- λ_O is the constant open mode failure rate of the three state device.
- λ_S is the constant short mode failure rate of the three state device.

The following system of equations is associated with Figure 8.10.

$$\frac{dP_N(t)}{dt} + (\lambda_O + \lambda_S)P_N(t) = 0 \tag{8.62}$$

$$\frac{dP_O(t)}{dt} - \lambda_O P_N(t) = 0 \tag{8.63}$$

$$\frac{dP_S(t)}{dt} - \lambda_S P_N(t) = 0 \tag{8.64}$$

At time t = 0, $P_N(0) = 1$ and $P_O(0) = P_S(0) = 0$.

By solving Equations (8.62) through (8.64) we get

$$P_N(t) = e^{-(\lambda_O + \lambda_S)t} \tag{8.65}$$

$$P_O(t) = q_O(t) = \frac{\lambda_O}{\lambda_O + \lambda_s}\left[1 - e^{-(\lambda_O + \lambda_S)t}\right] \tag{8.66}$$

Substituting Equations (8.66) and (8.67) into Equation (8.60) yields

$$R_{Sm}(t) = \prod_{i=1}^{m}\left[1 - \frac{\lambda_{Oi}}{\lambda_{Oi} + \lambda_{Si}}\left\{1 - e^{-(\lambda_{Oi} + \lambda_{Si})t}\right\}\right] - \prod_{i=1}^{m}\frac{\lambda_{Si}}{\lambda_{Oi} + \lambda_{Si}}\left\{1 - e^{-(\lambda_{Oi} + \lambda_{Si})t}\right\} \tag{8.68}$$

$$P_S(t) = q_S(t) = \frac{\lambda_S}{\lambda_O + \lambda_s}\left[1 - e^{-(\lambda_O + \lambda_S)t}\right] \tag{8.67}$$

where

$R_{Sm}(t)$ is the three state device series system reliability at time t.

For identical devices, Equation (8.68) becomes

$$R_{Sm}(t) = \left[1 - \frac{\lambda_O}{\lambda_O + \lambda_S}\left\{1 - e^{-(\lambda_O + \lambda_S)t}\right\}\right]^m - \left[\frac{\lambda_S}{\lambda_O + \lambda_S}\left\{1 - e^{-(\lambda_O + \lambda_S)t}\right\}\right]^m \tag{8.69}$$

Example 8.6

A system is composed of three independent and identical three state devices in series. Each device's open mode and short mode failure probabilities are 0.05 and 0.03, respectively. Calculate the system reliability.

Substituting the given data into Equation (8.61) yields

$$R_{S3} = (1 - 0.05)^3 - (0.03)^3$$
$$= 0.8573$$

Thus, the series system reliability is 0.8573.

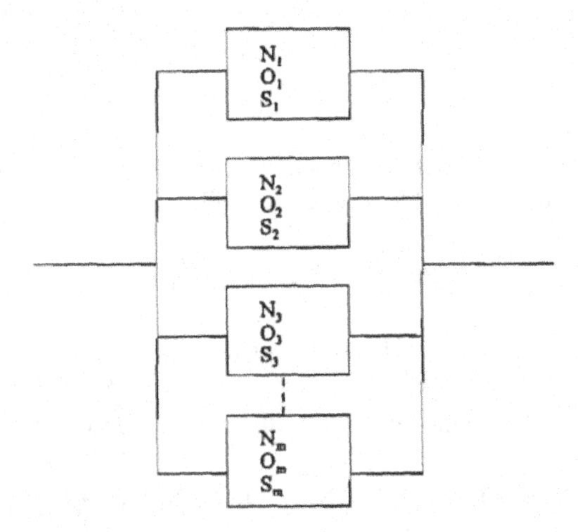

FIGURE 8.11 Block diagram of a three state device parallel network.

Example 8.7

A three state device network has two independent and identical units in series. Each unit's open mode and short mode failure rates are 0.002 and 0.004 failures/h, respectively. Calculate the network reliability for a 100-h mission.

Substituting the given data into Equation (8.69) yields

$$R_{S2}(100) = \left[1 - \frac{0.002}{0.002+0.004}\left\{1-e^{-(0.002+0.004)(100)}\right\}\right]^2$$

$$-\left[\frac{0.004}{0.002+0.004}\left\{1-e^{-(0.002+0.004)(100)}\right\}\right]^2$$

$$= 0.6314$$

Thus, the network reliability for a 100-h mission is 0.6314.

8.3.2 Reliability Evaluation of a Parallel Network

Figure 8.11 shows the block diagram of a parallel network having m three state devices. Each block denotes a three state device i that can operate normally (N_i) or fail either open (O_i) or short (S_i) circuited. The system functions normally when either all the devices work successfully or until at least one device operates normally and the remaining ones have failed in their open mode. In contrast, the system fails when all the devices fail in their open mode or any one of them fails in short mode.

As for the series system, the Figure 8.9 probability tree can be used to develop a reliability expression for the two unit parallel system. Thus, from Figure 8.9 the two non-identical device parallel system success paths are as follows:

- $N_1 N_2$
- $N_1 O_2$
- $O_1 N_2$

With the aid of the above paths and the relationships $P_1 + q_{O1} + q_{S1} = 1$ and $P_2 + q_{O2} + q_{S2} = 1$, the independent parallel network reliability is

$$R_{p2} = P_1 P_2 + P_1 q_{O2} + P_2 q_{O1} = (1 - q_{S1})(1 - q_{S2}) - q_{O1} q_{O2} \tag{8.70}$$

Similarly, the three non-identical unit parallel system success paths are as follows:

$$N_1 N_2 N_3, N_1 N_2 O_3, N_1 O_2 N_3, N_1 O_2 O_3, O_1 N_2 N_3, O_1 N_2 O_3, O_1 O_2 N_3$$

With the aid of the above paths, the reliability of the three independent non-identical unit parallel system is given by

$$\begin{aligned} R_{p3} &= P_1 P_2 P_3 + P_1 P_2 q_{O3} + P_1 q_{O2} P_3 + P_1 q_{O2} q_{O3} + q_{O1} P_2 P_3 \\ &+ q_{O1} P_2 q_{O3} + q_{O1} q_{O2} P_3 \\ &= (1 - q_{S1})(1 - q_{S2})(1 - q_{S3}) - q_{O1} q_{O2} q_{O3} \end{aligned} \tag{8.71}$$

Thus, for an m unit or device parallel system, we can generalize Equation (8.71) as follows:

$$R_{pm} = \prod_{i=1}^{m} (1 - q_{Si}) - \prod_{i=1}^{m} q_{Oi} \tag{8.72}$$

For identical units or devices, Equation (8.72) simplifies to

$$R_{pm} = (1 - q_S)^m - q_O^m \tag{8.73}$$

For a device's constant open and short mode failure rates, λ_O and λ_S, respectively, we substitute Equations (8.66) and (8.67) into Equation (8.72) to get

$$\begin{aligned} R_{pm}(t) &= \prod_{i=1}^{m} \left[1 - \frac{\lambda_{Si}}{\lambda_{Oi} + \lambda_{Si}} \left\{ 1 - e^{-(\lambda_{Oi} + \lambda_{Si})t} \right\} \right] \\ &- \prod_{i=1}^{m} \frac{\lambda_{Oi}}{\lambda_{Oi} + \lambda_{Si}} \left\{ 1 - e^{-(\lambda_{Oi} + \lambda_{Si})t} \right\} \end{aligned} \tag{8.74}$$

where

$R_{pm}(t)$ is the three state device parallel system reliability at time t.

For identical devices, Equation (8.74) simplifies to

$$R_{pm}(t) = \left[1 - \frac{\lambda_S}{\lambda_O + \lambda_S}\left\{1 - e^{-(\lambda_O + \lambda_S)t}\right\}\right]^m - \left[\frac{\lambda_S}{\lambda_O + \lambda_S}\left\{1 - e^{-(\lambda_O + \lambda_S)t}\right\}\right]^m \tag{8.75}$$

Example 8.8

Assume that in Example 8.6 the system is parallel instead of series and the other data are exactly the same. Calculate the parallel system reliability.

Inserting the given values into Equation (8.73) we get

$$R_{p3} = (1 - 0.03)^3 - (0.05)^3$$
$$= 0.9125$$

Thus, the parallel system reliability is 0.9125.

Example 8.9

Assume that in Example 8.7 the three state device network is parallel instead of series and the other given data are exactly the same. Determine the parallel network reliability for a 100-h mission.

Thus, substituting the specified values into Equation (8.75) yields

$$R_{p2}(100) = \left[1 - \frac{0.004}{0.002 + 0.004}\left\{1 - e^{-(0.002+0.004)(100)}\right\}\right]^2 - \left[\frac{0.002}{0.002 + 0.004}\left\{1 - e^{-(0.002+0.004)(100)}\right\}\right]^2$$
$$= 0.4663$$

Thus, the parallel network reliability for a 100-h mission is 0.4663.

8.3.3 Reliability Optimization of a Series Network

In this case for independent and identical devices, the series network reliability from Equation (8.61) is [10]

$$R_{Sm} = (1-q_o)^m - q_s^m \tag{8.76}$$

In order to obtain the optimum number of three state devices to be connected in series, we take partial derivatives of Equations (8.76) with respect to m and equate it to zero. Thus, we have

$$\frac{\partial R_{Sm}}{\partial m} = (1-q_o)^m \ell n(1-q_o) - q_s^m \ell n q_s = 0 \tag{8.77}$$

Solving Equation (8.77), we get

$$m^* = \frac{\ell n\left[\ell n q_s / \ell n(1-q_o)\right]}{\ell n\left[(1-q_o)/q_s\right]} \tag{8.78}$$

where

m* is the optimum number of devices in series.

Example 8.10

Assume that open and short mode failure occurrence probabilities of independent and identical three state devices are 0.02 and 0.05, respectively. Determine the number of devices to be arranged in series to maximize system reliability.

Thus, substituting the given data into Equation (8.78), we get

$$m^* = \frac{\ell n\left[\ell n(0.05)/\ell n(1-0.02)\right]}{\ell n\left[(1-0.02)/(0.05)\right]}$$

$$\simeq 2 \text{ devices}$$

It means two devices should be connected in series to maximize system reliability.

8.3.4 Reliability Optimization of a Parallel Network

From Equation (8.73), the parallel network, with independent and identical devices, reliability is

$$R_{pm} = (1-q_S)^m - q_O^m \tag{8.79}$$

In order to determine the optimum number of three state devices to be arranged in parallel, we take partial derivatives of Equation (8.79) with respect to m and equate it to zero. Thus, we get

$$\frac{\partial R_{pm}}{\partial m} = (1-q_s)^m \ell n(1-q_s) - q_o^m \ell n q_o = 0 \tag{8.80}$$

Solving Equation (8.80) yields

$$m^* = \frac{\ell n\left[\ell n q_o / \ell n(1-q_s)\right]}{\ell n\left[(1-q_s)/q_o\right]} \tag{8.81}$$

where

m* is the optimum number of devices in parallel.

Example 8.11

After examining the failure data of one type of identical three state devices, it is estimated that their open and short mode failure probabilities are 0.08 and 0.04, respectively. Calculate the number of independent devices to be placed in a parallel configuration to maximize system reliability.

Inserting the given data into Equation (8.81) yields

$$m^* = \frac{\ell n\left[\ell n(0.08)/\ell n(1-0.04)\right]}{\ell n\left[(1-0.04)/0.08\right]}$$

$$\simeq 2 \text{ devices}$$

Thus, for maximum system reliability two three state devices should be arranged in parallel.

8.4 PROBLEMS

1. Define the following three terms:
    - Three state device
    - Common-cause failure
    - Short mode failure
2. Discuss at least five causes for the occurrence of common cause failures.
3. Describe the methods that can be used to perform reliability analysis of redundant networks with common cause failures.
4. Using Equation (8.14), obtain an expression for mean time to failure of a k-out-of-m network with common cause failures.
5. Assume that in Figure 8.6 the failed unit is reparable from the system state 1 at a constant repair rate μ. Develop expressions for system reliability and mean time to failure.

6. A system is composed of two identical units in parallel and it can malfunction due to the occurrence of a common cause failure. The common cause failure rate is 0.0001 failures/h. Calculate the system reliability and common cause failure probability for a 200-h mission, if the unit failure rate is 0.001 failures/h.
7. An electronic device can fail either in its open mode or short mode or it simply operates normally. The open and short mode failure rates of the device are 0.007 and 0.005 failures/h, respectively. Calculate the probability of the device failing in its short mode during a 1000-h operational period.
8. A series system is composed of two independent and identical three state devices. The open and short mode failure rates of the device are 0.0003 and 0.007 failures/h, respectively. Calculate the system reliability for a 50-h mission period.
9. A parallel system is composed of three independent and non-identical three state devices. The open mode failure probabilities of devices A, B, and C are 0.02, 0.04, and 0.07, respectively. Similarly, the short mode failure probabilities of devices A, B, C are 0.03, 0.05, and 0.01, respectively. Calculate the open and short mode failure probabilities of the parallel system.
10. Assume that open and short mode failure occurrence probabilities of independent and identical three state devices are 0.12 and 0.10, respectively. Determine the number of devices to be connected in parallel to maximize system reliability.

8.5 REFERENCES

1. Taylor, J.R., A study of failure causes based on U.S. power reactor abnormal occurrence reports, *Reliab. Nucl. Power Plants*, IAEA-SM-195/16, 1975.
2. Epler, E.P., Common-mode considerations in the design of systems for protection and control, *Nuclear Safety*, 11, 323-327, 1969.
3. Ditto, S.J., Failures of systems designed for high reliability, *Nuclear Safety*, 8, 35-37, 1966.
4. Epler, E.P., The ORR emergency cooling failures, *Nuclear Safety*, 11, 323-327, 1970.
5. Jacobs, I.M., The common-mode failure study discipline, *IEEE Trans. Nuclear Sci.*, 17, 594-598, 1970.
6. Gangloff, W.C., Common-mode failure analysis is "in", *Electronic World*, October, 30-33, 1972.
7. Gangloff, W.C., Common mode failure analysis, *IEEE Trans. Power Apparatus Syst.*, 94, 27-30, 1975.
8. Fleming, K.N., A redundant model for common mode failures in redundant safety systems, *Proc. Sixth Pittsburgh Annu. Modeling Simulation Conf.*, 579-581, 1975.
9. Dhillon, B.S., On common-cause failures — Bibliography, *Microelectronics and Reliability*, 18, 533-534, 1978.
10. Dhillon, B.S. and Singh, C., *Engineering Reliability: New Techniques and Applications*, John Wiley & Sons, New York, 1981.

11. Dhillon, B.S. and Anude, D.C., Common-cause failures in engineering systems: a review, *Int. J. Reliability, Quality, Safety Eng.*, 1, 103-129, 1994.
12. Moore, E.F. and Shannon, C.E., Reliable circuits using less reliable relays, *J. Franklin Inst.*, 9, 191-208, 1956, and 281-297, 1956.
13. Creveling, C.J., Increasing the reliability of electronic equipment by the use of redundant circuits, *Proc. Inst. Radio Eng.* (IRE), 44, 509-515, 1956.
14. Dhillon, B.S., The Analysis of the Reliability of Multi-State Device Networks, Ph.D. Dissertation, 1975, available from the National Library of Canada, Ottawa.
15. Dhillon, B.S., Literature survey on three-state device reliability systems, 16, 601-602, 1977.
16. Lesanovsky, A., Systems with two dual failure modes — A survey, *Microelectronics and Reliability*, 33, 1597-1626, 1993.
17. WASH 1400 (NUREG-75/014), Reactor Safety Study, U.S. Nuclear Regulatory Commission, Washington, D.C., October 1975. Available from the National Technical Information Service, Springfield, VA.
18. Dhillon, B.S. and Proctor, C.L., Common mode failure analysis of reliability networks, *Proc. Annu. Reliability Maintainability Symp.*, 404-408, 1977.
19. Dhillon, B.S., *Reliability in Computer System Design*, Ablex Publishing, Norwood, NJ, 1987.
20. Dhillon, B.S. and Yang, N., Analysis of an engineering system with two types of common cause failures, *Proc. First Intl. Conf. Quality Reliability*, Hong Kong, 2, 393-397, 1995.
21. Dhillon, B.S. and Rayapati, S.N., Reliability evaluation of multi-state device networks with probability trees, *Proc. 6th Symp. Reliability Electron.*, Budapest, 27-37, 1985.

9 Mechanical Reliability

9.1 INTRODUCTION

Usually, the reliability of electronic parts is predicted on the assumption that their failure times are exponentially distributed (i.e., their failure rates are constant). The past experience also generally supports this assumption. Nonetheless, this assumption was derived from the bathtub hazard rate concept which states that during the useful life of many engineering items, the failure rate remains constant. However, in the case of mechanical items, the assumption of constant failure rate is not generally true. In fact, in many instances their increasing failure rate patterns can be represented by an exponential function.

The history of mechanical reliability may be traced back to World War II with the development of V1 and V2 rockets by the Germans. In 1951, Weibull published a statistical function to represent material strength and life length [1]. Today it is known as the Weibull distribution and it has played an important role in the development of mechanical reliability and reliability in general. Also, in the 1950s Freudenthal reported many advances to structural reliability [2-4].

In 1963 and 1964, the National Aeronautics and Space Administration (NASA) lost SYNCOM I and Mariner III due to busting of a high pressure tank and mechanical failure, respectively. As a result of these and other mechanical failures, researchers in the field called for the improvement in reliability and longevity of mechanical and electromechanical parts. In years to come, NASA spent millions of dollars to test, replace, and redesign various items, including mechanical valves, filters, actuators, pressure gauges, and pressure switches [5].

Some of the major projects initiated by NASA in 1965 were entitled "Designing Specified Reliability Levels into Mechanical Components with Time-Dependent Stress and Strength Distributions", "Reliability Demonstration Using Overstress Testing", and "Reliability of Structures and Components Subjected to Random Dynamic Loading" [6]. After the mid-1960s, many other people [7-9] contributed to mechanical reliability and a comprehensive list of publications on the subject is given in Reference 10.

This chapter presents various different aspects of mechanical reliability.

9.2 REASONS FOR THE DISCIPLINE OF MECHANICAL RELIABILITY AND MECHANICAL FAILURE MODES

There have been many reasons for the development of the mechanical reliability field. Five such reasons are presented in Figure 9.1 [11]. They are: influence from electronic reliability, stringent requirements and severe application environments, lack of design experience, cost and time constraints, and optimization of resources.

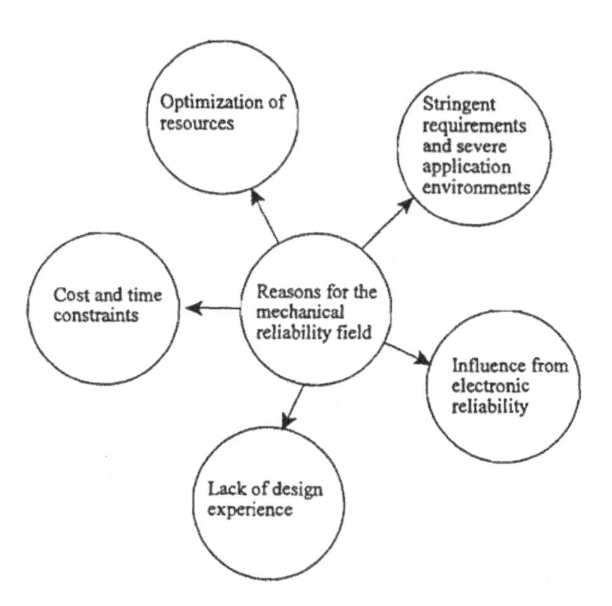

FIGURE 9.1 Basic reasons for the development of the mechanical reliability field.

Influence from electronic reliability for the stimulation of similar developments in mechanical engineering is an important factor because of vast advances made in predicting and analyzing reliability of electronic equipment. Stringent requirements and severe application environments are other important reasons because of many expensive systems to be used in military and space applications. Cost and time constraints make it difficult for designers to learn from past mistakes. Lack of design experience occurs because of rapid changes in technology; consequently, mechanical designers do not have the time to master the design, in particular design of complex equipment for use in aerospace or military environment. Optimization of resources is another good reason as the workable design is no longer considered satisfactory. Thus, a given design has to be optimized under such constraints as reliability, performance, weight, size, and cost.

There are many different types of failure modes associated with mechanical items. Good design practices help to reduce or eliminate the occurrence of these failure modes. Nonetheless, these failure modes include [12-14]:

- **Bending failure.** This may also be called a combined failure because it occurs when one outer surface is in tension and the other outer surface is in compression. The tensile rupture of the outer material is a representative of the bending failure.
- **Metallurgical failure.** This may also be called a material failure and is the result of extreme oxidation or operation in corrosive environment. The environmental conditions such as heat, nuclear radiation, erosion, and corrosive media accelerate the occurrence of metallurgical failures.

- **Tensile-yield-strength failure.** This failure occurs under pure tension and, more specifically, when the applied stress exceeds the yield strength of the material. Usually, its consequence is permanent set or permanent deformation in the structure and seldom catastrophic. The recommended value of safety factors for this kind of failure could be as low as 1 (i.e., design to yield) or a high of 4 or 5 for pressure vessels and bridges.
- **Instability failure.** This type of failure occurs in structural members such as columns and beams, particularly the ones manufactured using thin material where the loading is usually in compression. However, instability failure may also result because of torsion or by combined loading, i.e., including bending and compression. Usually this type of failure is crippling or leads to a total failure of the structure. The past experience indicates that a safety factor value of 2 is usually sufficient but avoid using less than 1.5.
- **Ultimate tensile-strength failure.** This leads to a complete failure of the structure at a cross-sectional point and it occurs (i.e., ultimate tensile-strength failure) when the applied stress is greater than the ultimate tensile strength. The recommended values of safety factors for this kind of failure could be as low as unity (i.e., design to fail) or a high of 5 or 6 for pressure vessels and bridges.
- **Material flaw failure.** This type of failure usually occurs due to factors such as poor quality assurance, fatigue cracks, weld defects, and small cracks and flaws. These factors reduce the allowable material strength and, as a consequence, lead to premature failure at the flawed spots.
- **Compressive failure.** This failure is similar to the tensile failures discussed earlier but with one exception, i.e., under compressive loads. The compressive failure causes permanent deformation/cracking/rupturing.
- **Fatigue failure.** This type of failure occurs because of repeated loading or unloading (or partial unloading) of an item. In order to prevent the occurrence of this kind of failure, make sure the proper materials are selected, e.g., steel outlasts aluminum under cyclic loading.
- **Bearing failure.** This type of failure is similar in nature to the compressive failure and it usually occurs due to a round cylindrical surface bearing on either a flat or a concave surface like roller bearings in a race. If repeated loading occurs, it is important to consider the possibility of the occurrence of fatigue failures.
- **Stress concentration failure.** This type of failure occurs in circumstances of uneven stress "flow" through a mechanical design. Usually, the concentration of stress occurs at sudden transitions from thick gauges to thin gauges, at sudden variations in loading along a structure, at various attachment conditions, or at right-angle joints.
- **Shear loading failure.** The occurrence of ultimate and yield failure takes place when the shear stress becomes greater than the strength of the material when applying high torsion or shear loads. Normally, these failures occur along a 45° axis in relation to the principal axis.

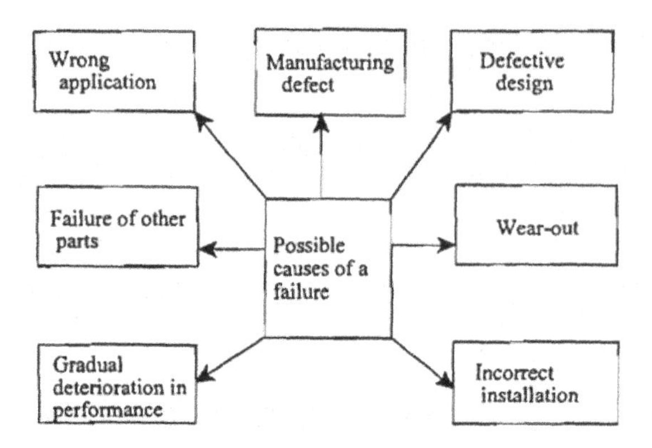

FIGURE 9.2 Seven possible causes of failure.

- **Creep/rupture failure.** Normally, long-term loads make elastic materials stretch even when they are less than the normal yield strength of the material. Nonetheless, material stretches (creeps) if the load is maintained continuously and usually it ultimately terminates in a rupture. It is to be noted that at elevated temperatures, creep accelerates.

9.3 GENERAL AND GEAR FAILURE CAUSES

There are many causes of failure. Seven of them are shown in Figure 9.2 [15].

A study conducted over a period of 35 years reported 931 gear failures and their major classifications were as follows [15]:

- Breakage: 61.2%
- Surface fatigue: 20.3%
- Wear: 13.2%
- Plastic flow: 5.3%

Furthermore, the failure causes were grouped into five distinct categories: (1) service related (74.7%), (2) heat-treatment related (16.2%), (3) design related (6.9%), (4) manufacturing related (1.4%), and (5) material related (0.8%). In turn, each of these five categories were further broken into elements as follows:

Service-Related (74.7%)

- Continual overloading: 25%
- Improper assembly: 21.2%
- Impact loading: 13.9%
- Incorrect lubrication: 11%
- Foreign material: 1.4%
- Abusive handling: 1.2%

- Bearing failure: 0.7%
- Operator errors: 0.3%

Heat-Treatment Related (16.2%)

- Incorrect hardening: 5.9%
- Inadequate case depth: 4.8%
- Inadequate core hardness: 2%
- Excessive case depth: 1.8%
- Improper tempering: 1%
- Excessive core hardness: 0.5%
- Distortion: 0.2%

Design-Related (6.9%)

- Wrong design: 2.8%
- Specification of suitable heat treatment: 2.5%
- Incorrect material selection: 1.6%

Manufacturing-Related (1.4%)

- Grinding burns: 0.7%
- Tool marks or notches: 0.7%

Material-Related (0.8%)

- Steel defects: 0.5%
- Mixed steel or incorrect composition: 0.2%
- Forging defects: 0.1%

9.4 SAFETY FACTOR AND SAFETY MARGIN

Safety factor and safety margin are used to ensure the reliability of mechanical items and, basically, are arbitrary multipliers. These indices can provide satisfactory design if they are established with care using considerable past experiences. Nonetheless, in today's environment as new designs involve new applications and new materials, more consistent approaches are required. Furthermore, the mechanical design totally based upon safety factors could be misleading and may be costly due to over-design or catastrophic because of under-design. Both safety factor and safety margin are described below.

9.4.1 Safety Factor

There are many different ways of defining a safety factor [9, 10, 16–19]. Two different definitions of the safety factor are presented below.

Definition I

The safety factor is defined by [20]:

$$SF = \frac{MFGST}{MFGS} \geq 1 \tag{9.1}$$

where

SF is the safety factor.
MFGST is the mean failure governing strength.
MFGS is the mean failure governing stress.

This safety factor is a good measure when both the stress and strength are described by a normal distribution. However, it is important to note that when the variation of stress and/or strength is large, this measure of safety becomes meaningless because of positive failure rate.

Definition II

The safety factor is defined by [12, 13]:

$$SF = \frac{SM}{MAWS} \tag{9.2}$$

where

SM is the strength of the material.
MAWS is the maximum allowable working stress.

The value of this safety factor is always greater than unity. In fact, its value less than unity indicates that the item under consideration will fail because the applied stress is greater than the strength.

The standard deviation of the safety factor is expressed by [12, 13]:

$$\sigma = \left[(\sigma_{th}/MAWS)^2 + \{SM/(MAWS)^2\}\sigma_s^2\right]^{1/2} \tag{9.3}$$

where

σ is the standard deviation of the safety factor.
σ_{th} is the standard deviation of the strength.
σ_s is the standard deviation of the stress.

All in all, there are many factors that must be carefully considered in selecting an appropriate value of the safety factor: uncertainty of load, cost, failure consequence, uncertainty of material strength, degree of uncertainty involved in relating applied stress to strength, etc. [21].

9.4.2 Safety Margin

This is defined as follows [9]:

$$\theta_m = SF - 1 \tag{9.4}$$

where

θ_m is the safety margin.
SF is the safety factor.

The value of this measure is always greater than zero and its negative value indicates that the item under consideration will fail. For normally distributed stress and strength, the safety margin is expressed by [9]:

$$\theta_m = \left(\mu_{th} - \mu_{ms}\right)/\sigma_{th} \tag{9.5}$$

where

μ_{th} is the average strength.
μ_{ms} is the maximum stress.
σ_{th} is the standard deviation of the strength.

The maximum stress, μ_{ms}, is given by

$$\mu_{ms} = \mu_s + k\sigma_s \tag{9.6}$$

where

μ_s is the mean value of the stress.
σ_s is the standard deviation of the stress.
k is a factor which takes values between 3 and 6.

It is to be noted that just like the safety factor, the safety margin is a random variable.

Example 9.1

Assume that the following data are known for a mechanical item under design:

$\mu_{th} = 30{,}000$ psi, $\mu_s = 15{,}000$ psi
$\sigma_{th} = 1200$ psi, $\sigma_s = 300$ psi

For k = 3 and 6, calculate the values of the safety margin.

Substituting the above data into Equation (9.6) yields

For k = 3

$$\mu_{ms} = 15,000 + 3\ (300)$$
$$= 15,900 \text{ psi}$$

For k = 6

$$\mu_{ms} = 15,000 + 6\ (300)$$
$$= 16,800 \text{ psi}$$

By inserting the above calculated values and the given data into Equation (9.5), we get

For k = 3

$$\sigma_m = \frac{30,000 - 15,900}{1200}$$
$$= 11.75$$

For k = 6

$$\sigma_m = \frac{30,000 - 16,800}{1200} = 11$$

Thus, the values of the safety margin for k = 3 and 6 are 11.75 and 11, respectively.

9.5 "DESIGN BY RELIABILITY" METHODOLOGY AND STRESS-STRENGTH MODELS

In order to design a reliable mechanical item, a systematic series of steps are required. The "design by reliability" methodology is made up of such steps [8, 22]:

- Define the design problem to be solved.
- Highlight and list all the relevant design variables and parameters.
- Conduct failure modes, effect and criticality analysis (FMECA) as per MIL-STD-1629 [23].
- Verify selection of critical design parameters.
- Develop relationship between the critical parameters and the failure governing criteria.

- Determine the failure-governing stress and strength functions and then the failure-governing stress and strength distributions.
- Calculate the reliability using the failure-governing stress and strength distributions for each critical failure mode.
- Iterate the design until the set reliability goals are satisfied.
- Optimize design with respect to factors such as reliability, maintainability, cost, safety, performance, volume, and weight.
- Repeat the design optimization for all associated critical parts.
- Calculate item reliability.
- Iterate the design until the reliability goals of the item are effectively satisfied.

Before we discuss stress-strength models, let us define stress and strength:

- **Stress.** This is any factor that tends to produce a failure. Some examples are mechanical load, electric current, environment, and temperature.
- **Strength.** This is the ability of an item to accomplish its identified goal adequately without a failure when subject to the external loading and environment.

When the probability distributions of both stress and strength of an item are known, the item's reliability may be determined through analytical means. Under such circumstances, the reliability of an item may be defined as the probability that the failure governing stress will not exceed the failure governing strength. Mathematically, it is expressed as follows:

$$R = P(x > y) = P(y < x) = P(x - y > 0) \tag{9.7}$$

where

R is the item reliability.
P is the probability.
y is the stress random variable.
x is the strength random variable.

Using Reference 10, we rewrite Equation (9.7) as follows:

$$R = \int_{-\infty}^{\infty} f(y) \left[\int_{y}^{\infty} f(x)\,dx \right] dy \tag{9.8}$$

where

f (y) is the probability density function of the stress.
f (x) is the probability density function of the strength.

Equation (9.8) can also be written in different forms and three such forms are as follows:

$$R = \int_{-\infty}^{\infty} f(x)\left[\int_{-\infty}^{x} f(y)\,dy\right]dx \tag{9.9}$$

$$R = \int_{-\infty}^{\infty} f(y)\left[1 - \int_{-\infty}^{y} f(x)\,dx\right]dy \tag{9.10}$$

$$R = \int_{-\infty}^{\infty} f(x)\left[1 - \int_{x}^{\infty} f(y)\,dy\right]dx \tag{9.11}$$

It is to be noted that Equations (9.10) and (9.11) were written using the following two relationships in Equations (9.8) and (9.9), respectively:

$$\int_{-\infty}^{y} f(x)\,dx + \int_{y}^{\infty} f(x)\,dx = 1 \tag{9.12}$$

$$\int_{-\infty}^{x} f(y)\,dy + \int_{x}^{\infty} f(y)\,dy = 1 \tag{9.13}$$

Using the above reliability equations, the following stress-strength models were developed for defined stress and strength probability density functions of an item:

Model I

In this case, stress and strength associated with an item are exponentially distributed. Thus, we have

$$f(y) = \lambda e^{-\lambda y}, \quad 0 \le y < \infty \tag{9.14}$$

and

$$f(x) = \alpha e^{-\alpha x}, \quad 0 \le x < \infty \tag{9.15}$$

where

λ and α are the reciprocals of the mean values of stress and strength, respectively.

Inserting Equations (9.14) and (9.15) into Equation (9.8) yields

$$R = \int_0^{\infty} \lambda e^{-\lambda y}\left[\int_y^{\infty} \alpha e^{-\alpha x}\right] dy$$

$$= \int_0^{\infty} \lambda e^{-(\alpha+\lambda)y} dy \tag{9.16}$$

$$= \frac{\lambda}{\lambda+\alpha}$$

For $\lambda = \frac{1}{\bar{y}}$ and $\alpha = \frac{1}{\bar{x}}$, Equation (9.16) becomes

$$R = \frac{\bar{x}}{\bar{x}+\bar{y}} \tag{9.17}$$

where

$\bar{x}$ and $\bar{y}$ are the mean strength and stress, respectively.

Dividing the top and bottom of the right-hand side of Equation (9.17) by $\bar{x}$, we get

$$R = \frac{1}{1+\bar{y}/\bar{x}} = \frac{1}{1+\beta} \tag{9.18}$$

where

$\beta = \bar{y}/\bar{x}$.

For reliable design $\bar{x} \geq \bar{y}$, thus $\beta \leq 1$. For given values of β, using Equation (9.18), we obtained the tabulated values for item reliability as shown in Table 9.1.

Table 9.1 values indicate that as the value of β increases from 0 to 1, the item reliability decreases from 1 to 0.5. More specifically, as the mean stress increases, the item reliability decreases, accordingly.

Example 9.2

Assume that stress and strength associated with an item are exponentially distributed with mean values of 10,000 and 35,000 psi, respectively. Compute the item reliability.

Substituting the given data into Equation (9.17) yields

$$R = \frac{35{,}000}{35{,}000+10{,}000} = 0.7778$$

Thus, the item reliability is 0.7778.

TABLE 9.1
Values for Item Reliability

β	Item reliability, R
0	1
0.1	0.9091
0.2	0.8333
0.3	0.7692
0.4	0.7143
0.5	0.6667
0.6	0.6250
0.7	0.5882
0.8	0.5556
0.9	0.5263
1	0.5

Model II

In this case, an item's stress and strength are described by normal and exponential probability density functions, respectively, as

$$f(y) = \frac{1}{\sigma\sqrt{2\pi}} \exp\left[-\frac{1}{2}\left(\frac{y-\mu}{\sigma}\right)^2\right] \qquad -\infty < y < \infty \tag{9.19}$$

$$f(x) = \alpha e^{-\alpha x}, \qquad x \geq 0 \tag{9.20}$$

where

- μ is the mean stress.
- σ is the standard deviation associated with stress.
- α is the reciprocal of the mean strength.

Substituting Equations (6.19) and (6.20) into Equation (9.8), we get

$$\begin{aligned} R &= \int_{-\infty}^{\infty} \left[\frac{1}{\sigma\sqrt{2\pi}} \exp\left[-\frac{1}{2}\left(\frac{y-\mu}{\sigma}\right)^2\right]\right] \left[\int_y^{\infty} \alpha e^{-\alpha x}\, dx\right] dy \\ &= \int_{-\infty}^{\infty} \frac{1}{\sigma\sqrt{2\pi}} \exp\left\{-\left[\frac{1}{2}\left(\frac{y-\mu}{\sigma}\right)^2 + \alpha y\right]\right\} dy \end{aligned} \tag{9.21}$$

since

$$\frac{(y-\mu)^2}{2\,\sigma^2}+\alpha y=\frac{2\mu\alpha\sigma^2-\alpha^2\sigma^4+\left(y-\mu+\alpha\sigma^2\right)^2}{2\,\sigma^2} \tag{9.22}$$

Equation (6.21) yields

$$R=\exp\left[-\frac{1}{2}\left(2\mu\alpha-\alpha^2\,\sigma^2\right)\right] \tag{9.23}$$

Example 9.3

Assume that the strength of an item follows an exponential distribution with a mean of 35,000 psi. The stress acting upon the item is normally distributed with mean 10,000 psi and standard deviation 5,000 psi. Calculate the item's reliability.

By inserting the given data into Equation (9.23), we get

$$R=\exp\left[-\frac{1}{2}\left\{\frac{2(10{,}000)}{35{,}000}-\frac{(5{,}000)^2}{(35{,}000)^2}\right\}\right]$$

$$=\exp\left[-\frac{1}{2}\{0.5714-0.0204\}\right]$$

$$=0.7592$$

Thus, the item's reliability is 0.7592.

Model III

In this case, the strength of the item is normally distributed and the stress acting upon it follows the exponential distribution. Thus, we have

$$f(x)=\frac{1}{\sigma_x\sqrt{2\,\pi}}\exp\left[-\frac{1}{2}\left(\frac{x-\mu_x}{\sigma_x}\right)^2\right],\qquad -\infty<x<\infty \tag{9.24}$$

$$f(y)=\lambda e^{-\lambda y},\quad y>0 \tag{9.25}$$

where

λ is the reciprocal of the mean stress.
μ_x is the mean strength.
σ_x is the strength standard deviation.

Using Equations (9.24) and (9.25) in Equation (9.9) yields

$$R = \int_{-\infty}^{\infty} \frac{1}{\sigma_x \sqrt{2\pi}} \exp\left[-\frac{1}{2}\left(\frac{x-\mu_x}{\sigma_x}\right)^2\right] dx \left[\int_0^x \lambda e^{-\lambda y} dy\right]$$
$$= 1 - \int_{-\infty}^{\infty} \frac{1}{\sigma_x \sqrt{2\pi}} \exp\left\{-\left[\frac{1}{2}\left(\frac{x-\mu_x}{\sigma_x}\right)^2 + \lambda x\right]\right\} dx \tag{9.26}$$

since

$$\frac{1}{2}\left(\frac{x-\mu_x}{\sigma^2}\right)^2 + \lambda x = \frac{2\mu_x \lambda \sigma_x^2 + \left(\sigma_x^2 \lambda - \mu_x + x\right)^2 - \sigma_x^4 \lambda^2}{2\sigma_x^2} \tag{9.27}$$

Equation (9.26) is rewritten to the following form:

$$R = 1 - \exp\left[-\frac{1}{2}\left(2\mu_x \lambda - \sigma_x^2 \lambda^2\right)\right] G \tag{9.28}$$

where

$$G \equiv \int_{-\infty}^{\infty} \frac{1}{\sigma_x \sqrt{2\pi}} \exp\left[-\frac{1}{2}\left(\frac{\sigma_x^2 \lambda - \mu_x + x}{\sigma_x}\right)^2\right] dx \tag{9.29}$$

Since G = 1 [24], Equation (9.28) simplifies to

$$R = 1 - \exp\left[-\frac{1}{2}\left(2\mu_x \lambda - \sigma_x^2 \lambda^2\right)\right] \tag{9.30}$$

Example 9.4

An item's stress and strength are described by exponential and normal distributions, respectively. The mean of the stress is 10,000 psi and the mean and standard deviation of the strength are 30,000 and 3,000 psi, respectively. Calculate the item's reliability.

Inserting the specified data into Equation (9.30), we get

$$R = 1 - \exp\left[-\frac{1}{2}\left\{\frac{2(30{,}000)}{10{,}000} - \frac{(3{,}000)^2}{(10{,}000)^2}\right\}\right]$$

$$= 1 - \exp\left[-\frac{1}{2}(6 - 0.09)\right]$$

$$= 0.9479$$

Thus, the item's reliability is 0.9479.

Other stress-strength models are presented in Reference 10.

9.6 MELLIN TRANSFORM METHOD

This is a useful method for estimating an item's reliability when its associated stress and strength distributions cannot be assumed, but there is an adequate amount of empirical data. This approach can also be used when stress and strength distributions associated with an item are known. Obviously, the method is based on Mellin transforms and for Equation (9.8) they are defined as follows:

$$X = \int_{y}^{\infty} f(x)\,dx = 1 - F_1(y) \tag{9.31}$$

and

$$Y = \int_{0}^{y} f(y)\,dy = F_2(y) \tag{9.32}$$

where

$F_1(y)$ and $F_2(y)$ are the cumulative distribution functions.

Taking the derivative of Equation (9.32) with respect to y, we have

$$\frac{dY}{dy} = f(y) \tag{9.33}$$

Thus, from Equation (9.33), we get

$$dY = f(y)\,dy \tag{9.34}$$

Obviously, it can be easily seen from Equation (9.32) that Y takes values of 0 to 1 (i.e., at $y = 0, Y = 0$ and at $y = \infty, Y = 1$). Thus, inserting Equations (9.31) and (9.34) into Equation (9.8) yields

$$R = \int_{0}^{1} X \, dY \tag{9.35}$$

The above equation indicates that the area under the X vs. Y plot represents item reliability. This area can be estimated by using the Simpson's rule, expressed below.

$$\int_{j}^{k} f(z)dz \simeq \frac{k-j}{3n}\left(W_0 + 4W_1 + 2W_2 + 4W_3 + \ldots + 2W_{n-2} + 4W_{n-1} + W_n\right) \tag{9.36}$$

where

$f(z)$ is a function of z defined over interval (j, k), i.e., $j \leq z \leq k$.
n is the even number of equal subdivided parts of interval (j, k).
$\frac{k-j}{n} = g$ is the width of the subdivided part.

At $j = z_0, z_1, z_2, z_3, \ldots, z_n = k$; the corresponding values of W are $W_0 = f(z_0)$, $W_1 = f(z_1)$, $W_2 = f(z_2)$, ..., $W_n = f(z_n)$.

Example 9.5

Use the Mellin transform method to estimate the item reliability in Example 9.2. Comment on the end result in comparison to the one obtained in Example 9.2.

Thus, substituting Equation (9.14) and the given relevant data of Example 9.2 into relationship (9.32) yields

$$\begin{aligned} Y &= \int_{0}^{y} \frac{1}{10,000} e^{-\left(\frac{1}{10,000}\right)y} dy \\ &= 1 - e^{-\left(\frac{1}{10,000}\right)y} \end{aligned} \tag{9.37}$$

Inserting Equation (9.15) and the given relevant data of Example 9.2 into Equation (9.31), we have

$$\begin{aligned} X &= 1 - \left[1 - e^{-\left(\frac{1}{35,000}\right)}\right] \\ &= e^{-\left(\frac{1}{35,000}\right)y} \end{aligned} \tag{9.38}$$

Table 9.2 presents values of Y and X using Equations (4.37) and (4.38), respectively, for assumed values of stress y.

TABLE 9.2
Computed Values of Y and X for the Assumed Values of Stress y

y (psi)	Y	X
0	0	1
4,000	0.33	0.89
8,000	0.55	0.80
12,000	0.70	0.71
16,000	0.80	0.63
20,000	0.87	0.57
24,000	0.91	0.50
28,000	0.94	0.45
32,000	0.96	0.40
36,000	0.97	0.36
40,000	0.98	0.32
44,000	0.99	0.29
48,000	0.992	0.25
∞	1	0

Figure 9.3 shows the plot of X vs. Y for data given in Table 9.2. The area under the Figure 9.3 curve is estimated using Equation (9.36) as follows:

$$R \simeq \frac{(1-0)}{3(4)}\left[W_0 + 4W_1 + 2W_2 + 4W_3 + W_4\right]$$

$$\simeq \frac{1}{12}\left[1 + 4(0.95) + 2(0.84) + 4(0.68) + 0\right]$$

$$\simeq 0.7667$$

The above result, i.e., $R = 0.7667$, is quite close to the one obtained using the analytical approach in Example 9.2, i.e., $R = 0.7778$.

9.7 FAILURE RATE MODELS

Over the years, many failure rate estimation mathematical models have been developed to determine failure rates of various mechanical items including brakes, clutches, pumps, filters, compressors, seals, and bearings [12, 13, 25–27]. This section presents failure rate calculation models for such items.

9.7.1 Break System Failure Rate Model

The brake system failure rate is expressed by [25]:

$$\lambda_{bs} = \lambda_{bfm} + \lambda_s + \lambda_b + \lambda_{se} + \lambda_a + \lambda_h \quad (9.39)$$

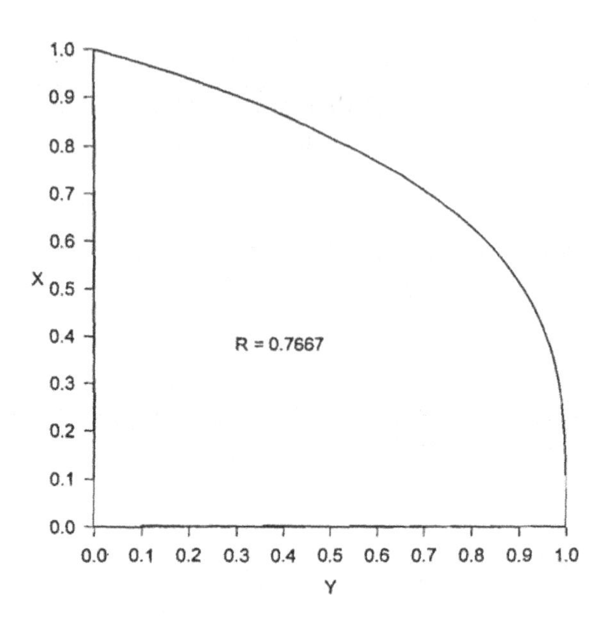

FIGURE 9.3 Plot of X vs. Y.

where

λ_{bs} is the break system failure rate, expressed in failures/10^6 h.
λ_{bfm} is the failure rate of brake friction materials.
λ_s is the failure rate of springs.
λ_b is the failure rate of bearings.
λ_{se} is the failure rate of seals.
λ_a is the failure rate of actuators.
λ_h is the failure rate of brake housing.

The values of λ_{bs}, λ_{bfm}, λ_s, λ_b, λ_{se}, λ_a, and λ_h are obtained through various means [25, 28]. For example, λ_{bfm} can be calculated by using the following equation [25]:

$$\lambda_{bfm} = \lambda_{bbfm} \prod_{i=1}^{4} f_i \tag{9.40}$$

where

λ_{bbfm} is the base failure rate associated with the brake friction material.
f_i is the ith multiplying factor that considers the effects on the base failure rate of items such as dust contaminants (i = 1), ambient temperature (i = 2), brake type (i = 3), and counter-surface roughness (i = 4).

In turn, the base failure rate of the brake friction material is given by [29]:

$$\lambda_{bbfm} = 3 \times 10^{-3}.R_W.W.n.\alpha.(AC)^2 \theta/(LT).A \tag{9.41}$$

where

LT is the lining thickness, expressed in inches.
A is the lining area, expressed in inches^2.
R_W is the friction material's specific wear rate, expressed in in.^{3}/ft-lbf.
W is the vehicle/aircraft weight, expressed in lbf.
n is the miles traveled per year.
α is the number of brake applications per miles.
AC is the average change in velocity per brake action, expressed in ft/s.
θ is the proportion of weight carried by lining during braking action. For example, in the case of a 4-wheel vehicle, typically each front brake carries 3/10 of the breaking load.

9.7.2 Clutch System Failure Rate Model

The clutch system failure rate is expressed by [25]:

$$\lambda_{cs} = \lambda_{cfm} + \lambda_s + \lambda_b + \lambda_{se} + \lambda_a \tag{9.42}$$

where

λ_{cs} is the clutch system failure rate, expressed in failure/10^6 h.
λ_{cfm} is the failure rate of clutch friction material.

Other symbols used in Equation (9.42) are the same as the ones used in Equation (9.39).
The failure rate of clutch friction materials, λ_{cfm}, is expressed by

$$\lambda_{cfm} = \lambda_{bcfm} f_1 f_2 \tag{9.43}$$

where

λ_{bcfm} is the base failure rate of clutch friction material.
f_1 is the factor that considers the effects on the base failure rate of multiple plates.
f_2 is the factor that considers the effects on the base failure rate of ambient temperature.

The clutch friction material base failure rate, λ_{bcfm}, is given by [25, 30]:

$$\lambda_{bcfm} = k E_a / 2 A W_m CFM_a LT \tag{9.44}$$

where

- k is the number of applications per hour.
- LT is the lining thickness, expressed in inches.
- AW_m is the average wear of the material.
- CFM_a is the area of the clutch friction material on each disc, expressed in inches^2.
- E_a is the average energy dissipated per engagement, expressed in ft–lbf.

9.7.3 Pump Failure Rate Model

The pump failure rate is expressed by [26]:

$$\lambda_p = \lambda_{ps} + \lambda_{pse} + \lambda_{pb} + \lambda_{pc} + \lambda_{pfd} \tag{9.45}$$

where

- λ_p is the pump failure rate, expressed in failures/10^6 cycles.
- λ_{ps} is the failure rate of the pump shaft.
- λ_{pse} is the failure rate of all pump seals.
- λ_{pb} is the failure rate of all pump bearings.
- λ_{pc} is the failure rate of the pump casing.
- λ_{pfd} is the failure rate of the pump fluid driver.

The failure rate of the pump shaft, λ_{ps}, can be calculated by using the following equation:

$$\lambda_{ps} = \lambda_{bps} \prod_{i=1}^{6} f_i \tag{9.46}$$

where

- λ_{bps} is the base failure rate of the pump shaft.
- f_i is the ith modifying factor; for i = 1 is associated with material temperature, i = 2 pump displacement, i = 3 casing thrust load, i = 4 shaft surface finish, i = 5 contamination, and i = 6 material endurance limit.

The pump seal failure rate, λ_{pse}, is expressed by

$$\lambda_{pse} = \lambda_{bpse} \prod_{i=1}^{7} f_i \tag{9.47}$$

where

λ_{bpse} is the pump seal base failure rate.
f_i is the ith modifying factor; for i = 1 is for effects of casing thrust load, i = 2 surface finish, i = 3 seal smoothness, i = 4 fluid viscosity, i = 5 pressure/velocity factor, i = 6 temperature, and i = 7 contaminates.

Similarly, the values of λ_{pb}, λ_{pc}, and λ_{pfd} can be calculated. Reference 26 presents procedures to calculate these failure rates.

9.7.4 Filter Failure Rate Model

The following equation can be used to predict filter failure rate [26]:

$$\lambda_f = \lambda_{bf} \prod_{i=1}^{6} f_i \tag{9.48}$$

where

λ_f is the filter failure rate, expressed in failures/10^6 h.
λ_{bf} is the filter base failure rate.
f_i is the ith modifying factor; for i = 1 is for temperature effects, i = 2 vibration effects, i = 3 water contamination effects, i = 4 cyclic flow effects, i = 5 cold start effects, and i = 6 differential pressure effects.

9.7.5 Compressor System Failure Rate Model

The compressor system failure rate is expressed by the following equation [27]:

$$\lambda_{com} = \lambda_{cs} + \lambda_{cse} + \lambda_{cva} + \lambda_{cdc} + \lambda_{cc} + \lambda_{cb} \tag{9.49}$$

where

λ_{com} is the compressor system failure rate, expressed in failures/10^6 h.
λ_{cs} is the failure rate of all compressor shafts.
λ_{cse} is the failure rate of all compressor seals.
λ_{cva} is the failure rate of valve assy (if any).
λ_{cdc} is the failure rate due to design configuration.
λ_{cc} is the failure rate of the compressor casing.
λ_{cb} is the failure rate of all compressor bearings.

The compressor shaft failure rate is expressed by

$$\lambda_{cs} = \lambda_{bcs} \prod_{i=1}^{5} f_i \tag{9.50}$$

where

λ_{bcs} is the compressor shaft base failure rate.
f_i is the ith modifying factor; for i = 1 is for material temperature, i = 2 material endurance limit, i = 3 displacement, i = 4 shaft surface finish, and i = 5 contamination.

The compressor seal failure rate can be predicted by using the following equation:

$$\lambda_{cse} = \lambda_{bcse} \prod_{i=1}^{10} f_i \tag{9.51}$$

where

λ_{bcse} is the compressor seal base failure rate.
f_i is the ith modifying factor; for i = 1 is for fluid pressure, i = 2 seal face pressure and gas velocity, i = 3 seal smoothness, i = 4 seal size, i = 5 temperature, i = 6 allowable leakage, i = 7 contaminants, i = 8 flow rate, i = 9 fluid viscosity, and i = 10 surface finish and other conductance parameters.

The compressor valve assembly failure rate is expressed by

$$\lambda_{cva} = \lambda_{so} + \lambda_{sa} + \lambda_{pa} \tag{9.52}$$

where

λ_{so} is the failure rate of the solenoid (if any).
λ_{sa} is the failure rate of sliding action valve assy (if any).
λ_{pa} is the failure rate of poppet assy (if any).

Procedures for calculating λ_{cdc}, λ_{cc}, and λ_{cb} are presented in Reference 27.

9.7.6 Bearing Failure Rate Model

The bearing failure rate is expressed by [12]:

$$\lambda_{br} = \lambda_{pbs} (AL/SL)^x (AE/0.006)^{2.36} (SV/OV)^{0.54} L_c^{2/3} \tag{9.53}$$

where

λ_{br} is the failure rate of bearing (using actual conditions), expressed in failure/10^6 h of operation.
x is a factor; x = 4.0 for ball bearings and x = 3.33 for roller bearings.

L_c is the actual contamination level, μg/m³/60 μg/m³.
AE is the alignment error, radians.
SV is the specification lubricant viscosity, lb.min/in².
OV is the operating lubricant viscosity, lb.min/in².
AL is the actual load, psi.
SL is the specification load, psi.
λ_{pbs} is the pseudo failure rate of the bearing and is expressed as follows:

$$\lambda_{pbs} = 10^6 M^{0.9}/(5.45)\, RL \tag{9.54}$$

where

M is the number of bearings in system.
RL is the rated life in revolutions and is given by:

$$RL = 16700/\text{rpm}\,(BLR/ERL)^{\alpha} \tag{9.55}$$

where

rpm is the shaft rotating velocity, expressed in rev/min.
α is a constant; $\alpha = 10/3$ for roller bearings and $\alpha = 3$ for ball bearings.
BLR is the basic load rating, lb.
ERL is the equivalent radial, load, lb.

9.8 FAILURE DATA SOURCES FOR MECHANICAL PARTS

Just like in the case of electronic parts, the availability of failure data for mechanical parts is very important. Usually, it is more difficult to obtain useful failure data for mechanical parts than for the electronic parts. Reference 10 lists over 50 sources for obtaining failure data for various mechanical items. Some of the important sources for obtaining mechanical failure related information are as follows:

- RADC Non-Electronic Reliability Notebook [31].
- Government Industry Data Exchange Program (GIDEP) [32].
- AVCO Corporation Data [33].
- Book: Mechanical Reliability [10].
- Martin Marietta Corporation Data [34].

Table 9.3 presents failure rates for selected mechanical items. Chapter 4 discusses the subject of failure data in more depth.

TABLE 9.3
Failure Rates for Selected Mechanical Parts [10, 12]

No.	Mechanical part	Use environment	Failure rate (failures/10^6 h)
1	Hose, pneumatic	Not applicable (N/A)	29.3
2	Pressure regulator	Ground, fixed	2.4
3	Fuel pump	Ground, fixed	176.4
4	Compressor (general)	Ground, mobile	33.6
5	Clutch (friction)	Ground, mobile	38.2
6	Filter, gas (air)	Ground, mobile	3.2
7	Heavy-duty ball bearing	N/A	14.4
8	Seal, O-ring	N/A	0.2
9	Brake assembly	N/A	16.8
10	Guide pin	N/A	13

9.9 OPTIMUM MODELS FOR MECHANICAL EQUIPMENT REPLACEMENT OR MAINTENANCE

During the mechanical equipment use, its maintenance and replacement are two important considerations. Over the years, professionals in the field have developed various types of mathematical models to determine optimum time intervals for equipment replacement or maintenance. This section presents two such models [10].

9.9.1 Equipment Replacement Model

This mathematical model is concerned with determining the optimum time interval between replacements by minimizing the mean annual cost associated with the equipment with respect to the time between replacements or the equipment life expressed in years. The model assumes that the equipment mean annual cost is made up of three components: (1) average operating cost, (2) average maintenance cost, and (3) average investment cost. Thus, the mean annual cost of the equipment is expressed by

$$MC = C_0 + C_m + \frac{CI}{x} + \frac{(x-1)}{2}(IOC + IMC) \tag{9.56}$$

where

- MC is the mean annual total cost of equipment.
- C_0 is the operational cost for the first year.
- C_m is the maintenance cost for the first year.
- x is the equipment life in years.
- IOC is the amount by which operational cost increases annually.

IMC is the amount by which maintenance cost increases annually.
CI is the cost of investment.

Taking the derivatives of Equation (9.56) with respect to x, we get

$$\frac{dMC}{dx} = \frac{1}{2}(IOC + IMC) - \frac{CI}{x^2} \tag{9.57}$$

Equating Equation (9.57) to zero and rearranging yields

$$x^* = \left[\frac{2\,CI}{IOC + IMC}\right]^{1/2} \tag{9.58}$$

where

x* is the optimum replacement time interval.

Example 9.6

Assume that the following data are specified for mechanical equipment:

$$IMC = \$500,\ IOC = \$600,\ \text{and}\ CI = \$15{,}000$$

Calculate the optimum time for the mechanical equipment replacement.
Using the above data in Equation (9.58) yields

$$x^* = \left[\frac{2(15{,}000)}{600 + 500}\right]^{1/2}$$

$$= 5.2 \text{ years}$$

Thus, the optimum time for the mechanical equipment replacement is 5.2 years.

9.9.2 Equipment Maintenance Model

This mathematical model can be used to calculate the optimum number of inspections per facility per unit time. This information is quite useful to decision makers because inspections are often disruptive but on the other hand such inspections normally reduce the equipment downtime because of fewer breakdowns. In this model, expression for the total equipment downtime is used to obtain the optimum number of inspections.

The total facility downtime per unit time is given by [35]:

$$TDT = n(DTI) + (DTB)k/n \tag{9.59}$$

where

TDT is the total facility downtime per unit time.
k is a constant for specific facility.
n is the number of inspections per facility per unit time.
DTI is the downtime per inspection for a facility under consideration.
DTB is the downtime per breakdown for a facility under consideration.

By differentiating Equation (9.59) with respect to n, we get

$$\frac{dTDT}{dn} = DTI - (DTB)\,k/n^2 \tag{9.60}$$

Setting Equation (9.60) equal to zero and rearranging yields

$$n^* = [k(DTB)/DTI]^{1/2} \tag{9.61}$$

where

n* is the optimum number of inspections per facility per unit time.

Inserting Equation (9.61) into Equation (9.59) yields

$$TDT^* = 2\,[(DTB)\,(DTI)k]^{1/2} \tag{9.62}$$

where

TDT* is the minimum total facility downtime per unit time.

Example 9.7

The following data are associated with certain mechanical equipment:

$$DTI = 0.02 \text{ month, } DTB = 0.1 \text{ month, and } k = 4. \tag{9.63}$$

Calculate the optimum number of inspections per month and the minimum value of the total equipment downtime by using Equations (9.61) and (9.62), respectively.

Thus, inserting the given data into Equation (9.61) yields

$$n^* = [4(0.1)/0.02]^{1/2}$$
$$= 4.5 \text{ inspections/month}$$

Similarly, substituting the specified data into Equation (9.62), we have

$$TDT^* = 2\left[(0.1)(0.02)\,4\right]^{1/2}$$

$$= 0.18 \text{ months}$$

Thus, the optimum number of inspections per month is 4.5 and the minimum value of the total equipment downtime is 0.18 month.

9.10 PROBLEMS

1. Give reasons for the development of the mechanical reliability field.
2. Describe the following mechanical failure modes:
   - Compressive failure
   - Stress concentration failure
   - Shear loading failure
   - Bending failure
3. What are the important causes of gear failures?
4. Give four different ways of defining a safety factor.
5. Define the terms "stress" and "strength".
6. An item's stress and strength are normally distributed. Develop an expression for the item reliability.
7. A component's stress and strength are distributed exponentially with mean values of 12,000 and 40,000 psi, respectively. Determine the item reliability.
8. Use the Mellin transform approach to estimate the item reliability in Problem 7. Comment on the end result in comparison to the one obtained for Problem 7.
9. List at least 10 important sources for obtaining failure data for mechanical items.
10. A mechanical component's stress and strength are Maxwellian distributed. Develop an expression for the component reliability.

9.11 REFERENCES

1. Weibull, W., A statistical distribution function of wide applicability, *J. Appl. Mech.*, 18, 293-297, 1951.
2. Freudenthal, A.M., Safety and probability of structural failures, *Trans. Am. Soc. Civil Eng.*, 121, 1337-1397, 1956.
3. Freudenthal, A.M., *Fatigue in Aircraft Structures*, Academic Press, New York, 1957.
4. Freudental, A.M. and Gumbel, E.J., Failure and survival in fatigue, *J. Appl. Phys.*, 25, 110-120, 1954.
5. Dhillon, B.S. and Singh, C., *Engineering Reliability: New Techniques and Applications*, John Wiley & Sons, New York, 1981.
6. Redler, W.M., Mechanical reliability research in the national aeronautics and space administration, *Proc. Reliability Maintainability Conf.*, New York, 763-768, 1966.

7. Lipson, C., Sheth, N.J., and Disney, R.L., Reliability Prediction-Mechanical Stress/Strength Inference, Rome Air Development Center, Griffiss Air Force Base, New York, 1967, Report No. RADC-TR-66-710.
8. Kececioglu, D., Reliability analysis of mechanical components and systems, *Nuclear Eng. Des.*, 19, 259-290, 1972.
9. Kececioglu, D. and Haugen, E.B., A unified look at design safety factors, safety margins and measures of reliability, *Proc. Annu. Reliability Maintainability Conf.*, 522–530, 1968.
10. Dhillon, B.S., *Mechanical Reliability: Theory, Models and Applications,* American Institute of Aeronautics and Astronautics, Washington, D.C., 1988.
11. Mechanical Reliability Concepts, American Society of Mechanical Engineers (ASME) Design Engineering Conference, ASME, New York, 1965.
12. Grant Ireson, W., Coombs, C.F., and Moss, R.Y., *Handbook of Reliability Engineering and Management,* McGraw-Hill, New York, 1996.
13. Doyle, R.L., Mechanical-system reliability, Tutorial Notes, *Annu. Reliability Maintainability Symp.*, 1992.
14. Collins, J.A., *Failure of Materials in Mechanical Design,* John Wiley & Sons, New York, 1981.
15. Lipson, C., Analysis and Prevention of Mechanical Failures, Course Notes No. 8007, University of Michigan, Ann Arbor, June 1980.
16. Howell, G.M., Factors of safety, *Machine Design,* July 12, 76-81, 1956.
17. Schoof, R., How much safety factor?, *Allis-Chalmers Elec. Rev.*, 21-24, 1960.
18. McCalley, R.B., Nomogram for selection of safety factors, *Design News,* Sept. 1957, pp. 138-141.
19. Shigley, J.E. and Mitchell, L.D., *Mechanical Engineering Design,* McGraw-Hill, New York, 610-611, 1983.
20. Bompass-Smith, J.H., *Mechanical Survival: The Use of Reliability Data,* McGraw-Hill, London, 1973.
21. Juvinall, R.C., *Fundamentals of Machine Component Design,* John Wiley & Sons, New York, 1983.
22. Dhillon, B.S. and Singh, C., *Engineering Reliability: New Techniques and Applications,* John Wiley & Sons, New York, 1981.
23. MIL-STD-1629, Procedures for Performing a Failure Mode, Effects, and Criticality Analysis (FMECA), Department of Defense, Washington, D.C.
24. Kececioglu, D. and Li, D., Exact solutions for the predication of the reliability of mechanical components and structural members, *Proc. Failure Prevention Reliability Conf.*, published by the American Society of Mechanical Engineers, ASME, New York, 1985, pp. 115-122.
25. Rhodes, S., Nelson, J.J., Raze, J.D., and Bradley, M., Reliability models for mechanical equipment, *Proc. Annu. Reliability Maintainability Symp.*, 127-131, 1988.
26. Raze, J.D., Nelson, J.J., Simard, D.J., amd Bradley, M., Reliability models for mechanical equipment, *Proc. Annu. Reliability Maintainability Symp.*, 130-134, 1987.
27. Nelson, J.J., Raze, J.D., Bowman, J., Perkins, G., and Wannamaker, A., Reliability models for mechanical equipment, *Proc. Annu. Reliability Maintainability Symp.*, 146-153, 1989.
28. Boone, T.D., Reliability Prediction Analysis for Mechanical Brake Systems, NAVA IR/SYSCOM Report, August 1981, Department of Navy, Department of Defense, Washington, D.C.

29. Minegishi, H., Prediction of Brake Pad Wear/Life by Means of Brake Severity Factor as Measured on a Data Logging System, Society of Automotive Engineers (SAE) Paper No. 840358, 1984, SAE, Warrendale, PA.
30. Spokas, R.B., Clutch Friction Material Evaluation Procedures, SAE Paper No. 841066, 1984, SAE, Warrendale, PA.
31. Shafer, R.E., Angus, J.E., Finkelstein, J.M., Yerasi, M., and Fulton, D.W., RADC Non-Electronic Reliability Notebook, Report No. RADC-TR-85-194, 1985, Reliability Analysis Center (RAC), Rome Air Development Center (RADC), Griffiss Air Force Base, Rome, NY.
32. GIDEP Operations Center, U.S. Department of Navy, Naval Weapons Station, Seal Beach, Corona Annex, Corona, CA.
33. Earles, D.R., *Handbook: Failure Rates,* AVCO Corporation, Massachussetts, 1962.
34. Guth, G., Development of Nonelectronic Part Cycle Failure Rates, Rept. No. 17D/A-050678, December 1977, Martin Marietta Corporation, Orlando, FL. Available from the National Technical Information Service (NTIS), Springfield, VA.

10 Human Reliability in Engineering Systems

10.1 INTRODUCTION

The failure of engineering systems is not only caused by hardware or software malfunctions, but also by human errors. In fact, the reliability of humans plays a crucial role in the entire system life cycle: design and development phase, manufacturing phase, and operation phase.

Even though in modern times the history of human factors may be traced back to 1898 when Frederick W. Taylor performed studies to determine the most appropriate designs for shovels [1], it was not until 1958 when Williams [2] recognized that human-element reliability must be included in the overall system-reliability prediction; otherwise, such a prediction would not be realistic. In 1960, the work of Shapero et al. [3] further stressed the importance of human reliability in engineering systems by pointing out that human error is the cause for 20 to 50% of all equipment failures.

In 1962, a database known as Data Store containing time and human performance reliability estimates for human engineering design features was established [4]. Also, in the 1960s two symposia concerning human reliability/error were held [5, 6]. In 1973, *IEEE Transactions on Reliability* published a special issue on human reliability [7]. In 1980, a selective bibliography on human reliability was published covering the period from 1958 to 1978 [8]. The first book on human reliability entitled *Human Reliability: With Human Factors* appeared in 1986 [9].

Over the years many professionals have contributed to the field of human reliability. A comprehensive list of publications on the subject is given in Reference 10.

This chapter presents different aspects of human reliability.

10.2 TERMS AND DEFINITIONS

There are numerous terms and definitions used in the human reliability area and some are presented below [9, 11-13].

- **Human factors:** This is a body of scientific facts pertaining to the characteristics of humans.
- **Human error:** This is the failure to perform a given task (or the performance of a forbidden action) that could result in the disruption of scheduled operations or damage to property and equipment.
- **Human performance:** This is a measure of human-functions/actions subject to specified conditions.

- **Continuous task:** This is a task that involves some sort of tracking activity; for example, monitoring a changing situation.
- **Human reliability:** This is the probability of accomplishing a task successfully by the human at any required stage in system operations within a stated minimum time (if such time requirement is stated).

10.3 HUMAN ERROR OCCURRENCE EXAMPLES AND STUDIES

In the past human error has been responsible for many accidents and equipment failures. Numerous studies have reported a significant proportion of equipment and other failures due to human error. Results of some of those studies are given below.

- Human operator accounted for over 90% of the documented air traffic control system errors [14].
- A study of 135 vessel failures occurring during the period from 1926–1988 discovered that 24.5% of failures were directly due to humans [15].
- Over 50% of all technical medical equipment problems are due to operator errors [16].
- Up to 90% of accidents both generally and in medical devices are caused by human mistakes [17, 28].
- A study of 23,000 defects in the production of nuclear components revealed that approximately 82% of the defects were due to human error [19].
- During the period from June 1, 1973 to June 30, 1975, 401 human errors occurred in U.S. commercial light-water nuclear reactors [20].

10.4 HUMAN ERROR OCCURRENCE CLASSIFICATION AND ITS TYPES AND CAUSES

Human errors can occur in many different ways and their occurrence may be classified into six distinct categories as shown in Figure 10.1. These are decision errors, action errors, transmission errors, checking errors, diagnostic errors, and retrieval errors.

Decision errors occur when the wrong decision is made after considering the situation. Action errors are the result of no action, incorrect action, or the performance of correct action on the wrong object when required. Transmission errors occur when information that must be passed to others is not sent, sent incorrectly, or sent to the wrong destination. Checking errors occur when the systems require checks, the incorrect checks are made, checks are omitted, or correct checks are made on the wrong objects. Diagnostic errors are the result of misinterpreting the actual situation when an abnormal event occurs. Retrieval errors occur when required information either from an individual, an individual's memory, or from any other reference source is not received or the incorrect information is received.

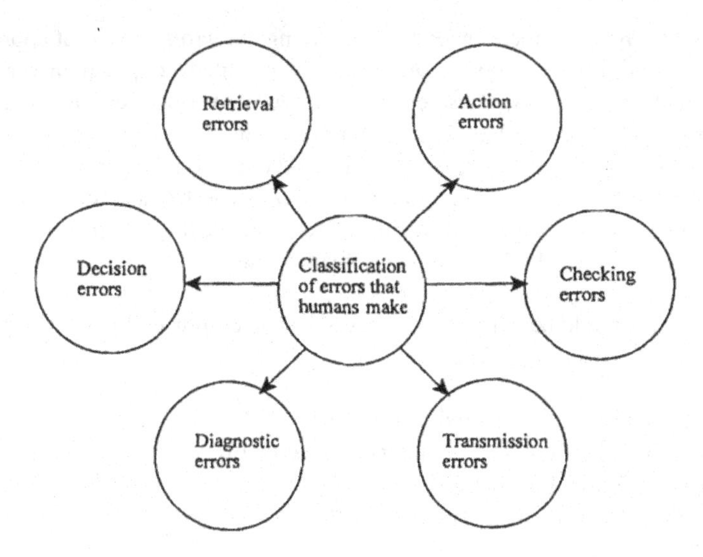

FIGURE 10.1 Categories of errors that humans make.

Human errors may be broken into many distinct types as follows [9, 21, 22]:

- **Design errors.** These types of errors are the result of inadequate design. The causes of these errors are assigning inappropriate functions to humans, failure to implement human needs in the design, failure to ensure the man-machine interaction effectiveness, and so on. An example of design errors is the placement of controls and displays so far apart that an operator is unable to use them in an effective manner.
- **Operator errors.** These errors are the result of operator mistakes and the conditions that lead to operator errors include lack of proper procedures, complex tasks, poor personnel selection and training, poor environment, and operator carelessness.
- **Assembly errors.** These errors occur during product assembly due to humans. Assembly errors may occur due to causes such as poorly designed work layout, inadequate illumination, excessive noise level, poor blueprints and other related material, excessive temperature in the work area, and poor communication of related information.
- **Inspection errors.** These errors occur because of less than 100% accuracy of inspectors. One example of an inspection error is accepting and rejecting out-of-tolerance and in-tolerance parts, respectively. Nonetheless, according to Reference 23, an average inspection effectiveness is close to 85%.
- **Maintenance errors.** These errors occur in the field due to oversights by the maintenance personnel. As the equipment becomes old, the likelihood of the occurrence of those errors may increase because of the increase in maintenance frequency. Some of the examples of maintenance errors are

calibrating equipment incorrectly, applying the wrong grease at appropriate points of the equipment, and repairing the failed equipment incorrectly.
- **Installation errors.** These errors occur due to various reasons including using the wrong installation related blueprints or instructions, or simply failing to install equipment according to the manufacturer's specification.
- **Handling errors.** These errors basically occur because of inadequate storage or transportation facilities. More specifically, such facilities are not as specified by the equipment manufacturer.

In general, there could be numerous causes for the occurrence of human error. Some of those are listed below [9, 21].

- Poor motivation of involved personnel.
- Poor training or skill of concerned personnel.
- Poor equipment design.
- Inadequate or poorly written equipment operating and maintenance procedures.
- Poor job environment: poor lighting, high/low temperature, high noise level, crowded work space, etc.
- Inadequate work tools.
- Complex tasks.
- Poor work layout.

10.5 HUMAN PERFORMANCE AND STRESS

Human performance depends on many different factors; thus, it may be said that humans perform differently under different conditions. More specifically, the important factors that will affect a person's performance include reaction to stress, time at work, fatigue, group interaction and identification, social pressure, repetitive work, supervisor's expectations, morale, social interaction, crew efficiency, and idle time [24].

In particular, stress is probably the most important factor that affects human performance. Over the years, many researchers have studied the relationship between human performance effectiveness and stress/anxiety. The resulting curve of their effort is shown in Figure 10.2 [25, 26]. This curve indicates that a moderate level of stress is necessary to increase human performance effectiveness to its maximum. At a very low stress level, the task becomes unchallenging and dull and most humans' performance effectiveness is not at the maximum. In contrast, as the stress bypasses its moderate level, the human performance effectiveness begins to decline. This decline is caused by factors such as fear, worry, and other types of psychological stress. In the highest stress region, human reliability is at its lowest.

Another important point to note from Figure 10.2 is that as the stress increases in Region A, human performance effectiveness improves. However, in Region B, human performance effectiveness decreases with the increasing stress.

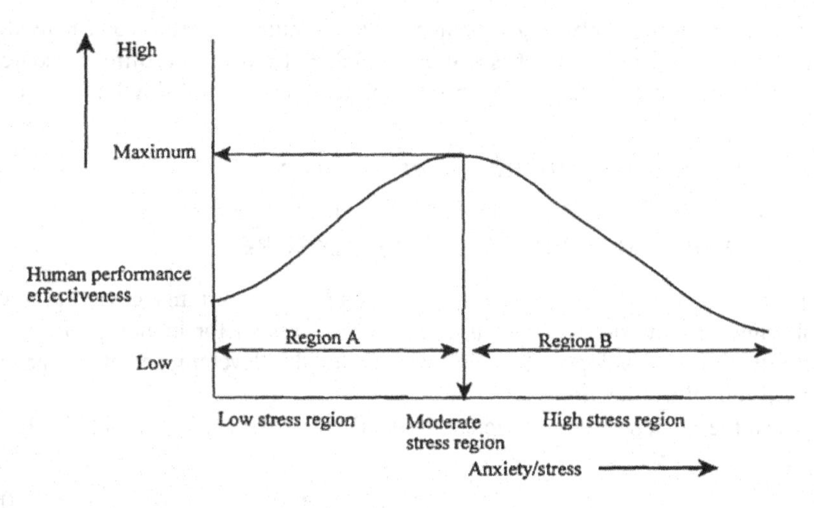

FIGURE 10.2 A hypothetical curve of human performance effectiveness vs. anxiety/stress.

10.5.1 Stress Factors and Operator Stress Characteristics

There are many factors that increase the stress on a human and in turn decrease his/her reliability in work and other environments. Some of these factors are listed below [25].

- Dissatisfied with the current job.
- Faced with serious financial problems.
- Working with individuals having unpredictable temperaments.
- Low chance of promotion from current position/job.
- Facing a possible work layoff.
- Current job/position below ability and experience.
- Conducting tasks under extremely tight time schedules.
- Having health problems.
- Excessive demands of superiors.
- Inadequate expertise to conduct the tasks required in the current job/position.
- Often takes work home to meet deadlines.
- Having difficulties with spouse/children.

A human operator has certain limitations in conducting certain tasks. The error occurrence probability increases when such limitations are exceeded. In order to improve operator reliability, such operator limitations or stress characteristics must be considered carefully during the design phase.

The operator stress characteristics include [12]: very short decision-making time, several displays difficult to discriminate, requirement to perform steps at high speed, poor feedback for the determination of accuracy of actions taken, requirement for prolonged monitoring, very long sequence of steps required to perform a task, requirement to make decisions on the basis of data obtained from various different sources, and requirement to operate at high speed more than one control simultaneously.

10.6 HUMAN PERFORMANCE RELIABILITY IN CONTINUOUS TIME AND MEAN TIME TO HUMAN ERROR (MTTHE) MEASURE

As humans perform time-continuous tasks such as aircraft maneuvering, scope monitoring, and missile countdown, a general expression for human performance reliability can be developed the same way as for the development of the general classical reliability function.

Thus, the human error occurrence probability in time interval Δt is [27-29]

$$P(X_2/X_1) = \lambda_h(t)\Delta t \tag{10.1}$$

where

X_1 is an error-free human performance event of duration time t.
X_2 represents an event that the human error will occur during time interval $[t, t + \Delta t]$.
$\lambda_h(t)$ is the human error rate at time t.

The joint probability of the error-free human performance is expressed by

$$P(\overline{X_2}/X_1)\, P(X_1) = P(X_1) - P(X_2/X_1)\, P(X_1) \tag{10.2}$$

where

$\overline{X}_2$ represents an event that human error will not occur during interval $[t, t + \Delta t]$.

More specifically, Equation (10.2) denotes an error-free human performance probability over intervals [o, t] and $[t, t + \Delta t]$.

Equation (10.2) may be rewritten to the following form:

$$R_h(t + \Delta t) = R_h(t) - R_h(t) P(X_1/X_2) \tag{10.3}$$

where

$R_h(t)$ is the human performance reliability at time t.
$R_h(t + \Delta t)$ is the human performance reliability at time $t + \Delta t$.

By inserting Equation (10.1) into Equation (10.3) we have

$$\frac{R_h(t+\Delta t)-R_h(t)}{\Delta t}=-R_h(t)\lambda_h(t) \tag{10.4}$$

In the limiting case, expression (10.4) becomes

$$\frac{dR_h(t)}{dt}=-R_h(t)\lambda_h(t) \tag{10.5}$$

By rearranging Equation (10.5) we get

$$-\lambda_h(t)dt=\frac{1}{R_h(t)}\cdot dR_h(t) \tag{10.6}$$

By integrating both sides of Equation (10.6) over the time interval [o, t] results in

$$-\int_0^t \lambda_h(t)dt=\int_0^t \frac{1}{R_h(t)}\cdot dR_h(t) \tag{10.7}$$

Since at t = 0, R_h (t) = 1 Equation (10.7) becomes

$$-\int_0^t \lambda_h(t)dt=\int_0^{R_h(t)} \frac{1}{R_h(t)}\cdot dR_h(t) \tag{10.8}$$

After evaluating the right-hand side of Equation (10.8), we get

$$\ln R_h(t)=-\int_0^t \lambda_h(t)dt \tag{10.9}$$

Thus, from Equation (10.10) we write

$$R_h(t)=e^{-\int_0^t \lambda_h(t)dt} \tag{10.10}$$

The above equation is the general human performance reliability function. It can be used to predict human reliability at time t when times to human error are described by any known statistical distribution. A study reported in Reference 27 collected human error data for time-continuous tasks under laboratory conditions. The Weibull, gamma, and log-normal distributions fitted quite well to these data.

In order to obtain a general expression for mean time to human error (MTTHE), we integrate Equation (10.10) over the interval [0, ∞]:

$$MTTHE = \int_0^\infty R_h(t)\,dt = \int_0^\infty \exp\left[-\int_0^t \lambda_h(t)\,dt\right]dt \tag{10.11}$$

The above equation can be used to obtain MTTHE when times to human error are governed by any known distribution function.

Example 10.1

Assume that the time to human error associated with a time-continuous task is exponentially distributed and the human error rate is 0.02 errors per hour. Calculate the following:

- Human performance reliability for a 5-h mission.
- MTTHE.

Thus, in this case we have

$$\lambda_h(t) = \lambda_h = 0.02 \text{ errors/h}$$

Inserting the above value into Equation (10.10) yields

$$\begin{aligned} R_h(t) &= e^{-\int_0^t (0.02)dt} \\ &= e^{-(0.02)t} \end{aligned} \tag{10.12}$$

At $t = 5$ h, from Equation (10.12) we get

$$\begin{aligned} R_h(5) &= e^{-(0.02)(5)} \\ &= 0.9048 \end{aligned} \tag{10.13}$$

By substituting Equation (10.12) into Equation (10.11) we get

$$\begin{aligned} MTTHE &= \frac{1}{(0.02)} \\ &= 50 \text{ h} \end{aligned}$$

It means the human performance reliability is 0.9048 and a human error can occur after every 50 h.

Example 10.2

A person is performing a time-continuous task and his/her time to human error is described by Weibull distribution. Obtain expressions for human reliability and MTTHE.

In this case, as the time to human error is Weibull distributed, the human error rate at time t is given by

$$\lambda_h(t) = (t/\theta)^{\alpha-1}\,\alpha/\theta \tag{10.14}$$

where

α is the shape parameter.
θ is the scale parameter.

By inserting Equation (10.14) into Equation (10.10) we have

$$\begin{aligned} R_h(t) &= e^{-\int_0^t \left[(t/\theta)^{\alpha-1}\alpha/\theta\right]dt} \\ &= e^{-(t/\theta)^{\alpha}} \end{aligned} \tag{10.15}$$

Using Equation (10.15) in Equation (10.11) yields

$$\begin{aligned} \text{MTTHE} &= \int_0^{\infty} \exp\left[-(t/\theta)^{\alpha}\right]dt \\ &= \theta\,\Gamma\left(\frac{1}{\alpha}+1\right) \end{aligned} \tag{10.16}$$

where

Γ (•) is the gamma function.

The human reliability and MTTHE are given by Equations (10.15) and (10.16), respectively.

10.7 HUMAN RELIABILITY EVALUATION METHODS

Over the years, many techniques to evaluate human reliability have been developed [9]. Each such technique has advantages and disadvantages. Some of these methods are presented below.

10.7.1 Probability Tree Method

This is used to perform task analysis by diagrammatically representing critical human actions and other events associated with the system. Usually, this method is utilized to perform task analysis in the technique for the Human Error Rate Prediction (THERP) approach [30]. Nonetheless, the branches of the probability tree represent diagrammatic task analysis. More specifically, the outcomes (e.g., success of failure) of each event are represented by the branching limbs of the tree and each tree branch is assigned an occurrence probability.

There are many benefits of this method including a visibility tool, simplified mathematical computations which in turn reduce the error occurrence probability due to computation, and estimate conditional probability readily, which may otherwise be obtained through complex probability equations.

This approach can also be used to calculate reliability of standard networks [31]. The method is demonstrated by solving the following example.

Example 10.3

A person performs two quality control tasks: α and β. Each of these two tasks can either be performed correctly or incorrectly. The task α is performed before task β and both the tasks are independent of each other. In other words, the performance of task α does not affect the performance of task β or vice-versa. Develop a probability tree for this example and obtain an expression for probability of not successfully completing the overall mission.

In this case, the person first performs task α correctly or incorrectly and then proceeds to task β which can also be performed correctly or incorrectly. This scenario is depicted by the probability tree shown in Figure 10.3. In this figure, α and β with bars denote unsuccessful events and without bars denote successful events. Other symbols used in the solution to the example are as follows:

P_s is the probability of success of the overall mission (i.e., performing both tasks α and β correctly).
P_f is the failure probability of the overall mission.
P_α is the probability of performing task α correctly.
$P_{\bar{\alpha}}$ is the probability of performing task α incorrectly.
P_β is the probability of performing task β correctly.
$P_{\bar{\beta}}$ is the probability of performing task β incorrectly.

Using Figure 10.3, the probability of success of the overall mission is

$$P_s = P_\alpha P_\beta \tag{10.17}$$

Similarly, using Figure 10.3, the failure probability of the overall mission is given by

$$P_f = P_{\bar{\alpha}} P_{\bar{\beta}} + P_{\bar{\alpha}} P_\beta + P_\alpha P_{\bar{\beta}} \tag{10.18}$$

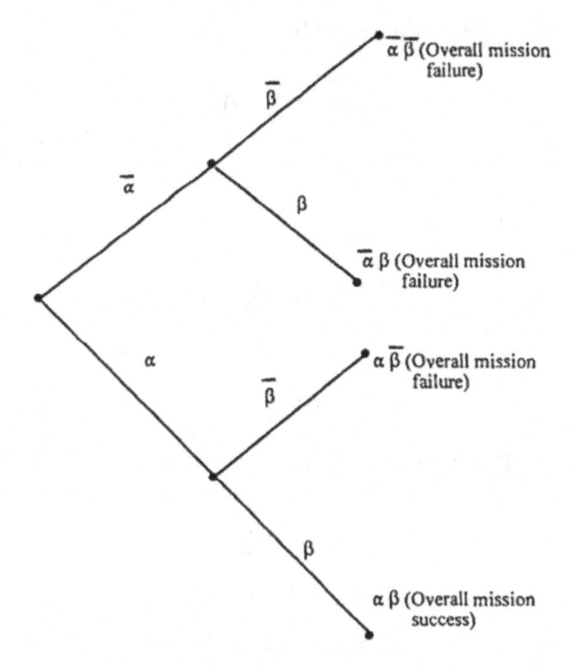

FIGURE 10.3 Probability tree for performing tasks α and β.

Since $P_\alpha + P_{\bar{\alpha}} = 1$ and $P_\beta + P_{\bar{\beta}} = 1$, by subtracting Equation (10.18) from unity yields the same result as Equation (10.17), i.e.,

$$P_s = 1 - P_f = 1 - \left[P_{\bar{\alpha}} P_{\bar{\beta}} + P_{\bar{\alpha}}\left(1 - P_{\bar{\beta}}\right) + P_{\bar{\beta}}\left(1 - P_{\bar{\alpha}}\right)\right] \tag{10.19}$$

$$P_s = 1 - \left[P_{\bar{\alpha}} + P_{\bar{\beta}} - P_{\bar{\alpha}} P_{\bar{\beta}}\right] \tag{10.20}$$

The above equation is exactly the same as Equation (10.17), if we write the right-hand term of Equation (10.17) in terms of failure probabilities. Nonetheless, the probability of not successfully completing the overall mission is given by Equation (10.18).

Example 10.4

Assume that in Example 10.3 the probabilities of performing tasks α and β incorrectly are 0.15 and 0.2, respectively. Determine the following:

- Probability of successfully completing the overall mission.
- Probability of not successfully completing the overall mission.
- The sum of the above two results is equal to unity.

Inserting the given data into Equation (10.20) yields

$$P_s = 1 - [0.15 + 0.2 - (0.15)(0.2)]$$
$$= 0.68$$

Since $P_\alpha + P_{\bar{\alpha}} = 1$, we get

$$P_\alpha = 1 - P_{\bar{\alpha}}$$
$$= 1 - 0.15$$
$$= 0.85$$

Similarly, since $P_\beta + P_{\bar{\beta}} = 1$, we have

$$P_\beta = 1 - P_{\bar{\beta}}$$
$$= 1 - 0.2$$
$$= 0.8$$

By substituting the above two values and the specified data into Equation (10.18) we get

$$P_f = (0.15)(0.2) + (0.15)(0.8) + (0.85)(0.2)$$
$$= 0.32$$

Thus,

$$P_s + P_f = 0.68 + 0.32 = 1$$

The probabilities of successfully/not successfully completing the overall missior are 0.68/0.32, respectively, and the sum of these two values is equal to unity.

10.7.2 Fault Tree Method

This approach is described in detail in Chapter 7 and it can also be used to perform human reliability analysis. The following examples demonstrate the application of this method in human reliability work.

Example 10.5

A person is required to do a certain job, J, composed of two independent tasks X and Y. These two tasks must be performed correctly for the successful completion

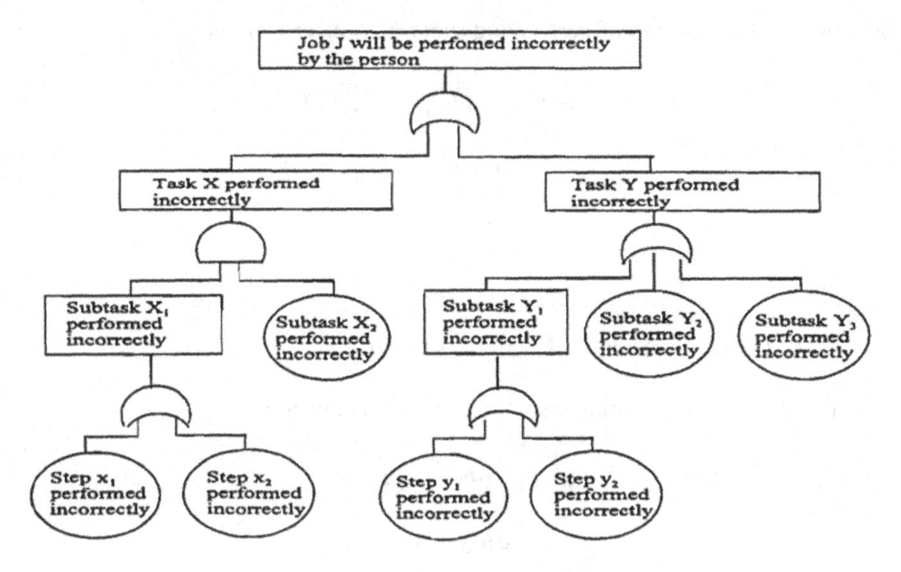

FIGURE 10.4 Fault tree for the performance of Job J incorrectly.

of the job. Task X is made up of two subtasks X_1 and X_2. If any one of these two subtasks is performed correctly, Task X can be completed successfully. Task Y is composed of subtasks Y_1, Y_2, and Y_3. All of these three subtasks must be performed correctly for the success of Task Y. Both subtasks, X_1 and Y_1, are composed of two steps each, i.e., x_1, x_2 and y_1, y_2, respectively. Both steps for each of these two subtasks must be completed correctly for subtask success. Develop a fault tree for the event that job J will be performed incorrectly by the person.

Figure 10.4 presents the fault tree for the example.

Example 10.6

Assume that the probability of occurrence of the basic events in Figure 10.4 is 0.04. Calculate the probability of occurrence of the top event, Job J will be performed incorrectly by the person. Assume in your calculations that the given fault tree is independent.

The probability of performing subtask X_1 incorrectly is given by

$$\begin{aligned} P(X_1) &= P(x_1) + P(x_2) - P(x_1)\,P(x_2) \\ &= 0.04 + 0.04 - (0.04)\,(0.04) \\ &= 0.0784 \end{aligned}$$

where

$P\,(x_i)$ is the probability of performing step x_i incorrectly, for i = 1, 2.

Similarly, the probability of performing subtask Y_1 incorrectly is

$$P(Y_1) = P(y_1) + P(y_2) - P(y_1)\,P(y_2)$$
$$= 0.04 + 0.04 - (0.04)\,(0.04)$$
$$= 0.0784$$

where

$P(y_i)$ is the probability of performing step y_i incorrectly, for i = 1, 2.

The probability of performing task X incorrectly is given by

$$P(X) = P(X_1) + P(X_2)$$
$$= (0.0784)\,(0.04)$$
$$= 0.0031$$

where

$P(X_i)$ is the probability of performing subtask X_i incorrectly, for i = 1, 2.

The probability of performing task Y incorrectly is expressed by

$$P(Y) = 1 - \left(1 - P(Y_1)\right)\left(1 - P(Y_2)\right)\left(1 - P(Y_3)\right)$$
$$= 1 - (1 - 0.0784)\,(1 - 0.04)\,(1 - 0.04)$$
$$= 0.1507$$

where

$P(Y_i)$ is the probability of performing subtask Y_i incorrectly, for i = 1, 2, 3.

The probability of performing job J incorrectly by the person is

$$P(J) = P(X) + P(Y) - P(X)\,P(Y)$$
$$= 0.0031 + 0.1507 - (0.0031)\,(0.1507)$$
$$= 0.1534$$

Figure 10.5 shows the fault tree with given and calculated probability values. The probability that job J will be performed incorrectly by the person is 0.1534.

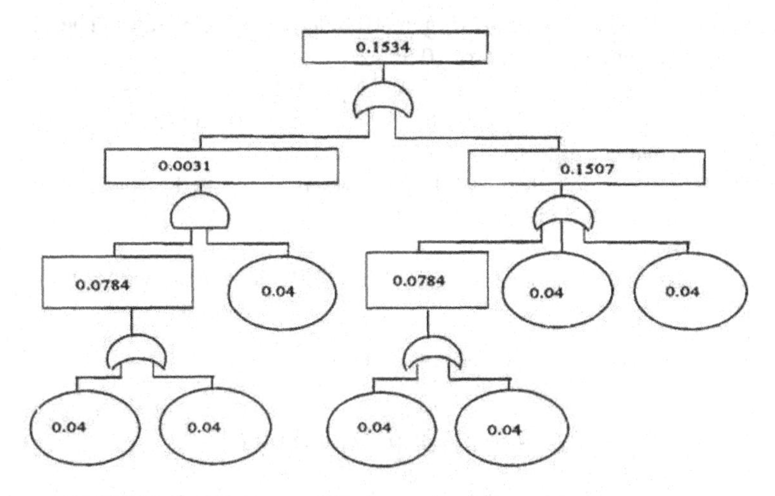

FIGURE 10.5 Fault tree with given and calculated probability values.

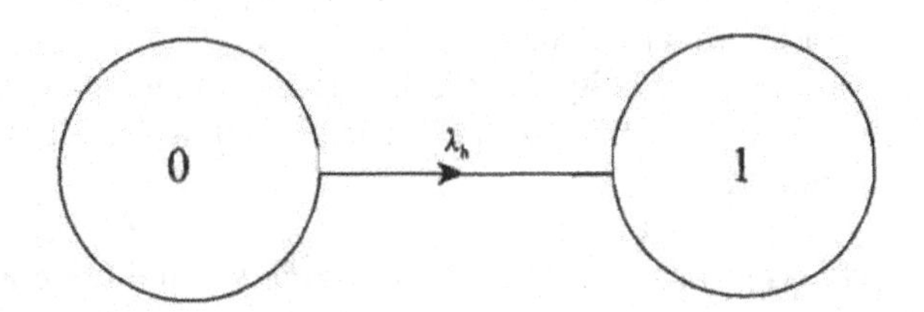

FIGURE 10.6 State space diagram for a human performing a time continuous task.

10.7.3 Markov Method

This is a powerful reliability engineering tool that can also be used to perform time continuous human reliability analysis. The following assumptions are associated with this method:

- The probability of a transition occurrence from one state to another in finite time Δt is $\lambda_h \Delta t$. The parameter λ_h in our case is the constant human error rate.
- The probability of two or more transitional occurrences in finite time Δt is negligible [i.e., $(\lambda_h \Delta t)(\lambda_h \Delta t) \rightarrow 0$].
- All occurrences are independent of each other.

The state space diagram of a human performing a time-continuous task is given in Figure 10.6. In this case, the system has two states (i.e., 0 and 1). The state 0 represents human performing the time continuous task normally and state 1 denotes human committed error. By using the Markov method, we can obtain the probabilities of the human being in states 0 and 1 at time t for known human error rate.

The numerals in the circles of Figure 10.6 denote system states. The following symbols are associated with Figure 10.6:

P_0 (t) is the probability that the human is performing his/her assigned task normally at time t.
P_1 (t) is the probability that the human has committed error at time t.
λ_h is the constant human error rate.

Using the Markov approach, we write down the following equations for Figure 10.6.

$$P_0(t+\Delta t) = P_0(t)(1-\lambda_h \Delta t) \tag{10.21}$$

$$P_1(t+\Delta t) = P_0(t)(\lambda_h \Delta t) + P_1(t) \tag{10.22}$$

where

P_0 (t + Δt) is the probability that the human is performing his/her assigned task normally at time t + Δt.
P_1 (t + Δt) is the probability that the human has committed error at time t.
$(1 - \lambda_h \Delta t)$ is the probability of the occurrence of no human error in finite time Δt.

Rearranging Equations (10.21) and (10.22) and taking the limits, we get

$$\frac{dP_0(t)}{dt} = -\lambda_h P_0(t) \tag{10.23}$$

$$\frac{dP_1(t)}{dt} = \lambda_h P_0(t) \tag{10.24}$$

At time t = 0, P_0 (0) = 1 and P_1 (0) = 0.

Solving Equations (10.23) and (10.24) using Laplace transforms [9] leads to

$$P_0(t) = e^{-\lambda_h t} \tag{10.25}$$

$$P_1(t) = 1 - e^{-\lambda_h t} \tag{10.26}$$

Thus, the human reliability, R_h (t), is given by

$$R_h(t) = P_0(t) = e^{-\lambda_h t} \tag{10.27}$$

By integrating Equation (10.27) over the time interval [0, ∞], we get the following equation for MTTHE:

$$\begin{aligned} \text{MTTHE} &= \int_0^\infty R_h(t)\,dt \\ &= \int_0^\infty e^{-\lambda_h t}\,dt \\ &= 1/\lambda_h \end{aligned} \tag{10.28}$$

Example 10.7

Assume that a person is performing a time continuous task and his/her human error rate is 0.006 errors per hour. Calculate the probability that the person will commit an error during an 8-h mission.

Substituting the given data into Equation (10.26) yields

$$\begin{aligned} P_1(8) &= 1 - e^{-(0.006)(8)} \\ &= 0.0469 \end{aligned}$$

There is an approximately 5% chance that the person will commit an error during the 8-h mission.

10.8 HUMAN RELIABILITY MARKOV MODELING

As demonstrated in the preceding section, the Markov method is a useful tool to perform various types of human reliability analysis. This section presents analysis of two mathematical models using the Markov method.

10.8.1 Reliability Analysis of a System with Human Error

This mathematical model represents a system which can fail either due to a hardware failure (i.e., a failure other than a human error) or to a human error. The system human/non-human failure rates are constant. The system state space diagram is shown in Figure 10.7. By using the Markov approach we can obtain equations for

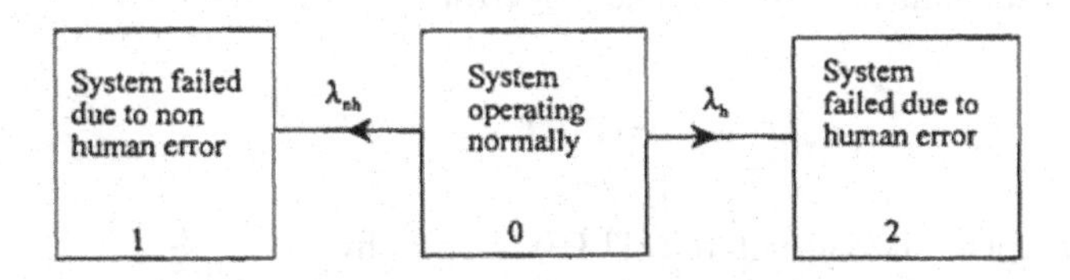

FIGURE 10.7 System transition diagram.

system reliability, system failure probability due to human error, system failure probability due to non-human error, and system mean time to failure. The numerals in the boxes of Figure 10.7 denote system state.

The following symbols are associated with this model:

λ_h is the constant human error rate.
λ_{nh} is the constant nonhuman error rate (i.e., hardware failure rate, etc.)
$P_i(t)$ is the probability that the system is in state i at time t; for i = 0 (means system operating normally), i = 1 (means system failed due to non-human error), and i = 2 (means system failed due to human error).

By applying the Markov method and using Figure 10.7, we get the following equations:

$$\frac{dP_0(t)}{dt} + (\lambda_h + \lambda_{nh}) P_0(t) = 0 \qquad (10.29)$$

$$\frac{dP_1(t)}{dt} - \lambda_{nh} P_0(t) = 0 \qquad (10.30)$$

$$\frac{dP_2(t)}{dt} - \lambda_h P_0(t) = 0 \qquad (10.31)$$

At time t = 0, $P_0(0) = 1$, $P_1(0) = P_2(0) = 0$.

Solving the above equations using Laplace transforms [9] results in

$$P_0(t) = e^{-(\lambda_h + \lambda_{nh})t} \qquad (10.32)$$

$$P_1(t) = \frac{\lambda_{nh}}{(\lambda_{nh} + \lambda_h)} \left[1 - e^{-(\lambda_h + \lambda_{nh})t}\right] \qquad (10.33)$$

$$P_2(t) = \frac{\lambda_h}{(\lambda_{nh} + \lambda_h)} \left[1 - e^{-(\lambda_h + \lambda_{nh})t}\right] \qquad (10.34)$$

The system reliability, $R_s(t)$, with human error is

$$R_s(t) = P_0(t) = e^{-(\lambda_h + \lambda_{nh})t} \qquad (10.35)$$

The system mean time to failure (MTTF) is given by

$$\begin{aligned} MTTF &= \int_0^{\infty} R_s(t)\,dt \\ &= \int_0^{\infty} e^{-(\lambda_h+\lambda_{nh})t}\,dt \\ &= 1/(\lambda_h+\lambda_{nh}) \end{aligned} \tag{10.36}$$

Example 10.7

A system can fail either due to a hardware failure or a human error and its hardware failure and human error rates are 0.0005 failures/h and 0.0001 errors/h, respectively. Calculate the system MTTF and failure probability due to human error for a 12-h mission.

Inserting the given data into Equation (10.36) yields

$$\begin{aligned} MTTF &= 1/(0.0001+0.0005) \\ &= 1666.7 \text{ h} \end{aligned}$$

Similarly, we substitute the specified data into Equation (10.34) to get

$$\begin{aligned} P_2(12) &= \frac{(0.0001)}{(0.0001+0.0005)}\left[1-e^{-(0.0001+0.0005)(12)}\right] \\ &= 0.0012 \end{aligned}$$

Thus, the system failure probability due to human error for a 12-h mission is 0.0012.

10.8.2 Reliability Analysis of a Human Performing a Time-Continuous Task Under Fluctuating Environment

This mathematical model represents a human performing a time continuous task under fluctuating environment. The environment fluctuates from normal to abnormal (stress) and vice versa [32]. Human error can occur under both environments. However, it is logical to assume that the human error rate under the abnormal environment will be greater than under the normal environment because of increased stress. In this model, it is also assumed that the rate of change from normal to abnormal environment and vice versa is constant along with human error rate under either environment. The model state space diagram is shown in Figure 10.8. The numerals in the boxes of this figure denote system state. The following symbols were used to develop equations for the model:

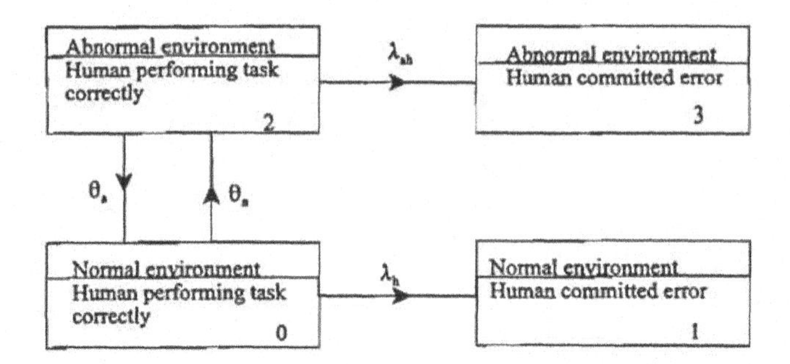

FIGURE 10.8 State space diagram of a human performing a time continuous task under fluctuating environment.

P_i (t) is the probability of the human being in state i at time t; for i = 0 (means human performing task correctly under normal environment), i = 1 (means human committed errors under normal environment), i = 2 (means human performing task correctly under abnormal environment), i = 3 (means human committed error under abnormal environment).
λ_h is the constant human error rate from state 0.
λ_{ah} is the constant human error rate from state 2.
θ_n is the constant transition rate from normal to abnormal environment.
θ_a is the constant transition rate from abnormal to normal environment.

By using the Markov method, we get the following system of equations from Figure 10.8:

$$\frac{dP_0(t)}{dt} + (\lambda_h + \theta_n) P_0(t) = P_2(t)\theta_a \tag{10.37}$$

$$\frac{dP_1(t)}{dt} - P_0(t)\lambda_h = 0 \tag{10.38}$$

$$\frac{dP_2(t)}{dt} + (\lambda_{ah} + \theta_a) P_2(t) = P_0(t)\theta_n \tag{10.39}$$

$$\frac{dP_3(t)}{dt} - P_2(t)\lambda_{ah} = 0 \tag{10.40}$$

At time t = 0, $P_0(0) = 1$, and $P_1(0) = P_2(0) = P_3(0) = 0$.

Solving the above equations using Laplace transforms we obtain

$$P_0(t) = (s_2 - s_1)^{-1} \left[(s_2 + \lambda_{ah} + \theta_n) e^{s_2 t} - (s_1 + \lambda_{ah} + \theta_n) e^{s_1 t} \right] \tag{10.41}$$

where

$$s_1 = \left[-a_1 + \left(a_1^2 - 4\, a_2\right)^{1/2}\right] \Big/ 2 \tag{10.42}$$

$$s_2 = \left[-a_1 - \left(a_1^2 - 4\, a_2\right)^{1/2}\right] \Big/ 2 \tag{10.43}$$

$$a_1 \equiv \lambda_h + \lambda_{ah} + \theta_n + \theta_a \tag{10.44}$$

$$a_2 \equiv \lambda_h \left(\lambda_{ah} + \theta_a\right) + \theta_n \lambda_{ah} \tag{10.45}$$

$$P_1(t) = a_4 + a_5\, e^{s_2 t} - a_6\, e^{s_1 t} \tag{10.46}$$

where

$$a_3 = 1/\left(s_2 - s_1\right) \tag{10.47}$$

$$a_4 = \lambda_h \left(\lambda_{ah} + \theta_a\right) / s_1 s_2 \tag{10.48}$$

$$a_5 = a_3 \left(\lambda_h + a_4 s_1\right) \tag{10.49}$$

$$a_6 = a_3 \left(\lambda_h + a_4 s_2\right) \tag{10.50}$$

$$P_2(t) = \theta_n\, a_3 \left(e^{s_2 t} - e^{s_1 t}\right) \tag{10.51}$$

$$P_3(t) = a_7 \left[\left(1 + a_3\right) \left(s_1 e^{s_2 t} - s_2 e^{s_1 t}\right)\right] \tag{10.52}$$

where

$$a_7 \equiv \lambda_{ah}\, \theta_n / s_1 s_2 \tag{10.53}$$

The reliability of the human is expressed by

$$R_h(t) = P_0(t) + P_2(t) \tag{10.54}$$

Thus, the MTTHE is

$$\begin{aligned} MTTHE &= \int_0^\infty R_h(t)\,dt \\ &= \int_0^\infty [P_0(t) + P_2(t)]\,dt \qquad (10.55) \\ &= (\lambda_{ah} + \theta_n + \theta_a)/a_2 \end{aligned}$$

Example 10.8

Assume that a person is performing a time continuous task under normal and abnormal environments with error rates 0.002 errors/h and 0.003 errors/h, respectively. The transition rates from normal to abnormal environment and vice versa are 0.02 per hour and 0.04 per hour, respectively. Calculate the MTTHE.

Substituting the given data into Equation (10.55) we get

$$\begin{aligned} MTTHE &= \frac{(0.003 + 0.02 + 0.04)}{0.002(0.003 + 0.04) + (0.02)(0.003)} \\ &= 431.51 \text{ h} \end{aligned}$$

Thus, the mean time to human error is 431.51 h.

10.9 HUMAN ERROR DATA

Human reliability predictions are only as effective as the body of quantitative data on which they are made. More specifically, the human error data are the backbone of any human reliability prediction. Thus, the collection and maintenance of human error data are at least as important as making various types of human reliability predictions. Human error data can be collected through several means including [33–35]:

- Experimental studies
- Expert judgments
- Self-made error reports
- Human data recorder
- Automatic data recorder
- Published literature

The data collected from experimental studies is usually generated under the laboratory conditions. These conditions may not be the true representatives of actual conditions. In addition, the method is time consuming and rather expensive. The main advantage of data collected through experimental means is that the data are probably the least influenced by the subjective elements that may procure some error. One example of data based on the experimental findings is the Data Store [4]. Expert judgments are another approach to obtain human reliability data. This approach is used quite often by human reliability methodologists and has two attractive features: (i) it is comparatively easy to develop because a large amount of data can be collected from a small number of expert respondents, and (ii) it is relatively cheaper to develop. Some of the drawbacks of this method are less reliable than data obtained through other means and frequent use of less experienced experts than required.

In the case of self-made error reports, the person who makes the error also reports that error. One of the drawbacks of this method is that people are generally reluctant to confess making an error. The human data recorder approach calls for the physical presence of a person to observe task performance and document events as necessary. Some of the disadvantages of this approach are that it is expensive and the observer may fail to recognize committed errors. In operational system testing of human-machine systems, the human data recorder approach is used quite frequently. In the case of the automatic data recorder method, the use of instrumentation permits the automatic recording of all operator actions. Two typical examples are the Operational Performance Recording and Evaluation Data System (OPREDS) [35] and the Performance Measurement System. The latter system was developed by the General Physics Corporation for recording actions in nuclear power plant simulators. The published literature approach calls for collecting the data from publications such as journals, conference proceedings, and books.

10.9.1 Specific Human Error Data Banks and Sources

Over the years many data banks for obtaining human error related information have been developed [9, 37, 38]. Nine such data banks listed in Table 10.1 are reviewed in Reference 39.

TABLE 10.1
Human Error Data Banks

No.	Data bank title
1	Data Store [4]
2	Operational Performance Recording and Evaluation Data System (OPREDS) [36]
3	Nuclear Plant Reliability Data System [40]
4	Aerojet General Method [41]
5	Aviation Safety Reporting System [42]
6	Bunker-Ramo Tables [43]
7	Air Force Inspection and Safety Center Life Sciences Accident and Incident Reporting System [44]
8	Safety Related Operator Action (SROA) Program [39]
9	Technique for Establishing Personnel Performance Standards (TEPPS) [39]

Other important sources for obtaining human error related data are References 45 through 48. Over 20 sources for obtaining human reliability data are listed in References 9 and 34.

10.9.2 Brief Description of Selected Human Error Data Banks

This section briefly describes three important data banks for obtaining human error related information [39].

- **Data Store.** It has served as an important source for obtaining human reliability data [49]. The Data Store was established in 1962 by the American Institute for Research, Pittsburgh [4] and it contains estimates for time and human performance reliability. More specifically, this data bank possesses data relating human error probabilities to design features. All in all, Data Store is probably the most developed data bank for application during the design process.
- **Aviation Safety Reporting System.** Originally this data bank was developed by the National Aeronautics and Space Administration (NASA) and contains information on civil aircraft accidents/incidents. However, the data is based on voluntary reporting and it receives roughly 400 human error reports per month.
- **Operation Performance Recording and Evaluation Data System.** This system was developed to collect data on operational human performance by the U.S. Navy Electronics Laboratory, San Diego. The system permits automatic monitoring of human operator actions and in the late 1960s and early 1970s it was used to collect data on various U.S. Navy ships. Nonetheless, the system was restricted to switch turning and button-pushing.

10.9.3 Human Error Data for Selective Tasks

In order to give examples of the types of available human reliability data, Table 10.2 presents human error data for selective tasks taken from published sources [34].

10.10 PROBLEMS

1. Define the following terms:
    - Human reliability
    - Human error
    - Human factors
2. Give some examples of well-publicized catastrophic disasters due to humans.
3. What are the important categories of errors that humans make?

TABLE 10.2
Human Error Data for Selective Tasks

Number	Error/task description	Performance reliability	Errors per plant-month (boiling water reactors)	Errors per million operations
1	Finding maintenance (scheduled) approaches in maintenance manual	0.997	—	—
2	Reading gauge incorrectly	—	—	5000
3	Requirement misinterpretation/ misunderstanding	—	0.0074	—
4	Adjusting incorrectly	—	0.026	—
5	Incorrect servicing	—	0.03	—
6	Connecting hose incorrectly	—	—	4700
7	Turning rotary selector switch to certain position	0.9996	—	—
8	Installation o-ring incorrectly	—	—	66700

4. Describe the following terms:
   - Operator errors
   - Maintenance errors
   - Assembly errors
5. Discuss the human performance effectiveness vs. stress relationship.
6. List at least 12 available sources for obtaining human error data.
7. Assume that a person performs three independent tasks associated with a maintenance job, i.e., tasks x, y, and z. Each task can either be performed correctly or incorrectly. The task x is performed before task y and task y is performed before task z. Obtain an expression for probability of not successfully completing the maintenance job by developing a probability tree.
8. A person is performing a time-continuous task with exponentially distributed time to human error. Calculate the following, if the person's error rate is 0.05 errors/h:
   - Person's reliability over an 8-h period.
   - Person's mean time to human error.
9. An engineering system can fail either due to a human error or a non-human error. Both the human error and non-human error rates are 0.006 and 0.009 failures per hour, respectively. Calculate the system failure probability due to a human error over a 100-h period.
10. A person is performing a time-continuous task with Rayleigh distributed time to human error. Develop an expression for mean time to human error.

10.11 REFERENCES

1. Chapnis, *Man-Machine Engineering*, Wadsworth Publishing Company, Belmont, CA, 1965.
2. Williams, H.L., Reliability evaluation of the human component in man-machine systems, *Elect. Manufact.*, April, 78-82, 1958.
3. Shapero, A., Cooper, J.I., Rappaport, M., Shaeffer, K.H., and Bates, C.J., Human Engineering Testing and Malfunction Data Collection in Weapon System Programs, WADD Technical Report, No. 60-36, Wright-Patterson Air Force Base, Dayton, OH, February 1960.
4. Munger, S.J., Smith, R.W., and Payne, D., An Index of Electronic Equipment Operability: Data Store, American Institute for Research, Pittsburgh, PA, Report No. AIR-C43-1/62 RP(1), 1962.
5. *Proc. Symp. Quantification of Human Performance*, sponsored by the Electronic Industries Association and the University of New Mexico, Albuquerque, NM, August 1964.
6. Askren, W.B., Ed., *Proc. Symp. Reliability of Human Performance in Work*, Report No. AMRL-TR-67-88, Aerospace Medical Research Laboratories, Wright-Patterson Air Force Base, Ohio, May 1967.
7. Regulinski, T.L., Ed., Special issue on human reliability, *IEEE Trans. Reliability*, 22, August 1973.
8. Dhillon, B.S., On human reliability-bibliography, *Microelectronics and Reliability*, 20, 371-373, 1980.
9. Dhillon, B.S., *Human Reliability: With Human Factors*, Pergamon Press, New York, 1986.
10. Dhillon, B.S. and Yang, N., Human reliability: a literature survey and review, *Microelectronics and Reliability*, 34, 803-810, 1994.
11. Hagen, E.W., Ed., Human reliability analysis, *Nuclear Safety*, 17, 315-326, 1976.
12. Meister, D., Human factors in reliability, in *Reliability Handbook*, Ireson, W.G., Ed., McGraw-Hill, New York, 1966, 12.2-12.37.
13. Definitions of Effectiveness Terms for Reliability, Maintainability, Human Factors and Safety, MIL-STD-721B, August 1966. Available from the Naval Publications and Forms Center, 5801 Tabor Avenue, Philadelphia, PA.
14. Kenney, G.C., Spahn, M.J., and Amato, R.A., The Human Element in Air Traffic Control: Observations and Analysis of Performance of Controllers and Supervisors in Providing Air Traffic Control Separation Services, METREK Div., MITRE Corp., Report No. MTR-7655, December 1977.
15. Organizational Management and Human Factors in Quantitative Risk Assessment, Report No. 33/1992 (Report 1), 1992, British Health and Safety Executive (HSE), London.
16. Dhillon, B.S., Reliability technology in health care systems, *Proc. IASTED Int. Symp. Comp. Adv. Technol. Med., Health Care, Bioeng.*, 84-87, 1990.
17. Nobel, J.L. Medical devices failures and adverse effects, *Pediat-Emerg. Care*, 7, 120-123, 1991.
18. Bogner, M.S., Medical devices and human error, in *Human Performance in Automated Systems: Current Research and Trends*, Mouloua, M. and Panasuraman, R., Eds., Lawrence Erlbaum Associates, Hillsdale, NJ, 1994, pp. 64-67.
19. Rook, L.W., Reduction of Human Error in Industrial Production, Report No. SCTM 93-63 (14), June 1962, Sandia Naitonal Laboratories, Albuquerque, NM.
20. Joos, D.W., Sabril, Z.A., and Husseiny, A.A., Analysis of gross error rates in operation of commercial nuclear power stations, *Nuclear Eng. Design*, 52, 265-300, 1979.

21. Meister, D., The problem of human-initiated failures, *Proc. Eighth Natl. Symp. Reliability and Quality Control,* 234-239, 1962.
22. Cooper, J.I., Human-initiated failures and man-function reporting, *IRE Trans. Human Factors,* 10, 104-109, 1961.
23. McCornack, R.L., Inspector Accuracy: A Study of the Literature, Report No. SCTM 53-61 (14), 1961, Sandia Corporation, Albuquerque, NM.
24. Lee, K.W., Tillman, F.A., and Higgins, J.J., A literature survey of the human reliability component in a man-machine system, *IEEE Trans. Reliability,* 37, 24-34, 1988.
25. Beech, H.R., Burns, L.E., and Sheffield, B.F., *A Behavioural Approach to the Management of Stress,* John Wiley & Sons, Chichester, 1982.
26. Hagen, E.W., Ed., Human reliability analysis, *Nuclear Safety,* 17, 315-326, 1976.
27. Regulinski, T.L. and Askern, B., Mathematical modeling of human performance reliability, *Proc. Annu. Symp. Reliability,* 5-11, 1969.
28. Askern, W.B. and Regulinski, T.L., Quantifying human performance for reliability analysis of systems, *Human Factors,* 11, 393-396, 1969.
29. Regulinski, T.L. and Askren, W.B., Stochastic modelling of human performance effectiveness functions, *Proc. Annu. Reliability Maintainability Symp.,* 407-416, 1972.
30. Swain, A.D., A Method for Performing a Human-Factors Reliability Analysis, Report No. SCR-685, Sandia Corporation, Albuquerque, NM, August 1963.
31. Dhillon, B.S. and Rayapati, S.N., Reliability evaluation of multi-state device networks with probability trees, *Proc. Sixth Symp. Reliability in Electron.,* Hungarian Academy of Sciences, Budapest, August 1985, pp. 27-37.
32. Dhillon, B.S., Stochastic models for predicting human reliability, *Microelectronics and Reliability,* 21, 491-496, 1982.
33. Meister, D., Human Reliaiblity, in *Human Factors Review,* Muckler, F.A., Ed., Human Factors Society, Santa Monica, CA, 1984, pp. 13-53.
34. Dhillon, B.S., Human error data banks, *Microelectronics and Reliability,* 30, 963-971, 1990.
35. Dhillon, B.S. and Singh, C., *Engineering Reliability: New Techniques and Applications,* John Wiley & Sons, New York 1981.
36. Urmston, R., Operational Performance Recording and Evaluation Data System (OPREDS), Descriptive Brochures, Code 3400, Navy Electronics Laboratory Center, San Diego, CA, November 1971.
37. Kohoutek, H.J., Human centered design, in *Handbook of Reliability Engineering and Management,* Grant Ireson, W., Coombs, C.F., and Moss, R.Y., Eds., McGraw-Hill, New York, 1996, pp. 9.1–9.30.
38. LaSala, K.P., Human reliability: An overview, tutorial notes, *Annu. Reliability Maintainability Symp.,* 1–45, 1992.
39. Topmiller, D.A., Eckel, J.S., and Kozinsky, E.J., Human Reliability Data Bank for Nuclear Power Plant Operations: A Review of Existing Human Reliability Data Banks, Report No. NUREG/CR2744/1, U.S. Nuclear Regulatory Commission, Washington, D.C., 1982.
40. Reporting Procedures Manual for the Nuclear Plant Reliability Data System (NPRDS), South-West Research Institute, San Antonio, TX, December 1980.
41. Irwin, I.A., Levitz, J.J., and Freed, A.M., Human Reliability in the Performance of Maintenance, Report No. LRP 317/TDR-63-218, Aerojet-General Corporation, Sacramento, CA, 1964.
42. Aviation Safety Reporting Program, FAA Advisory Circular No. 00-46B, Federal Aviation Administration (FAA), Washington, D.C., June 15, 1979.

43. Hornyak, S.J., Effectiveness of Display Subsystems Measurement Prediction Techniques, Report No. TR-67-292, Rome Air Development Center (RADC), Griffis Air Force Base, Rome, NY, September 1967.
44. Life Sciences Accident and Incident Classification Elements and Factors, AFISC Operating Instruction No. AFISCM, 127-6, U.S. Air Force, Washington D.C., December 1971.
45. Stewart, C., The Probability of Human Error in Selected Nuclear Maintenance Tasks, Report No. EGG-SSDC-5580, Idaho National Engineering Laboratory, Idaho Falls, ID, 1981.
46. Boff, K.R. and Lincoln, J.E., Engineering Data Compedium: Human Perception and Performance, Vols. 1-3, Armstrong Aerospace Medical Research Laboratory, Wright-Patterson Air Force Base, Ohio, 1988.
47. Gertman, D.I. and Blackman, H.S., *Human Reliability and Safety Analysis Data Handbook*, John Wiley & Sons, New York, 1994.
48. Swain, A.D. and Guttmann, H.E., Handbook of Human Reliability Analysis with Emphasis on Nuclear Power Plant Applications, Report No. NUREG/CR-1278, The United States Nuclear Regulatory Commission, Washington, D.C., 1983.
49. Meister, D., Human reliability data base and future systems, *Proc. Annu. Reliability Maintainability Symp.*, 276-280, 1993.

11 Reliability Testing and Growth

11.1 INTRODUCTION

Just like in the case of any other reliability activity, reliability testing is an important reliability task. In fact, it may be called one of the most important reliability activities of a reliability program. The main purpose of reliability testing is to obtain information regarding failures, in particular, the product/equipment tendency to fail as well as the failure consequences. This type of information is extremely useful in controlling failure tendencies along with their consequences. A good reliability test program may be classified as the one providing the maximum amount of information concerning failures from a minimal amount of testing [1]. Over the years, many important publications on reliability testing have appeared; in particular, two such publications are listed in References 2 and 3.

In the design and development of new complex and sophisticated systems, the first prototypes usually contain various design and engineering related deficiencies. In fact, according to References 4 and 5, the reliability of revolutionary design product/systems could be very low, i.e., 15 to 50% of their mature design capability. It means without the initiation of various corrective measures during the development stage to improve reliability, such products'/systems' reliability could very well remain at the low initial value. Nonetheless, correcting the weaknesses or errors in design, manufacturing methods, elimination of bad components, etc. leads to a product's reliability growth [6]. The term "reliability growth" may be defined as the positive improvement in a reliability parameter over a time period because of changes made to the product design or the manufacturing process [7]. Similarly, the term "reliability growth program" may be described as a structure process for discovering reliability related deficiencies through testing, analysis of such deficiencies, and implementation of corrective measures to lower their occurrence rate.

The serious thinking concerning reliability growth may be traced back to the late 1950s. In 1964, a popular reliability growth monitoring model was postulated by Duane [8]. Comprehensive lists of publications up to 1980 on reliability growth are given in Refences 9 and 10.

This chapter discusses reliability testing and growth.

11.2 RELIABILITY TESTING

A wide range of topics can be covered under this subject, including classification of reliability tests, success testing, confidence interval estimates for mean time between failures, and accelerated life testing. Each of these topics is presented below.

11.2.1 Reliability Test Classifications

Basically, reliability tests may be grouped under three classifications: (1) reliability development and demonstration testing, (2) qualification and acceptance testing, and (3) operational testing [11].

Reliability development and demonstration testing is concerned with meeting objectives such as to indicate if any design changes are required, to determine if the design is to be improved to satisfy the reliability requirement, and to verify improvements in design reliability. The nature of this type of testing depends on factors such as the level of complexity under consideration, and the type of system/subsystem being investigated. For example, in the case of electronic parts, the reliability development and demonstration testing could take the form of life tests to evaluate if the part can satisfy its reliability goals; if not, what measures are necessary?

In order to meet reliability development and demonstration objectives effectively, the accumulated test data must be of a kind that allows insight into the failure probabilities/failure effects for a certain design under consideration. Furthermore, these data serve as a good basis for reliability analysis and assessment for two particular items: design under consideration and subsequent related programs.

Qualification and acceptance testing is concerned with meeting two basic objectives: to arrive at decision if a part/assembly/end item is to be accepted or not, and to determine if a certain design is qualified for its projected use. These objectives differ from the objectives of other reliability tests, particularly with regard to the accept/reject mechanism. Qualification and acceptance testing incorporates the usual testing and screening the quality-control function of incoming parts. With respect to the materials and components to be used in the system/equipment under development, the qualification and acceptance testing starts early in the program.

Operational testing is concerned with objectives that include verifying the results of reliability analysis conducted during the system/equipment design and development, providing data indicating desirable changes to operating policies and procedures with respect to reliability/maintainability, and providing data for subsequent activities. All in all, the operational testing provides the feedback from practice to theory.

11.2.2 Success Testing

This type of testing is sometimes practiced in receiving inspection and in engineering test laboratories where a no-failure test is specified. Usually, the main goal for this test is to ensure that a certain reliability level has been achieved at a given confidence level.

In this case, for zero failures, the lower 100 $(1-\alpha)\%$ confidence limit on the desired reliability level can be written as [12]

$$R_{Low} = \alpha^{1/k} \tag{11.1}$$

where

k is the number items/units placed on test.
α is the level of significance or consumer's risk.

Thus, with 100 $(1 - \alpha)$% confidence, it may be stated that

$$R_{Low} \leq R_T \quad (11.2)$$

where

R_T is the true reliability.

Taking the natural logarithms of the both sides of Equation (11.1) leads to

$$\ell n R_{Low} = \frac{1}{k} \ell n \alpha \quad (11.3)$$

Thus, from Equation (11.3), we get

$$k = \frac{\ell n \alpha}{\ell n R_{Low}} \quad (11.4)$$

The desired confidence level, C, is expressed by

$$C = 1 - \alpha \quad (11.5)$$

Rearranging Equation (11.5), we get

$$\alpha = 1 - C \quad (11.6)$$

Using relationships (11.2) and (11.6) in Equation (11.4), we get

$$k = \frac{\ell n (1 - C)}{\ell n R_T} \quad (11.7)$$

Equation (11.7) can be used to determine the number of items to be tested for specified reliability and confidence level.

Example 11.1

Assume that 95% reliability of a television set is to be demonstrated at 85% confidence level. Determine the number of television sets to be placed on test when no failures are allowed.

Inserting the above given data into Equation (11.7) yields

$$k = \frac{\ln(1-0.85)}{\ln 0.95}$$

$$= 37$$

Thus, 37 television sets must be placed on test.

11.2.3 Confidence Interval Estimates for Mean Time Between Failures

In many practically inclined reliability studies, the time to item failure is assumed to be exponentially distributed. Thus, the item failure rate becomes constant and, in turn, the mean time between failures (MTBF) is simply the reciprocal of the failure rate (i.e., $1/\lambda$, where λ is the item constant failure rate).

In testing a sample of items with exponentially distributed times to failures, a point estimate of MTBF can be made but, unfortunately, this figure provides an incomplete picture because it fails to give any surety of measurement. However, it would probably be more realistic if we say, for example, that after testing a sample for t hours, m number of failures have occurred and the actual MTBF lies somewhere between specific upper and lower limits with certain confidence.

The confidence intervals on MTBF can be computed by using the χ^2 (chi-square) distribution. The general notation used to obtain chi-square values is as follows:

$$\chi^2(p, df) \tag{11.8}$$

where

p is a quantity, function of the confidence coefficient.
df is the degrees of freedom.

The following symbols are used in subsequent associated formulas [11, 13]:

γ is the mean life or mean time between failures.
θ is the acceptable error risk.
$C = 1-\theta$ is the confidence level.
k is the number of items that were placed on test at zero time (i.e., $t = 0$).
m is the number of failures accumulated to time t^*, where t^* denotes the life test termination time.
m^* is the number of preassigned failures.

There are two different cases for estimating confidence intervals:

- Testing is terminated at a preassigned time, t^*.
- Testing is terminated at a preassigned number of failures, m^*.

Thus, for the above two cases, to compute upper and lower limits, the formulas that can be used are as follows [11, 13]:

- Preassigned truncation time, t*

$$\left(\frac{2X}{\chi^2\left(\frac{\theta}{2}, 2m+2\right)}, \frac{2X}{\chi^2\left(1-\frac{\theta}{2}, 2m\right)}\right) \tag{11.9}$$

- Preassigned number of failures, m*

$$\left(\frac{2X}{\chi^2\left(\frac{\theta}{2}, 2m\right)}, \frac{2X}{\chi^2\left(1-\frac{\theta}{2}, 2m\right)}\right) \tag{11.10}$$

The value of X is determined by the test types: replacement test (i.e., the failed unit is replaced or repaired), non-replacement test.

Thus, for the replacement test, we have

$$X = kt^* \tag{11.11}$$

Similarly, for the non-replacement test, we get

$$X = (k-m)t^* + \sum_{i=1}^{m} t_i \tag{11.12}$$

where

t_i is the ith failure time.

In the case of censored items (i.e., withdrawal or loss of unfailed items) the value of X becomes as follows:

- For replaced failed units but non-replacment of censored items

$$X = (k-s)t^* + \sum_{j=1}^{s} t_j \tag{11.13}$$

where

s is the number of censored items
t_j is the jth, censorship time.

- For non-replaced failed and censored items

$$X = (k - s - m)t^* + \sum_{j=1}^{s} t_j + \sum_{i=1}^{m} t_i \tag{11.14}$$

The tabulated values of χ^2 (p, df) are presented in Table 11.1.

Example 11.2

Assume that 16 identical medical devices were placed on test at time t = 0 and none of the failed units were replaced and the test was terminated after 100 h. Four medical devices failed after 10, 30, 50, and 90 h of operation. Estimate the medical devices' MTBF and their upper and lower limits with 90% confidence level.

Substituting the specified data into Equation (11.12) yields

$$X = (16 - 4)(100) + (10 + 30 + 50 + 90)$$
$$= 1380 \text{ h}$$

Thus, the medical devices' MTBF is given by

$$\gamma = \frac{1380}{4} = 345 \text{ h}$$

Inserting the given and other values into relationship (11.9) and using Table 11.1 yields the following upper and lower limit values for the MTBF:

$$\text{Upper limit} = \frac{2(1380)}{\chi^2(0.95, 8)}$$
$$= \frac{2(1380)}{2.73}$$
$$= 1010.9 \text{ h}$$

$$\text{Lower limit} = \frac{2(1380)}{\chi^2(0.05, 10)}$$
$$= \frac{2(1380)}{18.30}$$
$$= 150.8 \text{ h}$$

TABLE 11.1
Values of Chi-Square

Degrees of freedom	Probability						
	0.99	0.95	0.9	0.5	0.1	0.05	0.01
2	0.02	0.1	0.21	1.38	4.6	5.99	9.21
4	0.29	0.71	1.06	3.35	7.77	9.44	13.27
6	0.87	1.63	2.2	5.34	10.64	12.59	16.81
8	1.64	2.73	3.49	7.34	13.36	15.5	20.09
10	2.55	3.94	4.86	9.34	15.98	18.3	23.2
12	3.57	5.22	6.3	11.34	18.54	21.02	26.21
14	4.66	6.57	7.79	13.33	21.06	23.68	29.14
16	5.81	7.96	9.31	15.33	23.54	26.29	32
18	7.01	9.39	10.86	17.33	25.98	28.86	34.8
20	8.26	10.85	12.44	19.33	28.41	31.41	37.56
22	9.54	12.33	14.04	21.33	30.81	33.92	40.28
24	10.85	13.84	15.65	23.33	33.19	36.41	42.98
26	12.19	15.37	17.29	25.33	35.56	38.88	45.64
28	13.56	16.92	18.93	27.33	37.91	41.33	48.27
30	14.95	18.49	20.59	29.33	40.25	43.77	50.89

Thus, we can state with 90% confidence that the medical devices' true MTBF will lie within 150.8 h and 1010.9 h or $150.8 \leq \gamma \leq 1010.9$.

Example 11.3

Assume that 20 identical electronic parts were put on test at zero time and at the occurrence of the tenth failure, the testing was stopped. The tenth failure occurred at 150 h and all the failed parts were replaced.

Determine the MTBF of the electronic parts and upper and lower limits on MTBF at 80% confidence level.

Inserting the given data into Equation (11.11) we get

$$X = (20)(150) = 3000 \text{ h}$$

Thus, electronic parts' MTBF is

$$\gamma = \frac{3000}{10} = 300 \text{ h}$$

Substituting the specified and other values into relationship (11.10) and using Table 11.1, we get the following values of MTBF upper and lower limits:

$$\text{Upper limit} = \frac{2(3000)}{\chi^2(0.90, 20)}$$

$$= \frac{6000}{12.44}$$

$$= 482.3 \text{ h}$$

$$\text{Lower limit} = \frac{2(3000)}{\chi^2(0.1, 20)}$$

$$= \frac{6000}{28.41}$$

$$= 211.1 \text{ h}$$

At 80% confidence level, the electronic parts' true MTBF will lie within 211.1 h and 482.3 h or $211.1 \le \gamma \le 482.3$.

Example 11.4

On the basis of data given in Example 11.3, determine the following:

- Probability of an electronic part's success for a 200-h mission.
- Upper and lower limits on this probability at 80% confidence level.

Thus, the electronic part's reliability is

$$R(200) = e^{-\left(\frac{200}{300}\right)}$$

$$= 0.5134$$

The upper and lower limit reliability are as follows:

$$\text{Upper limit} = e^{-\left(\frac{200}{482.3}\right)}$$

$$= 0.6606$$

$$\text{Lower limit} = e^{-\left(\frac{200}{211.1}\right)}$$

$$= 0.3877$$

Thus, the reliability of the electronic part is 0.5134 and its upper and lower values at 80% confidence level are 0.6606 and 0.3877, respectively.

11.2.4 Accelerated Life Testing

This is used to obtain quick information on items' reliabilities, failure rates, life distributions, etc. by subjecting the test items to conditions such that the malfunctions occur earlier. Thus, the accelerated life testing is a useful tool to make long term reliability prediction within a short time span. The following two approaches are employed to perform an accelerated life test [14–17]:

- **Approach I.** This is concerned with accelerating the test by using the item under consideration more intensively than its normal usage. Usually, the products such as a high bulb of a telephone set and a crankshaft of a car used discretely or non-continuously can be tested by using this approach. However, it is not possible to use this approach for an item such as a mainframe computer in constant use. Under such a scenario, the next approach can be used.
- **Approach II.** This is concerned with conducting the test at very high stress (e.g., temperature, humidity, and voltage) levels so that failures can be induced in a very short time interval. Typical examples of items for application under this approach are air traffic control monitors, components of a power-generating unit, and communication satellites. Usually, accelerated failure time testing over the accelerated stress testing is preferred because there is no need to make assumptions regarding the relationship of time to failure distributions at both normal and accelerated conditions. Nonetheless, the accelerated-stress testing results are related to the normal conditions by using various mathematical models. One such model will be presented subsequently.

Relationship Between the Accelerated and Normal Conditions

This section presents relationships between the accelerated and normal conditions for the following items [14, 18]:

- **Probability density function**
  The normal operating condition failure probability density function is expressed by

$$f_n(t) = \frac{1}{\beta} f_s\left(\frac{t}{\beta}\right) \tag{11.15}$$

where

t is time.
$f_n(t)$ is the normal operating condition failure probability density function.
β is the acceleration factor.
$f_s\left(\frac{t}{\beta}\right)$ is the stressful operating condition failure probability density function.

- **Cumulative distribution function**
  The normal operating condition cumulative distribution function is

$$F_n(t) = F_s\left(\frac{t}{\beta}\right) \tag{11.16}$$

where

$F_s\left(\frac{t}{\beta}\right)$ is the stressful operating condition cumulative distribution function.

- **Time to failure**
  The time to failure at normal operating condition is given by

$$t_n = \beta t_s \tag{11.17}$$

where

t_n is the time to failure at normal operating condition.
t_s is the time to failure at stressful operating condition.

- **Hazard rate**
  The normal operating condition hazard rate is given by

$$h_n(t) = \frac{f_n(t)}{1 - F_n(t)} \tag{11.18}$$

Substituting Equations (11.15) and (11.16) into Equation (11.18) yields

$$h_n(t) = \frac{\frac{1}{\beta} f_n\left(\frac{t}{\beta}\right)}{1 - F_s\left(\frac{t}{\beta}\right)} \tag{11.19}$$

Thus from Equation (11.19), we have

$$h_n(t) = \frac{1}{\beta} h_s\left(\frac{t}{\beta}\right) \tag{11.20}$$

where

$h_s\left(\frac{t}{\beta}\right)$ is the stressful operating condition hazard rate.

Acceleration Model

For an exponentially distributed time to failure at an accelerated stress, s, the cumulative distribution function is

$$F_s(t) = 1 - e^{-\lambda_s t} \tag{11.21}$$

where

λ_s is the parameter or the constant failure rate at the stressful level.

Thus, from Equation (11.16) and (11.21), we get

$$F_n(t) = F_s\left(\frac{t}{\beta}\right) = 1 - e^{-\lambda_s t/\beta} \tag{11.22}$$

Similarly, using relationship (11.20), we have

$$\lambda_n = \lambda_s/\beta \tag{11.23}$$

where

λ_n is the constant failure rate at the normal operating condition.

For both non-censored and censored data, the failure rate at the stressful level can be estimated from the following two equations, respectively [14]:

- **Noncensored data**

$$\lambda_s = k \Big/ \sum_{j=1}^{k} t_j \tag{11.24}$$

where

k is the total number of items under test at a certain stress.
t_j is the jth failure time; for j = 1, 2, 3, 4, ..., k.

- **Censored data**

$$\lambda_s = q \Big/ \left(\sum_{j=1}^{q} t_j + \sum_{j=1}^{k-q} t'_j \right) \tag{11.25}$$

where

q is the number of failed items at the accelerated stress.
t'_j is the jth censoring time.

Example 11.5

Assume that a sample of 40 integrated circuits were accelerated life tested at 135°C and their times to failure were exponentially distributed with a mean value of 7500 h. If the value of the acceleration factor is 30 and the integrated circuits' normal operating temperature is 25°C, calculate the integrated circuits, operating at the normal conditions, failure rate, mean time to failure, and reliability for a 5000-h mission.

In this case, the failure rate of the integrated circuits at the accelerated temperature is given by

$$\lambda_s = 1/(\text{integrated circuits' mean life under accelerated testing})$$

$$= \frac{1}{7500} \simeq 0.000133 \text{ failure/h}$$

Inserting the above result and the specified data into Equation (11.23), we get

$$\lambda_n \simeq \frac{0.000133}{30}$$

$$\simeq 4.4333 \times 10^{-6} \text{ failure/h}$$

Thus, the integrated circuits' mean time to failure ($MTTF_n$) at the normal operating condition is

$$MTTF_n = \frac{1}{\lambda_n}$$

$$\simeq 225,563.9 \text{ h}$$

The integrated circuits' reliability for a 5000-h mission at the normal operating condition is

$$R(5,000) = \exp\left(-\left\{4.4333 \times 10^{-6}\right\}(5,000)\right)$$

$$\simeq 0.9781$$

Thus, the integrated circuits' failure rate, mean time to failure and reliability at the normal operation are 4.4333×10^{-6} failure/h, 225,563.9 h, and 0.9781, respectively.

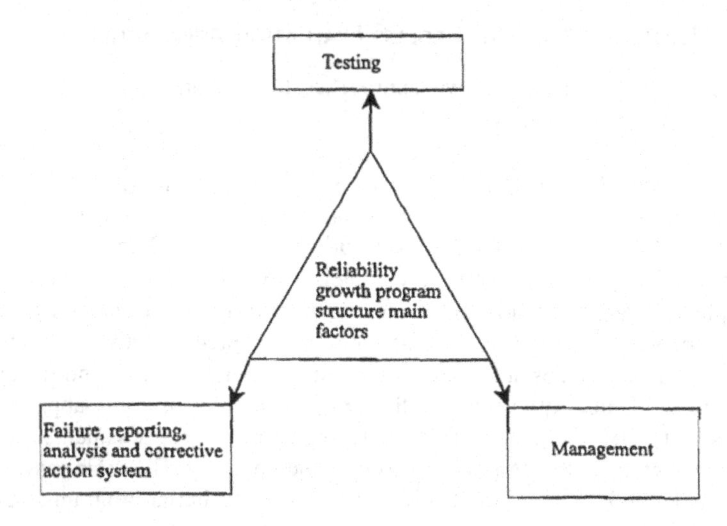

FIGURE 11.1 Reliability growth program basic structure main factors.

11.3 RELIABILITY GROWTH

Over the years many advances have been made in the area of reliability growth. These are in management, modeling, technical standards, etc. This section presents different aspects of reliability growth.

11.3.1 Reliability Growth Program

As described earlier, the reliability growth program is a structured process used to discover reliability deficiencies through testing, analyzing such deficiencies, and implementation of corrective measures to lower the rate of occurrence. Some of the important advantages of the reliability growth program include assessments of achievement and projecting the product reliability trends. Nonetheless, the fundamental structure of a reliability growth program consists of three basic factors as shown in Figure 11.1. These are management, testing, and failure reporting, analysis, and corrective action system (FRACAS) [5]. The management is concerned with making it happen and the testing provides opportunity to discover deficiencies in areas such as design, components, manufacturing, and procedures. FRACAS is primarily a process used for determining failure root causes and recommending appropriate corrective measures.

Some of the major management considerations concerning a reliability growth program are as follows:

- The level of the estimated initial reliability without reliability growth testing and its acceptability. Obviously, this information is related to how much of the system/product is new technology.
- Determination of the reliability growth slot within the system/product development program.
- Reliability growth program's impact on cost and schedule.

11.3.2 Reliability Growth Process Evaluation Approaches

A manager can basically use two approaches to evaluate the reliability growth process. Each of these is described below [7].

- **Approach I.** This approach makes use of assessments (i.e., quantitative evaluations of the current reliability status) based on data from the detection of failure sources; thus, it is a results oriented approach. In order to evaluate reliability growth progress, a comparison between the assessed and the planned values is made. Subsequently, the appropriate decisions are taken.
- **Approach II.** This approach monitors various process activities to ensure that such activities are being performed in a timely manner and the effort and work quality are as per the program plan. Thus, this approach is activity oriented and is practiced to supplement the assessments. Sometimes early in the program it is relied upon entirely. The main reason for this is the lack of adequate objective data in the initial program stages.

11.4 RELIABILITY GROWTH MODELS

Over the years many mathematical models for reliability growth have been developed. Ten such models are described in detail in Reference 10. Furthermore, Reference 7 presents an overview of 17 reliability growth models by grouping them into two categories: discrete and continuous. Two such models are presented below.

11.4.1 Duane Model

In 1964, J.T. Duane was the first person to report the most commonly accepted pattern for reliability growth. The Duane model basically is a graphical approach to perform analysis of reliability growth data and is simple and straightforward to understand. Nonetheless, the two important benefits of this approach are as follows [2]:

- The straight line used by the Duane plot can be fitted by eye to the data points.
- Various facts can be depicted by the Duane plot which otherwise could be hidden by statistical analysis. For example, even though the application of a goodness-of-fit test may conclude the rejection of a certain reliability growth model, it will not provide any possible reasons for the rejection. On the other hand, a plot of the same data might provide some possible reasons for the problem.

In contrast, the drawbacks of this model are the reliability parameters cannot be estimated as well in comparison to a statistical model and no interval estimates can be calculated. Nonetheless, with respect to the development of this model Duane collected data for several systems developed by General Electric to determine if any systematic changes in reliability improvement took place during their development. After performing analysis of these data, he concluded that a plot of cumulative hours vs. cumulative failure rate fell close to a straight line on log-log paper, under the maintenance of a continued reliability effort. Thus, he defined the cumulative failure rate of his model as follows:

$$\lambda_k = \frac{f}{T} = \theta T^{-\alpha} \tag{11.26}$$

where

λ_k is the cumulative failure rate.
α is a parameter denoting the growth rate.
θ is a parameter determined by circumstances such as product complexity, design margin, and design objective.
T is the total test hours.
f denotes failures during T.

In order to estimate the values of the parameters, we take the logarithms of both sides of Equation (11.26) to get

$$\log \lambda_k = \log \theta - \alpha \log T \tag{11.27}$$

Equation (11.27) is the equation for a straight line. Thus, the plot of the logarithm of the cumulative failure rate, λ_k, against the logarithm of cumulative operating hours, T, can be used to estimate the values of α and θ. The slope of the straight line is equal to α and at $T = 1$, θ is equal to the corresponding cumulative failure rate.

The least-squares method can be used to have a more accurate straight line fit in estimating α and θ [10].

11.4.2 Army Material System Analysis Activity (AMSAA) Model

This is another model that can be used to track reliability growth within test phases [7, 19, 20]. The model allows, for the purpose of reliability growth tracking, the development of rigorous statistical procedures. The following two assumptions are associated with this model:

- Reliability growth can be modeled as a nonhomogeneous Poisson process (NHPP) within a test phase.
- On the basis of failures and test time within a test phase, the cumulative failure rate is linear on log-log scale.

The following symbols are associated with this model:

t is the test time from the start of the test phase.
M(t) is the number of system failures by time t.

E [M(t)] is the expected value of M (t).
θ and β are the parameters.

Thus, the expected value of M(t) is expressed by

$$E[M(t)] = \theta\, t^{\beta} \tag{11.28}$$

By differentiating Equation (11.28) with respect to t, we get the following expression for the intensity function:

$$\lambda(t) = \theta \beta t^{\beta-1} \tag{11.29}$$

where

$\lambda(t)$ is the intensity function.

The instantaneous mean time between failures is expressed by

$$m(t) = 1/\lambda(t) \tag{11.30}$$

where

m (t) is the instantaneous mean time between failures.

The model is described in detail in References 2 and 7.

11.5 PROBLEMS

1. Discuss the following:
   - Demonstration testing
   - Qualification and acceptance testing
   - Success testing
2. What is the difference between reliability demonstration testing and reliability growth modeling?
3. Assume that 99% reliability of a medical device is to be demonstrated at 60% confidence level. Calculate the total number of medical devices to be placed on test when no failures are allowed.
4. What are the two approaches used to perform an accelerated life test? Describe them in detail.
5. What are the benefits of reliability growth testing?
6. Twenty-five identical electrical devices were put on test at zero time, none of the failed units were replaced, and the test was stopped after 200 h. Seven devices malfunctioned after 14, 25, 30, 65, 120, 140, and 150 h of operation. Calculate the electrical devices' mean time between failures and their associated upper and lower limits with 80% confidence level.

7. Define the following two terms:
   - Reliability growth
   - Reliability growth program
8. A sample of 60 identical electronic parts were accelerated life tested at 140°C and their times to failure were represented by an exponential distribution with a mean value of 4000 h. The normal operating temperature of these parts is 30°C and the value of the acceleration factor is 28. Determine the mean time to failure, failure rate, and reliability for a 100-h mission under the normal operating conditions.
9. What are the major management considerations concerning a reliability growth program?
10. A sample of 30 identical mechanical devices were placed on test at time $t = 0$ and at the occurrence of the seventh failure, the testing was terminated. The last or the seventh failure occurred at 200 h and all the failed devices were replaced. Calculate the MTBF of the mechanical devices and the upper and lower limits on MTBF at 70% confidence level.

11.6 REFERENCES

1. AMC Pamphlet AMCP 702-3, Quality Assurance, U.S. Army Material Command, Washington, D.C., 1968.
2. MIL-HDBK-781, Reliability Test Methods, Plans and Environments for Engineering Development, Qualification and Production, Department of Defense, Washington, D.C.
3. MIL-STD-781, Reliability Design Qualification and Production Acceptance Test: Exponential Distribution, U.S. Department of Defense, Washington, D.C.
4. Benton, A.W. and Crow, L.H., Integrated reliability growth testing, *Proc. Annu. Reliability Maintainability Symp.*, 160-166, 1989.
5. Crow, L.H., Reliability growth management, models, and standards, in tutorial notes, *Annu. Reliability Maintainability Symp.*, 1-12, 1994.
6. Mead, P.H., Reliability growth of electronic equipment, *Microelectronics and Reliability*, 14, 439-443, 1975.
7. MIL-HDBK-189, Reliability Growth Management, Department of Defense, Washington, D.C.
8. Duane, J.T., Learning curve approach to reliability monitoring, *IEEE Trans. Aerospace*, 563-566, 1964.
9. Dhillon, B.S., Reliability growth: A survey, *Microelectronics and Reliability*, 20, 743-751, 1980.
10. Dhillon, B.S., *Reliability Engineering in Systems Design and Operation*, Van Nostrand Reinhold Company, New York, 1983.
11. Von Alven, W.H., Ed., *Reliability Engineering*, Prentice-Hall, Englewood Cliffs, NJ, 1964.
12. Grant Ireson, W., Coombs, C.F., and Moss, R.Y., *Handbook of Reliability Engineering and Management*, McGraw-Hill, New York, 1996.
13. Dhillon, B.S., *Systems Reliability, Maintainability and Management*, Petrocelli Books, New York, 1983.
14. Elsayed, E.A., *Reliability Engineering*, Addison Wesley Longman, Reading, MA, 1996.

15. Nelson, W., *Accelerated Testing*, John Wiley & Sons, New York, 1980.
16. Bain, L.J. and Engelhardt, M., *Statistical Analysis of Reliability and Life-Testing Models: Theory*, Marcel Dekker, New York, 1991.
17. Meeker, W.Q. and Hahn, G.J., *How to Plan an Accelerated Life Test: Some Practical Guidelines*, American Society for Quality Control (ASQC), Milwaukee, WI, 1985.
18. Tobias, P.A. and Trindade, D., *Applied Reliability*, Van Nostrand Reinhold Company, New York, 1986.
19. Crow, L.H., Estimation procedures for the Duane model, *Proc. U.S. Army Mater. Syst. Anal. Act. (AMSAA) Reliability Growth Symp.*, Aberdeen Proving Ground, Maryland, September 1972.
20. Crow, L.H., Reliability analysis for complex repairable systems, in *Reliability and Biometry*, Proschan, F. and Serfling, R.J., Eds., Society for Industrial and Applied Mathematics (SIAM), Philadelphia, PA, 1974, pp. 379-410.

12 Reliability in Computer Systems

12.1 INTRODUCTION

In recent years, the applications of computers have increased quite dramatically ranging from personal use to control space systems. Today's computers are much more complex and powerful than their infant versions. As the computers are made up of both the software and hardware components, the percentage of the total computer cost spent on software has changed quite remarkably from the days of the first generation computers. For example, in 1955 software (i.e., including software maintenance) accounted for 20% of the total computer cost and in 1985, the software cost increased to 90% [1].

For an effective performance of a computer, both its hardware and software must function with considerable reliability. In fact, according to various past studies, the software is many times more likely to fail than hardware. Nonetheless, as computers are becoming more complex and sophisticated, the demand on their reliability has increased exponentially. For example, NASA's Saturn V Launch computer (circa 1964) had a mean time to failure (MTTF) goal of 25,000 h and the SIFT and FTMP avionic computers designed and developed in the late 1970s to control dynamically unstable aircraft were expected to have a MTTF of 1 billion hours [2, 3]. The early history of computer hardware reliability may be traced back to the works of Shannon [4], Hamming [5], Von Neumann [6], and Moore and Shannon [7]. For example, the triple modular redundancy (TMR) scheme was first proposed by Von Neumann [6] to improve system reliability in 1956. Since then, many other people have contributed to the computer hardware reliability and a comprehensive list of publications on the subject is given in Reference 8.

The serious effort on software reliability appeared to have started at Bell Laboratories in 1964 [9]. An evidence of this effort is a histogram of problems per month concerning switching system software. In 1967, Floyd [10] considered approaches for formal validation of software programs and Hudson [11] proposed Markov birth–death models. Since those years, a large number of publications on the subject have appeared [8].

This chapter presents important aspects of computer hardware and software reliability.

12.2 TERMS AND DEFINITIONS

As there are certain terms specifically concerned with computer hardware and software reliability, this section presents some of those terms along with their definitions [8, 12-15].

- **Software error.** This is a clerical, conceptual, or syntactic discrepancy which leads to one or more faults in the software.
- **Fault.** This is an attribute which adversely affects an item's reliability.
- **Covered fault.** This is a fault from which the concerned item can recover automatically.
- **Uncovered fault.** This is a fault from which the concerned item cannot recover automatically.
- **Debugging.** This is the process of eradicating and isolating errors.
- **Fault-tolerant computing.** This is the ability to execute specified algorithms successfully irrespective of computer hardware failures and software errors [16].
- **Software reliability.** This is the probability of a given software operating for a stated time interval, without an error, when used under the designed conditions on the stated machine.
- **Software testing.** This is the process of executing software to determine if the results it produces are correct [15].

12.3 HARDWARE RELIABILITY VS. SOFTWARE RELIABILITY

In order to have a better understanding of the differences between computer hardware and software reliability, Table 12.1 presents comparisons of the important concerned areas [8, 17–19].

12.4 COMPUTER FAILURE CAUSES

Computers fail due to many different causes. The major sources of computer failures are shown in Figure 12.1 [20, 21]. These are human errors, communication network failures, peripheral device failures, environmental and power failures, processor and memory failures, mysterious failures, saturation, and gradual erosion of the data base. The first six of these system failure sources are described below.

Human errors are, in general, the result of operator oversights and mistakes. Often operator errors occur during starting up, running, and shutting down the system. Communication network failures are associated with inter module communication and, usually, most of them are of a transient nature. The use of "vertical parity" logic can help to detect over two-thirds of errors in communication lines. Peripheral device failures are important but they seldom cause a system shutdown. The frequently occurring errors in peripheral devices are transient or intermittent and the usual reason for their occurrence is the electromechanical nature of the devices. The causes, such as electromagnetic interference, air conditioning equipment failure, earthquakes, and fires, are responsible for the occurrence of environmental failures. Power failures can occur due to factors such as transient fluctuations in voltage or frequency and total power loss from the local utility company.

Usually, processor errors are catastrophic but they occur quite rarely, as there are occasions when the central processor fails to execute instructions properly due

TABLE 12.1
Comparison of Hardware and Software Reliability

No.	Hardware reliability	Software reliability
1	Wears out.	Does not wear out.
2	Many hardware parts fail according to the bathtub hazard rate curve.	Software does not fail according to the bathtub hazard rate curve.
3	A hardware failure is mainly due to physical effects.	A software failure is caused by programming error.
4	The failed system is repaired by performing corrective maintenance.	Corrective maintenance is really redesign.
5	Interfaces are visual.	Interfaces are conceptual.
6	The hardware reliability field is well established, particularly in the area of electronics.	The software reliability field is relatively new.
7	Obtaining good failure data is a problem.	Obtaining good failure data is a problem.
8	Usually redundancy is effective.	Redundancy may not be effective.
9	Potential for monetary savings.	Potential for monetary savings.
10	Preventive maintenance is performed to inhibit failures.	Preventive maintenance has no meaning in software.
11	It is possible to repair hardware by using spare modules.	It is impossible to repair software failures by using spare modules.
12	Hardware reliability has well-established mathematical concepts and theory.	Software reliability still lacks well-established mathematical concepts and theory.
13	Has a hazard function.	Has an error rate function.
14	Mean time to repair has significance.	Mean time to repair does not have any significance.
15	Reliability is time-related with failures occurring as a function of operational/storage time.	Reliability is not time-related and failures occur when a program step/path containing the fault is executed.

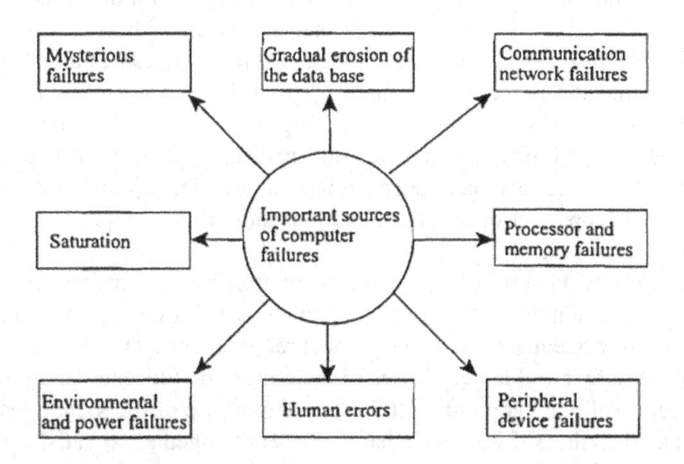

FIGURE 12.1 Major sources of computer failures.

to a "dropped bit". Nowadays, memory parity errors are very rare because of an impressive improvement in hardware reliability and also they are not necessarily fatal. Mysterious failures occur unexpectedly and thus in real-time systems such failures are never properly categorized. For example, when a normally functioning system stops operating suddenly without indicating any problem (i.e., software, hardware, etc.), the malfunction is called the mysterious failure.

12.5 SOFTWARE LIFE CYCLE PHASES AND ASSOCIATED ERROR SOURCES

A software life cycle may be divided into many different phases: requirements definition and analysis, preliminary design, detailed design, code and unit testing, integration and system testing, acceptance testing, maintenance and operation, and retirement. Within each of these phases, there are various sources of error, thus requiring careful consideration. All of these life cycle phases are described below [12].

The requirements definition and analysis phase defines and documents the engineering requirements for the software to be developed and its end product is a software requirement specification that allows the development organization to proceed with the design. Some of the sources that may generate errors in this phase include incomplete requirements, conflicting requirements, inadequately described user needs, and nonfeasible requirements.

In the preliminary design phase, the software product is decomposed into a hierarchal structure consisting of computer software configuration items. In turn, these configuration items are decomposed further into computer software components and modules. During the detailed design phase, the computer software component design specifications are progressively described in more detail until the identification of a set of primitive (i.e., non-decomposable) computer software units. Some of the sources that may induce errors in the detailed design phase are inconsistent data definitions, input data range errors, inadequate validity checking, unsatisfactory error recovery, and wrong error analysis for algorithms.

In the code and unit testing phase, the programmer translates the detailed software design into the programming language and the unit testing is performed to verify unit functionality. As coding is a fault-prone process, a number of standard bugs that may occur during this phase include missing code, incomplete initialization or resetting of variables and parameters, unreachable code, improper logic for branch conditions and loops, out-of-range calculations and infinite loops, and improper or incomplete validity checks.

At the completion of the unit tests, the integration testing starts and the separately tested units are assembled into logical combinations and tested again to demonstrate that the assembly adequately satisfies product requirements. The process is repeated by combining larger and larger classes of units until the full integration of all units of a computer software configuration item. The purpose of the system testing is to demonstrate that the software and hardware work together to satisfy all system functions as well as that the end product is ready for release to production. The errors/faults introduced in the test phase significantly influence the test program's capability to uncover faults in the software and those errors/faults include test plans/procedures that have wrongly interpreted software requirements and errors/faults in code written for the test program.

In the acceptance testing phase, the test team determines if the product under consideration meets its original requirements. The acceptance test plan is developed and the phase terminates when all the tests in the acceptance plan are successfully executed.

In the maintenance and operation phase, primarily the attention is focused on rectifying errors that appear in the exercise of the software or on fine-tuning the software to improve its performance.

In the retirement phase, the user may decide not to use the software any more and discard it altogether. The primary reason is that the software may have become difficult to maintain because of recurrent changes.

12.6 SOFTWARE RELIABILITY IMPROVEMENT METHODS

There are a number of methods to improve reliability of software products. The classifications of these methods include reliable software design, fault-tolerant design, formal methods, and testing.

12.6.1 Reliable Software Design Methods

Under this classification, there are a number of techniques that can help programmers to systematically derive the software design from its detailed specification [22]. These include structured programming, top-down programming, and data structure design.

The clarity of the design is the direct aim of the *structured programming*. More specifically, the structured programming calls for easy to understand design and a need to develop a program structure in which the vital software elements and their inter-relationships are easily identifiable. Some of the easy to use rules associated with structure programming include restricted use of GO TO statements in modules, ensuring that each module has only one entry and one exit, using a language and compiler with structured programming capabilities for the implementation purpose, and ensuring that all statements, including subroutine calls, are commented [23].

There are many advantages associated with the structured programming: increase in the productivity of programmers with respect to instructions coded per man-hour, a useful tool in localizing an error, maximum amount of code can be reused in redesign work, and an effective tool in understanding the program design by the designer and others. The *top-down programming* or design is also sometimes referred to as hierarchic decomposition because of the cascading nature of the process [24]. Nonetheless, the top-down design is a decomposition process concerned with directing attention to the program control structure or to the program flow of control. Furthermore, the top-down design is also concerned with the production of code at a later stage.

The top-down programming begins with a module representing the program in its entirety and it decomposes the module into subroutines. In turn, each subroutine is broken down further and this process continues until all the broken down elements are easy to comprehend and work with. The program structure is analogous to a tree; the trunk of the tree represents the module containing the entire program and its leaves represent the computer language statements. Some of the benefits of the top-down structured programming are reduction in the cost of testing, better quality software, software in a more readable form, lower cost of software maintenance, increase in confidence in software, etc.

The *data structure design* is inclined toward the development of a detailed program design and is also referred to as Jackson Structured Programming. The structure of the input files drive the methodology and its resulting design and from these files software is read and by the structure of the output files it is generated. According to this approach, an effective software design reproduces the data structure in the input/output files. Currently, this is the most systematically available design method and is simple, teachable, consistent (e.g., two totally different programmers will end up writing the same program), and anti-inspirational. The downside is that the method is not applicable to scientifically based programs because their contents are essentially algorithmic (i.e., functional).

12.6.2 Fault-Tolerant Software Design Methods

It would never be possible to produce error-free software, irrespective of the degree of testing or other measures [25, 26]. Under such circumstances, one approach to improve software reliability is to create a fault-tolerant architecture to recover from software failures. The expensiveness of the approach usually limits applications to high-risk areas, for example, embedded software for controlling satellites. Three methods of fault-tolerant software design are recovery-block design, N-version programming, and consensus recovery block. The *recovery-block design* is concerned with developing versions of software under the same set of requirements on the assumption that the pieces of software being developed are independent, and thus the likelihood of their failing simultaneously is negligible. The versions are ordered from the most reliable to the least reliable. The most reliable software is executed first and after the acceptance test run, if its output is rejected then a recovery block is entered. The recovery is conducted in the following three steps:

- Recover input.
- Execute most reliable software and restore input.
- Submit the output to the same acceptance test as before. In the event of rejection, the next recovery block is entered and the process continues.

In the case of *N-version programming*, many versions of the software are developed independently, N programs are executed in parallel or simultaneously. Subsequently the resulting outputs are compared and when at least programs have the same output, it is taken as the correct result and the process continues. This approach is not suitable where multiple correct outputs can be generated and also it discriminates correct results or solutions affected by rounding errors.

The *consensus recovery block method* combines attributes of the preceding two approaches; thus, it attempts to discard the weaknesses of those two methods. Nonetheless, this approach requires the development of N-versions of a software and of an acceptance test voting procedure. The reliability factor is used to rank the different versions of the software after execution of all software versions. The resulting outputs are submitted to the voting mechanism. If no agreement is reached, the order of reliability is used in submitting each output successively to the acceptance test. As soon as one of the resulting outputs passes the test, the process terminates and the software under consideration continues with its operation. All in all, it may be said that this method is more reliable than recovery-block design and N-version programming approaches.

12.6.3 Formal Methods

These include formal specification approaches and formal verification [12, 27]. The formal specification approaches consist of a detailed specification of the software requirements using a formal language. More specifically, the nature language of traditional programming is not used, but rather a "formal" substitute having exactly defined semantics and syntax. A typical example of such a language is mathematics. Nonetheless, the formal specification approaches help to enforce clarity in requirement definitions and leave little room for ambiguity and misinterpretation, which is usually a concern when using a natural language.

The formal verification makes use of inference on the formal specification in order to demonstrate that the software meets its objectives [28]. All in all, it may be said that the formal approaches are not properly understood by the involved professionals because of their complexity and need for considerable training.

12.6.4 Testing

This may simply be described as the process of executing a software program to uncover errors. There are many different types of software testing including module testing, top-down testing, bottom-up testing, and sandwich testing [8, 29–31]. Each of these testing methods is discussed below.

- **Module Testing.** This is concerned with testing each module usually in isolation from the rest of the system, but subject to the environments to be experienced in the system. Usually, a general module test program is developed instead of writing an entirely new program for each module to be tested. Because it is easy to correct uncovered errors and more grave consequences of errors can be discovered at a later stage, it is advisable to conduct module testing thoroughly [32].
- **Top-Down Testing.** This is concerned with integrating and testing the software program from the top end to the bottom end and in the program structure, the top module is the only module that is unit tested in isolation. At the conclusion of the top module testing, this module calls the remaining modules one by one to merge with it. In turn, each combination is tested and the process continues until the completion of combining and

testing of all the modules [31]. The advantages of the top-down testing are efficiency in locating problem areas at the higher program levels, easy representation of test cases after adding the input/output functions, and early demonstrations because of the early skeletal program. Similarly, some of the drawbacks of the top-down testing are that stub modules must be written, stub modules may be complex, test conditions could be impossible or difficult to create, and the representation of test cases in stubs can be difficult prior to adding the input/output functions [12].

- **Bottom-Up Testing.** This is concerned with integrating and testing the software program from the bottom end to the top end and the terminal modules are module (unit) tested in isolation. These modules do not call any other modules and at the conclusion of the terminal module testing, the modules that directly call these tested modules are the ones in line for testing. The modules are not tested in isolation; in fact, the testing is conducted together with the earlier tested lower level modules. The process continues until reaching the top of the program. Some of the benefits of the bottom-up testing are ease in creating test conditions, efficiency in locating problem areas at the lower levels of the program, and ease in observating test results. Similarly, the bottom-up testing drawbacks include: driver modules must be produced, and the program does not exist as an entity until the adding of the final module [12].
- **Sandwich Testing.** This method is the result of combining the top-down testing and bottom-up testing methods. Obviously, the idea behind developing this approach was to extract benefits of both top-down testing and bottom-up testing and eliminate some of their drawbacks. As both top-down testing and bottom-up testing are performed simultaneously, the program is integrated from both sides (i.e., top and bottom). Even though the resulting meeting point from both sides depends upon the program being tested, it can be predetermined by reviewing the structure of the program.

12.7 SOFTWARE RELIABILITY ASSESSMENT METHODS

There are many qualitative and quantitative software reliability assessment methods available in the published literature. Such methods may be grouped into three distinct classifications [12]:

1. Analytical approaches
2. Software metrics
3. Software reliability models

12.7.1 Analytical Approaches

Two important techniques belonging to the analytical approaches are failure mode and effect analysis (FMEA) and fault tree analysis (FTA). Both of these methods are widely used to assess hardware reliability and they can equally be applied to assess reliability of software. FMEA and FTA are described in detail in Chapters 6 and 7, respectively.

12.7.2 Software Metrics

These may be described as quantitative indicators of the degree to which a software item or process possesses a specified attribute. Often, the application of the software metrics is made to determine the status of a trend in a software development activity as well as to determine risk of going from one phase to another. Two software measures are described below.

Defect Density Measure

This is an important measure and can be applied to determine the reliability growth during the design phase. The measure requires establishing defect severity categories and possesses some ambiguity, since its low value may mean either an inadequate review process or a good product. Nonetheless, the cumulative defect ratio for design is expressed by

$$\theta_{cd} = \sum_{i=1}^{\alpha} N_i / L \qquad (12.1)$$

where

θ_{cd} is the cumulative defect ratio for design.
α is the total number of reviews.
N_i is the total number of unique defects, at or above a given severity level, discovered in the ith design review.
L is the number of source lines of design statement expressed in thousands, in the design phase.

In the event of having estimates of defect density greater than the ones for comparable projects, review the development process to determine if poor training/practices are responsible or if the requirements are ambiguous or incomplete. Under such circumstance, it may be the correct course to delay development until such time the corrective measures can be taken. In contrast, if estimates of defect density are less than for comparable projects, review the methodology and the review process itself. If the assessed review process is considered to be satisfactory, it is quite reasonable to conclude that the development phases are generating low-defect software products.

Code and Unit Test Phase Measure

For this phase, another form of the defect density measure is more appropriate and again this form requires the establishment of defect severity classifications. Thus, the cumulative defect ratio for code is expressed by

$$\gamma_{cd} = \sum_{i=1}^{\alpha} M_i / SL \tag{12.2}$$

where

γ_{cd} is the cumulative defect ratio for code.
α is the total number of reviews.
M_i is the total number of unique defects, at or above a given severity level, discovered in the ith code review.
SL is the number of source lines of code reviewed, expressed in thousands.

12.7.3 Software Reliability Models

Even though over the years many software reliability models have been developed [33-35], the first one appeared in 1967 [36]. Most of the current software reliability models may be grouped into four classifications as presented in Table 12.2 [12]. The two important assumptions associated with classification I models are indigenous and seeded faults have equal probability of detection and the seeded faults are distributed randomly in the software program under consideration. Mills seeding model falls under this category [37].

The key assumptions associated with the classification II models include independent test intervals, homogeneously distributed testing during intervals, and independent faults discovered during non-overlapping time intervals. Examples of the models belonging to this classification are Musa model [38] and the Shooman model [39].

There are many key assumptions associated with the classification III models: independent embedded faults, equal probability of exposure of each fault, correction process does not introduce faults, and independent times between failures. Two

TABLE 12.2
Classification of Software Reliability Models

No.	Classification	Description
I	Fault seeding	This incorporates those models that determine the number of faults in the program at zero time via seeding of extraneous faults.
II	Failure count	This includes models counting the number of failures/faults occurring in given time intervals.
III	Times between failures	This incorporates models providing the time between failure estimations.
IV	Input domain based	This incorporates models that determine the program/software reliability under the circumstance the test cases are sampled randomly from a known operational distribution of inputs to the program/software.

examples of the models belonging to this classification are the Jelinski and Moranda model [40] and the Schick and Wolverton model [21].

The classification IV models, i.e., input domain based models, have three key assumptions: (1) inputs selected randomly, (2) input domain can be partitioned into equivalence groups, and (3) known input profile distribution. Two examples of this classification model are the Nelson model [42] and the Ramamoorthy and Bastani model [43].

Some of the software reliability models are presented below.

Mills Model

This is a different and more pragmatic approach to software reliability prediction proposed by Mills [37] in 1972. Mills argues that an assessment of the faults remaining in a given software program can be made through a seeding process that makes an assumption of a homogeneous distribution of a representative category of faults. Prior to starting the seeding process, a fault analysis is required to determine the expected types of faults in the code and their relative occurrence frequency. Nonetheless, an identification of both seeded and unseeded faults is made during reviews or testing and the discovery of seeded and indigenous faults allows an assessment of remaining faults for the fault type under consideration. It is to be noted that the value of this measure can only be estimated, provided the seeded faults are discovered.

The maximum likelihood of the unseeded faults is expressed by [12]

$$N_1 = [N_{sf}\, m_{fu}]/m_{sf} \tag{12.3}$$

where

N_1 is the maximum likelihood of the unseeded faults.
N_{sf} is the number of seeded faults.
m_{fu} is the number of unseeded faults uncovered.
m_{sf} is the number of seeded faults discovered.

Thus, the number of unseeded faults remaining in a given program is

$$N = N_1 - m_{fu} \tag{12.4}$$

Example 12.1

A software program under consideration was seeded with 35 faults and, during testing, 60 faults of the same type were found. The breakdowns of the faults uncovered were 25 seeded faults and 35 unseeded faults. Estimate the number of unseeded faults remaining in the program.

Substituting the given data into Equation (12.3) yields

$$N_1 = (35)(35)/25 = 49 \text{ faults}$$

Inserting the above result and the other specified data value into Equation (12.4), we get

$$N = 49 - 35 = 14 \text{ faults}$$

It means 14 unseeded faults still remain in the program.

Musa Model

This model belongs to classification II models of Table 12.2 and is based on the premise that reliability assessments in the time domain can only be based upon actual execution time, as opposed to calender or elapsed time. The reason for this is that only during execution does the software program become exposed to failure-provoking stress. Nonetheless, this model should only be used after the completion of integration or, more specifically, when all the relevant modules are in place [44].

Some of the important assumptions associated with the model are as follows:

- Execution time between failures is piecewise exponentially distributed.
- Failure intervals are statistically independent and follow Poisson distribution.
- Failure rate is proportional to the remaining defects.

A comprehensive list of assumptions may be found in Reference 8. Musa developed the following simplified equation to obtain the net number of corrected faults:

$$m = M\left[1 - \exp\left(-kt/MT_s\right)\right] \tag{12.5}$$

where

- m is the net number of corrected faults.
- M is the initial number of faults.
- T_s is the mean time to failure at the beginning of the test.
- t is time.
- k is the testing compression factor defined as the mean ratio of detection rate of failures during test to the rate during normal use of the software program.

Mean time to failure, T, increases exponentially with execution time and is expressed by

$$T = T_s \exp\left(kt/MT_s\right) \tag{12.6}$$

Thus, the reliability at operational time t is

$$R(t) = \exp(-t/T) \tag{12.7}$$

From the above relationships, we obtain the number of failures that must occur to improve mean time to failure from, say, T_1 to T_2 [44]:

$$\Delta m = MT_s\left(\frac{1}{T_1} - \frac{1}{T_2}\right) \tag{12.8}$$

The additional execution time required to experience Δm is given by

$$\Delta t = \left(\frac{MT_s}{k}\right) \ln\left(T_2/T_1\right) \tag{12.9}$$

Example 12.2

A newly developed software is estimated to have approximately 450 errors. Also, at the beginning of the testing process, the recorded mean time to failure is 4 h. Determine the amount of test time required to reduce the remaining errors to 15, if the value of the testing compression factor is 5. Estimate reliability over a 100-h operational period.

Substituting the given data into Equation (12.8), we get

$$(450 - 15) = (450)(4)\left(\frac{1}{4} - \frac{1}{T_2}\right) \tag{12.10}$$

Rearranging Equation (12.10) yields

$$T_2 = 120 \text{ h}$$

Inserting the above result and the other specified data into Equation (12.9), we get

$$\Delta t = \left[\frac{(450)(4)}{5}\right] \ln(120/4)$$

$$= 1224.4 \text{ h}$$

Thus, for the given and calculated values from Equation (12.7), we get

$$R(100) = \exp(-100/120) = 0.4346$$

Thus, the required testing time is 1224.4 h and the reliability of the software for the specified operational period is 0.4346.

Shooman Model

This is one of the earliest software reliability models and has influenced the development of several others over the time period [43, 44]. The model does not require fault collection during debugging on a continuous basis and it can be used for software programs of all sizes. The key assumptions associated with this model are as follows [8]:

- Total machine instructions remain constant.
- Debugging does not introduce new errors.
- The hazard function is proportional to residual or the remaining software errors.
- The residual errors are obtained by subtracting the number of cumulative rectified errors from the total errors initially present.
- The total number of errors in the software program remains constant at the start of integration testing and they decrease directly as errors are corrected.

The model's hazard function is expressed by

$$\lambda(t) = C F_r(x) \tag{12.11}$$

where

$\lambda(t)$ is the hazard rate.
t is the system operating time.
C is the proportionality constant.
x is the elapsed or debugging time since the start of system integration.
$F_r(x)$ is the number of software faults remaining in the program at time x.

In turn, $F_r(x)$ is defined by

$$F_r(x) = \frac{F_z}{I} - F_c(x) \tag{12.12}$$

where

F_z is the number of initial faults at time $x = 0$.
I is the total number of machine language instructions.
$F_c(x)$ is the cumulative number of faults corrected in interval x.

Inserting Equation (12.12) into Equation (12.11) yields

$$\lambda(t) = C\left[\frac{F_z}{I} - F_c(x)\right] \tag{12.13}$$

In reliability theory, the reliability of an item at time t is expressed by [8]

$$R(t) = \exp\left[-\int_0^t \lambda(t)\,dt\right] \tag{12.14}$$

Thus, inserting Equation (12.13) into Equation (12.14), we get

$$R(t) = \exp\left[-C\left\{\frac{F_z}{I} - F_c(x)\right\}t\right] \tag{12.15}$$

By integrating Equation (12.15) over the interval [0, ∞], the following expression for mean time to failure (MTTF) results:

$$\begin{aligned} MTTF &= \int_0^{\infty} \exp\left[-C\left\{\frac{F_z}{I} - F_c(x)\right\}t\right] dt \\ &= 1 \Big/ C\left\{\frac{F_z}{I} - F_c(x)\right\} \end{aligned} \tag{12.16}$$

In order to estimate the constants, C and F_z, we use the maximum likelihood estimation approach to get [44-46]

$$\hat{C} = \left(\sum_{i=1}^{N} M_i\right) \Big/ \sum_{i=1}^{N} \left[\frac{\hat{F}_z}{I} - F_c(x_i)\right] W_i \tag{12.17}$$

and

$$\hat{C} = \left[\left(\sum_{i=1}^{N} M_i\right) \Big/ \left\{\frac{\hat{F}_z}{I} - F_c(x_i)\right\}\right] \Big/ \sum_{i=1}^{N} W_i \tag{12.18}$$

where

- N ($N \geq 2$) is the number of tests following the debugging intervals $(0, x_1)$, $(0, x_2)$, …, $(0, x_N)$.
- W_i is the total time of successful and unsuccessful (i.e., all) runs in the ith test.
- M_i is the total number of runs terminating in failure in the ith test.

Power Model

This model is also known as the Duane model because Duane [47] originally proposed it for hardware reliability in 1964. For software products, the same behavior has been observed. The reason for calling it a power model is that the mean value function, m(t), for the cumulative number of failures by time t is taken as a power of t, i.e.,

$$m(t) = \theta t^{\alpha}, \text{ for } \theta > 0, \alpha > 0 \tag{12.19}$$

For $\alpha = 1$, we get the homogeneous Poisson process model. The key assumption associated with this model is that the cumulative number of failures by time t, N (t), follows a Poisson process with value function described by Equation (12.19). In order to implement the model, the data requirement could be either of the following [48]:

- Elapsed times between failures, i.e., $y_1, y_2, y_3, \ldots, y_n$, where $y_i = t_i - t_{i-1}$ and $t_0 = 0$.
- Actual times the software program failed, i.e., $t_1, t_2, t_3, \ldots, t_n$.

If T is the time for which the software program was under observation, then we can write

$$\frac{m(T)}{T} = \frac{\theta T^{\alpha}}{T} = \frac{\text{Expected number of faults by T}}{\text{Total testing time, T}} \tag{12.20}$$

Taking the natural logarithms of Equation (12.20), we get

$$D = \ln \theta + (\alpha - 1) \ln T \tag{12.21}$$

The above equation plots as a straight line and it is the form fitted to given data.

Differentiating Equation (12.19) with respect to t we get the following expression for the failure intensity function:

$$\frac{dm(t)}{dt} = \theta \alpha t^{\alpha - 1} = \lambda(t) \tag{12.22}$$

For $\alpha > 1$, Equation (12.22) is strictly increasing; thus, there can be no growth in reliability [48].

Using the maximum likelihood estimation method, we get [49]

$$\hat{\theta} = n / t_n^{\hat{\alpha}} \tag{12.23}$$

and

$$\hat{\alpha} = n \Big/ \sum_{i=1}^{n-1} \ln(t_n / t_i) \tag{12.24}$$

The maximum likelihood estimation for the MTTF [i.e., for the (n + 1)st failure] is given by [48, 49]

$$\hat{\text{MTTR}} = t_n / n\hat{\alpha} \tag{12.25}$$

Air Force Model

This model is the result of the U.S. Air Force Rome Laboratory's [50] effort to predict software reliability in the earlier phases of the software life cycle. In this model, the predictions of fault density are developed and subsequently they can be transformed into other reliability measures such as failure rates [48]. The factors related to fault density at the initial phases include development environment, nature of application, traceability, program size, complexity, modularity, language type, extent of reuse, quality review results, standards review results, and anomaly management. Thus, the initial fault density is expressed by [48, 50]

$$\beta = (DA)(NA)(T)(PS)(C)(M)(LT)(ER)(QR)(SR)(AM) \quad (12.26)$$

where

- β is the initial fault density.
- DA is the development environment.
- NA is the nature of application.
- T is traceability.
- PS is the program size.
- C is complexity.
- M is modularity.
- LT is the language type.
- ER is the extent of reuse.
- QR is the quality review results.
- SR is the standards review results.
- AM is the anomaly management.

The initial failure rate, λ_I, is expressed by

$$\begin{aligned} \lambda_I &= (LEF)(FER)(NIF) \\ &= (LEF)(FER)(\beta \times LSC) \end{aligned} \quad (12.27)$$

where

- LSC is the number of lines of source code.
- LEF is the program's linear execution frequency.
- FER is the fault expose ratio ($1.4 \times 10^{-7} \leq FER \leq 10.6 \times 10^{-7}$).

The linear execution frequency, LEF, of the program is expressed by

$$LEF = (MIR)/(OIP) \quad (12.28)$$

where

- MIR is the mean instruction rate.
- OIP is the number of object instructions in the program.

The object instructions in the program, OIP, are given by

$$OIP = (SI)(CER) \tag{12.29}$$

where

SI is the number of source instructions.
CER is the code expansion ratio. More specifically, this is the ratio of machine instructions to source instructions and, normally, its average value is taken as 4.

The number of inherent faults, IF, is expressed by

$$IF = (\beta)(LSC) \tag{12.30}$$

Inserting Equations (12.28) through (12.30) into Equation (12.27) yields

$$\lambda_1 = (MIR)(FER)(IF)/(SI)(CER) \tag{12.31}$$

12.8 FAULT MASKING

In fault tolerant computing the term fault masking is used in the sense that a system having redundancy can tolerate a number of failures prior to its failure. The implication of the term masking is that a problem has surfaced somewhere within a digital system, but because of the design, the problem does not effect the overall system operation. Probably, the best known fault masking technique is modular redundancy.

12.8.1 Triple Modular Redundancy (TMR)

In this scheme, three identical redundant units or modules perform the same task simultaneously and the voter compares the outputs of all the tree modules and sides with the majority. The TMR system only experiences a failure when more than one module/unit fails or the voter fails. In other words, this type of redundancy can tolerate failure of a single module/unit. Figure 12.2 shows a block diagram of the TMR scheme.

The TMR scheme was first proposed by Von Neumann [6] in 1956 and one important example of its application was the SATURN V launch vehicle computer. This computer used TMR with voters in the central processor and duplication in the main memory [51]. The reliability of the TMR system with voter and the independent units is

$$R_{TMRV} = \left(3R_m^2 - 2R_m^3\right)R_V \tag{12.32}$$

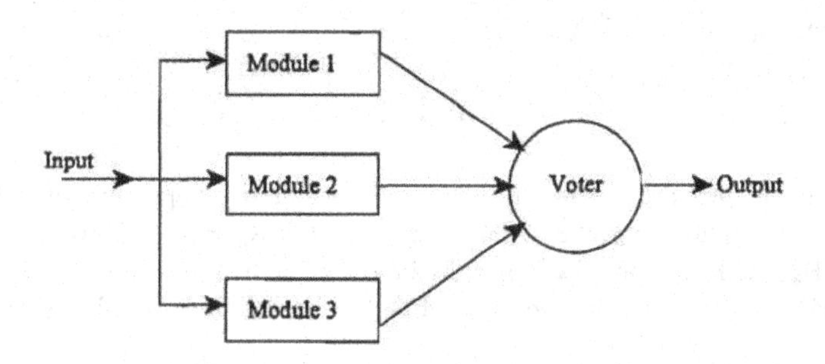

FIGURE 12.2 TMR system with voter.

where

R_{TMRV} is the reliability of the TMR system with voter.
R_V is the voter reliability.
R_m is the module/unit reliability.

With perfect voter, i.e., $R_V = 1$, Equation (12.32) reduces to

$$R_{TMR} = 3\,R_m^2 - 2\,R_m^3 \tag{12.33}$$

where

R_{TMR} is the reliability of the TMR system with perfect voter.

The improvement in reliability of the TMR system over a single unit system is determined by the single unit's reliability and the reliability of the voter. In the case of perfect voter, the reliability of the TMR system given by Equation (12.33) is only better than the single unit system when the reliability of the single unit or module is greater than 0.5. At $R_V = 0.9$, the reliability of the TMR system is only marginally better than the single unit system reliability when the reliability of the single unit is approximately between 0.667 and 0.833 [12]. Furthermore, when $R_V = 0.8$, the TMR system reliability is always less than a single unit's reliability.

For the single unit system's reliability, $R_S = R_m$, and the perfect voter, i.e., $R_V = 1$, the ratio of R_{TMR} to R_c is expressed by [3]

$$r = \frac{R_{TMR}}{R_S} = \frac{3\,R_m^2 - 2\,R_m^3}{R_m} = 3\,R_m - 2\,R_m^2 \tag{12.34}$$

Differentiating Equation (12.34) with respect to R_m and equating it to zero, we get

$$\frac{dr}{dR_s} = 3 - 4\,R_m = 0 \tag{12.35}$$

Thus, from Equation (12.35), we have

$$R_m = 0.75$$

This means that the maximum value of the reliability improvement ratio, r, and the maximum reliability of the TMR system occur at $R_m = 0.75$. Thus, inserting this value for R_m into Equations (12.33) and (12.34) we get the following maximum values of the TMR system reliability and the reliability improvement ratio, respectively:

$$R_{TMR} = 3(0.75)^2 - 2(0.75)^3$$

$$\simeq 0.84$$

and

$$r = 3(0.75) - 2(0.75)^2$$

$$= 1.125$$

Example 12.3

The reliability of the TMR system with perfect voter is given by Equation (12.33). Determine the point where both the single unit or simplex system reliability is equal to the TMR system reliability. Assume that the simplex system reliability is given by

$$R_s = R_m \tag{12.36}$$

In this case, we equate Equations (12.33) and (12.36) to get

$$3\,R_m^2 - 2\,R_m^3 = R_m \tag{12.37}$$

Therefore,

$$2\,R_m^2 - 3\,R_m + 1 = 0 \tag{12.38}$$

The roots of the above quadratic equation are

$$R_m = \frac{3 + [9 - 4(2)(1)]^{1/2}}{2(2)} = 1$$

and

$$R_m = \frac{3 - [9 - 4(2)(1)]^{1/2}}{2(2)} = 1/2$$

Thus, at $R_m = 1$, ½ the reliability of the simplex and TMR systems is the same. This means that the TMR system reliability will only be better than the reliability of the simplex system when $R_m > 0.5$.

Example 12.4

Assume that the reliability of the single unit in Example 12.3 is time dependent and is expressed by

$$R_m(t) = e^{-\lambda_m t} \tag{12.39}$$

where

t is time.
λ_m is the unit/module constant failure rate.

Determine the point where both the single unit or simplex system reliability is equal to the TMR system reliability.

Since from Example 12.3 we have $R_m = 1$, ½ we can write

$$e^{-\lambda_m t} = 1 \tag{12.40}$$

and

$$e^{-\lambda_m t} = 1/2 \tag{12.41}$$

that is

$$\lambda_m t = 0.6931$$

Equations (12.40) and (12.41) indicate that at $t = 0$ or $\lambda_m = 0$ and at $\lambda_m t = 0.6931$, respectively, the reliability of the simplex and TMR systems is the same. The reliability of the TMR system will only be better then the simplex system reliability when the value of the λt is less than 0.6931.

TMR System Time Dependent Reliability and Mean Time to Failure (MTTF)

For constant failure rates of units, the TMR system with voter reliability using Equation (12.32) is

$$\begin{aligned} R_{TMRV}(t) &= \left[3\,e^{-2\lambda_m t} - 2\,e^{-3\lambda_m t}\right] e^{-\lambda_v t} \\ &= 3\,e^{-(2\lambda_m + \lambda_v)t} - 2\,e^{-(3\lambda_m + \lambda_v)t} \end{aligned} \tag{12.42}$$

where

t is time.
R_{TMRV} is the TMR system with voter reliability at time t.
λ_m is the module/unit constant failure rate.
λ_V is the voter constant failure rate.

Integrating Equation (12.42) over the interval from 0 to ∞, we get the following expression for the MTTF of the TMR system with voter:

$$MTTF_{TMRV} = \int_0^\infty \left[3\, e^{-(2\lambda_m+\lambda_v)t} - 2\, e^{-(3\lambda_m+\lambda_v)t}\right] dt$$
$$= \frac{3}{(2\lambda_m+\lambda_V)} - \frac{2}{(3\lambda_m+\lambda_V)} \tag{12.43}$$

For the perfect voter, i.e., $\lambda_V = 0$, Equations (12.42) and (12.43) simplify to

$$R_{TMR}(t) = 3\, e^{-2\lambda_m t} - 2\, e^{-3\lambda_m t} \tag{12.44}$$

and

$$MTTF_{TMR} = \frac{3}{2\lambda_m} - \frac{2}{3\lambda_m} = \frac{5}{6\lambda_m} \tag{12.45}$$

where

$R_{TMR}(t)$ is the TMR system with perfect voter reliability at time t.
$MTTF_{TMR}$ is the MTTF of the TMR system with perfect voter.

Example 12.5

Assume that the failure rate of a unit/module belonging to an independent TMR system with voter is $\lambda_m = 0.0001$ failures/10^6 h. Calculate the TMR system MTTF, if the voter failure rate is $\lambda_V = 0.00002$ failures/10^6 h. Also, compute the system reliability for a 100-h mission.

Inserting the specified data into Equation (12.43) yields

$$MTTF_{TMRV} = \frac{3}{2(0.0001)+0.00002} - \frac{2}{3(0.0001)+0.00002}$$
$$\simeq 7{,}386 \text{ h}$$

Similarly, substituting the given data into Equation (12.42) yields

$$R_{TMRV}(100) = 3\,e^{-[2(0.0001)+0.00002](100)} - 2\,e^{-[3(0.0001)+0.00002](100)}$$

$$\simeq 0.9977$$

Thus, the reliability and MTTF of the TMR system with voter are 0.9977 and 7386 h, respectively.

Example 12.6

Repeat the Example 12.5 calculations for a TMR system with perfect voter. Comment on the end results.

Thus, in this case we have $\lambda_V = 0$.

Substituting the remaining given data into Equations (12.44) and (12.45), we get

$$R_{TMR}(100) = 3\,e^{-(2)(0.0001)(100)} - 2\,e^{-(3)(0.0001)(100)}$$

$$\simeq 0.9997$$

and

$$MTTF_{TMR} = \frac{5}{6(0.0001)}$$

$$\simeq 8{,}333 \text{ h}$$

In this case, the reliability and MTTF of the TMR system with perfect voter are 0.9997 and 8333 h, respectively. The perfect voter helped to improve the TMR system reliability and MTTF, as can be observed by comparing Examples 12.5 and 12.6 results.

TMR System with Repair and Perfect Voter Reliability Analysis

In the preceding TMR system reliability analysis, all units were considered non-repairable. This is quite true in certain applications like space exploration but in others it is possible to repair the failed units. Thus, in this case we consider an independent and identical unit repairable TMR system with perfect voter. As soon as any one of the TMR system units fails, it is immediately repaired. When more than one unit fails, the TMR system is not repaired. The TMR system state space diagram is shown in Figure 12.3.

Markov technique is used to develop state probability equations [8] for the state space diagram shown in Figure 12.3. The following assumptions are associated with this TMR system model:

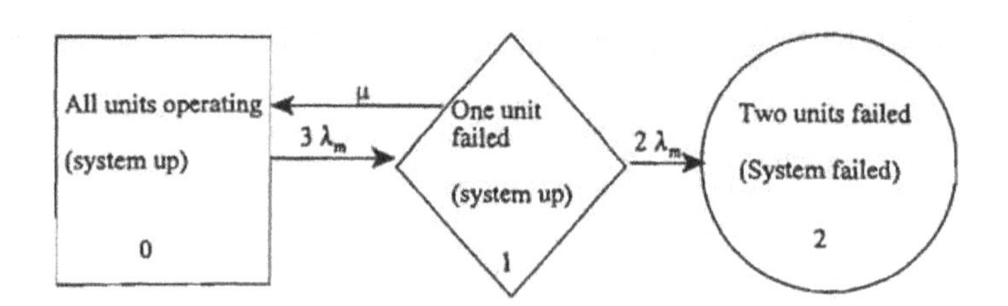

FIGURE 12.3 TMR system with repair state space diagram.

- The system is made up of three independent and identical units with a perfect voter.
- All system units are active and the system fails when more than one unit fails.
- Failure and repair rates of a unit are constant.
- The system is only repaired when one unit fails and a repaired unit is as good as new.
- The numerals in the box, diamond, and circle of Figure 12.3 identify specific system states.

The following symbols were used to develop equations for this model:

- j is the jth state of the TMR system shown in Figure 12.3; i = 0 (three units up), i = 1 (one unit failed — system up), i = 2 (two units failed — system failed).
- P_j (t) is the probability that the TMR system is in state j at time t; for j = 0, 1, and 2.
- λ_m is the unit/module failure rate.
- μ is the unit repair rate.

The following differential equations are associated with Figure 12.3:

$$P'(t) = \mu P_1(t) - 3\lambda_m P_0(t) \tag{12.46}$$

$$P_1'(t) = 3\lambda_m P_0(t) - (2\lambda + \mu) P_1(t) \tag{12.47}$$

$$P_2'(t) = 2\lambda_m P_1(t) \tag{12.48}$$

where the prime denotes differentiation with respect to time t. At time t = 0, P_0 (0) = 1 and P_1 (0) = P_2 (0) = 0.

By solving Equations (12.46) through (12.48), we get the following expression for the TMR system reliability, with repair and perfect voter:

$$R_{TMRr}(t) = P_0(t) + P_1(t)$$
$$= \frac{1}{x_1 - x_2}\left[(5\lambda_m + \mu)\left(e^{x_1 t} - e^{x_2 t}\right) + x_1 e^{x_1 t} - x_2 e^{x_2 t}\right] \quad (12.49)$$

where

$$x_1, x_2 = \left[-(5\lambda_m + \mu) \pm \left(\lambda^2 + \mu^2 + 10\,\lambda\,\mu\right)^{1/2}\right] / 2 \quad (12.50)$$

Integrating Equation (12.49) over the interval [0, ∞] yields

$$MTTF_{TMRr} = \int_0^\infty R_{TMRr}(t)\,dt$$
$$= \frac{5}{6\,\lambda_m} + \frac{\mu}{6\,\lambda_m^2} \quad (12.51)$$

where

$MTTF_{TMRr}$ is the TMR system, with repair and perfect voter, mean time to failure.

For no repair facility, i.e., $\mu = 0$, Equation (12.51) becomes the same as Equation (12.45).

Example 12.7

A unit/module of an independent TMR system, with repair and perfect voter, has a failure rate of 0.005 failures/h. The unit repair rate is 0.6 repairs/h. Calculate the following:

- System with repair mean time to failure.
- System mean time to failure, if there is no repair.

Comment on the end results.

Substituting the given data into Equation (12.51) yields

$$MTTF_{TMRr} = \frac{5}{6(0.005)} + \frac{0.6}{6(0.006)^2}$$
$$\simeq 4167 \text{ h}$$

Similarly, for no repair, inserting the given data into Equation (12.45) yields

$$MTTF_{TMR} = \frac{5}{6(0.005)}$$
$$\simeq 167 \text{ h}$$

It means the introduction of repair helped to increase the TMR system MTTF from 167 h to 4167 h.

12.8.2 N-Modular Redundancy (NMR)

This is a logical extension of TMR and it contains N identical modules/units. The number N is any odd number and is defined by N = 2 m + 1. The NMR system will be successful if at least (m+1) units operate normally. As the voter acts in series with the N-module system, the total system fails if the voter fails. For independent units, the reliability of the NMR system is

$$R_{NMRV} = R_V \left[\sum_{i=0}^{m} \binom{N}{i} R_m^{N-i} (1 - R_m)^i \right] \quad (12.52)$$

$$\binom{N}{i} \equiv N!/(N-i)!\, i! \quad (12.53)$$

where

R_{NMRV} is the reliability of the NMR system with voter.
R_V is the voter reliability.
R_m is the module reliability.

The time dependent and other reliability analysis of this system can be performed in a manner similar to the TMR system analysis. Additional redundancy schemes may be found in Reference 8.

12.9 PROBLEMS

1. Make a comparison of hardware reliability and software reliability.
2. What are the major sources of computer failures? Describe them in detail.
3. Define the following terms:
   - Software error
   - Debugging
   - Software reliability
   - Fault-tolerant computing

4. Discuss software life cycle phases and associated error sources.
5. Describe the following software design methods:
   - Structure programming
   - Top-down programming
   - Jackson structured programming
6. Describe the following techniques associated with software reliability improvement:
   - N-version programming
   - Top-down testing
   - Bottom-up testing
7. What are the major classifications of software reliability models? Describe them briefly.
8. A software under development was seeded with 30 faults and during the testing process 70 faults of the same kind were discovered. The breakdowns of the faults uncovered were 25 seeded faults and 45 unseeded faults. Estimate the number of unseeded faults remaining in the software under consideration.
9. The failure rate of a unit/module belonging to an independent TMR system with voter is 0.0001 failures/h and the voter failure rate is 0.00003 failures/h. Calculate the TMR system MTTF and its reliability for a 200-h mission.
10. Develop state probability expressions for the TMR system state space diagram shown in Figure 12.3. Also, prove that the system reliability is given by Equation (12.49).

12.10 REFERENCES

1. Keene, S.J., Software reliability concepts, *Annu. Reliability and Maintainability Symp. Tutorial Notes*, 1-21, 1992.
2. Pradhan, D.K., Ed., *Fault-Tolerant Computing Theory and Techniques*, Vols. 1 and 2, Prentice-Hall, Englewood Cliffs, NJ, 1986.
3. Shooman, M.L., Fault-tolerant computing, *Annu. Reliability and Maintainability Symp. Tutorial Notes*, 1-25, 1994.
4. Shannon, C.E., A mathematical theory of communications, *Bell System Tech. J.*, 27, 379-423 and 623-656, 1948.
5. Hamming, W.R., Error detecting and error correcting codes, *Bell System Tech. J.*, 29, 147-160, 1950.
6. Von Neumann, J., Probabilistic logics and the synthesis of reliable organisms from reliable components, in *Automata Studies*, Shannon, C.E. and McCarthy, J., Eds., Princeton University Press, Princeton, NJ, 1956, pp. 43-98.
7. Moore, E.F. and Shannon, C.E., Reliable circuits using less reliable relays, *J. Franklin Inst.*, 262, 191-208, 1956.
8. Dhillon, B.S., *Reliability in Computer System Design*, Ablex Publishing, Norwood, NJ, 1987.
9. Haugk, G., Tsiang, S.H., and Zimmerman, L., System testing of the no. 1 electronic switching system, *Bell System Tech. J.*, 43, 2575-2592, 1964.
10. Floyd, R.W., Assigning meanings to program, *Math. Aspects Comp. Sci.*, XIX, 19-32, 1967.

11. Hudson, G.R., Programming Errors as a Birth-and-Death Process, Report No. SP-3011, System Development Corporation, 1967.
12. Pecht, M., Ed., *Product Reliability, Maintainability, and Supportability Handbook*, CRC Press, Boca Raton, FL, 1995.
13. Lipow, M., Prediction of software failures, *J. Syst. Software*, 1, 71-75, 1979.
14. Anderson, R.T., *Reliability Design Handbook*, Rome Air Development Center, Griffiss Air Force Base, Rome, NY, 1976.
15. Glass, R.L., *Software Reliability Guidebook*, Prentice-Hall, Englewood Cliffs, NJ, 1979.
16. Avizienis, A., Fault-tolerant computing: An overview, *Computer*, January/February, 5-8, 1971.
17. Kline, M.B., Software and hardware reliability and maintainability: What are the differences?, *Proc. Annu. Reliability and Maintainability Symp.*, 179-185, 1980.
18. Grant Ireson, W., Coombs, C.F., and Moss, R.Y., *Handbook of Reliability Engineering and Management*, McGraw-Hill, New York, 1996.
19. Dhillon, B.S., *Reliability Engineering in Systems Design and Operation*, Van Nostrand Reinhold Company, New York, 1983.
20. Yourdon, E., The causes of system failures — Part II, *Modern Data*, 5, 50-56, 1972.
21. Yourdon, E., The causes of system failures — Part III, *Modern Data*, 5, 36-40, 1972.
22. Bell, D., Morrey, L., and Pugh, J., *Software Engineering: A Programming Approach*, Prentice-Hall, London, 1992.
23. Wang, R.S., Program with measurable structure, *Proc. Am. Soc. Quality Control Conf.*, 389-396, 1980.
24. Koestler, A., *The Ghost in the Machine*, Macmillan, New York, 1967.
25. Neufelder, A.M., *Ensuring Software Reliability*, Marcel Dekker, New York, 1993.
26. Scott, R.K., Gault, J.W., and McAllister, D.G., Fault tolerant software reliability modeling, *IEEE Trans. Software Eng.*, 13(5), 1987.
27. Neuhold, E.J. and Paul, M., Formal description of programming concepts, *Int. Fed. Info. Process. (IFIP) Conf. Proc.*, 310-315, 1991.
28. Galton, A., Logic as a formal method, *Comp. J.*, 35(5), 213-218, 1992.
29. Beizer, B., *Software System Testing and Quality Assurance*, Van Nostrand Reinhold Company, New York, 1984.
30. Myers, G.J., *The Art of Software Testing*, John Wiley & Sons, New York, 1979.
31. Myers, G.J., *Software Reliability: Principles and Practices*, John Wiley & Sons, New York, 1976.
32. Kopetz, H., *Software Reliability*, Macmillan, London, 1979.
33. Musa, J.D., Iannino, A., and Okumoto, K., *Software Reliability*, McGraw-Hill, New York, 1987.
34. Sukert, A.N., An investigation of software reliability models, *Proc. Annu. Reliability Maintainability Symp.*, 478-484, 1977.
35. Schick, G.J. and Wolverton, R.W., An analysis of competing software reliability models, *IEEE Trans. Software Eng.*, 4, 104-120, 1978.
36. Hudson, G.R., Program Errors as a Birth and Death Process, Report No. SP-3011, System Development Corporation, December 4, 1967.
37. Mills, H.D., On the Statistical Validation of Computer Programs, Report No. 72-6015, 1972. IBM Federal Systems Division, Gaithersburg, MD.
38. Musa, J.D., A theory of software reliability and its applications, *IEEE Trans. Software Eng.*, 1, 312-327, 1975.
39. Shooman, M.L., Software reliability measurement and models, *Proc. Annu. Reliability Maintainability Symp.*, 485-491, 1975.

40. Jelinski, Z. and Moranda, P.B., Software reliability research, in *Proceedings of the Statistical Methods for the Evaluation of Computer System Performance*, Academic Press, 1972, pp. 465-484.
41. Schick, G.J. and Wolverton, R.W., Assessment of software reliability, Proceedings of the Operations Research Physica-Verlag, Wurzburg-Wien, 1973, pp. 395-422.
42. Nelson, E., Estimating software reliability from test data, *Microelectronics and Reliability*, 17, 67-75, 1978.
43. Ramamoorthy, C.V. and Bastani, F.B., Software reliability: status and perspectives, *IEEE Trans. Software Eng.*, 8, 354-371, 1982.
44. Dunn, R. and Ullman, R., *Quality Assurance for Computer Software*, McGraw-Hill, New York, 1982.
45. Craig, G.R., Software Reliability Study, Report No. RADC-TR-74-250, Rome Air Development Center, Griffiss Air Force Base, Rome, NY, 1974.
46. Thayer, T.A., Lipow, M., and Nelson, E.C., *Software Reliability*, North-Holland Publishing Company, New York, 1978.
47. Duane, J.T., Learning curve approach to reliability monitoring, *IEEE Trans. Aerospace*, 2, 563-566, 1964.
48. Lyu, M.R., Ed., *Handbook of Software Reliability Engineering*, McGraw-Hill, New York, 1996.
49. Crow, L.H., in *Reliability Analysis for Complex Repairable Systems, Reliability and Biometry*, Proschan, F. and Serfling, R.J., Eds., Society for Industrial and Applied Mathematics (SIAM), Philadelphia, PA, 1974, pp. 379-410.
50. Methodology for Software Reliability Prediction and Assessment, Report No. RL-TR-92-52, Volumes I and II, Rome Air Development Center, Griffiss Air Force Base, Rome, NY, 1992.
51. Mathur, F.P. and Avizienis, A., Reliability analysis and architecture of a hybrid redundant digital system: Generalized triple modular redundancy with self-repair, *1970 Spring Joint Computer Conference, AFIPS Conf. Proc.*, 36, 375-387, 1970.

13 Robot Reliability and Safety

13.1 INTRODUCTION

Robots are increasingly being used in the industry to perform various types of tasks: material handling, arc welding, spot welding, etc.

The history of robots/automation may be traced back to the ancient times when the Egyptians built water-powered clocks and the Chinese and Greeks built water- and steam-powered toys. Nonetheless, the idea of the functional robot originated in Greece in the writings of Aristotle (4th century B.C.), the teacher of Alexander The Great, in which he wrote: "If every instrument could accomplish its own work, obeying or anticipating the will of others ..." [1].

In 1920, Karl Capek (1890–1938), a Czechoslovak science-fiction writer first coined the word "robot" and used it in his play entitled "Rossums Universal Robots" which opened in London in 1921. In the Czechoslovakian language, robot means "worker". In 1939, Isaac Asimov wrote a series of stories about robots and in 1942, he developed the following three laws for robots [2]:

- A robot must not harm a human, nor, through inaction, permit a human to come to harm.
- Robots must obey humans unless their orders conflict with the preceding law.
- A robot must safeguard its own existence unless it conflicts with the preceding two laws.

In 1954, George Devol [3] developed a programmable device that could be considered the first industrial robot. Nonetheless, in 1959, the first commercial robot was manufactured by the Planet Corporation and Japan, today the world leader in the use of robots, imported its first robot in 1967 [3]. In 1970, the first conference on industrial robots was held in Chicago, Illinois, and five years later (i.e., in 1975), the Robot Institute of America (RIA) was founded. Today, there are many journals and a large number of publications on robots.

In 1983, the estimated world robot population was around 30,000 [4] and by the end of 1998, it is forecasted to be around 820,000 [5].

A robot may be defined as a reprogrammable multi-functional manipulator designed to move material, parts, tools, or specialized items through variable programmed motions for the performance of various tasks [6]. As a robot has to be safe and reliable, this chapter discusses both of these topics.

13.2 ROBOT SAFETY

As in the case of general system safety, robot safety is an important factor in the effectiveness of robot systems. It is expected that the rapidly increasing robot population will further emphasize the importance of robot safety in a significant manner.

The review of published literature indicates that in the 1970s, robot safety problems became quite apparent to the professionals working in the area. Consequently, 1980 witnessed a flood of publications on the topic. In fact, some of the important publications of the 1980s and the early 1990s were as follows:

- In 1985, the Japanese Industrial Safety and Health Association (JISHA) developed a document entitled "An Interpretation of the Technical Guidance on Safety Standards, etc. of Industrial Robots" [7].
- In 1985, a book entitled *Robot Safety*, edited by Benny and Yong [8], appeared and it contained many useful articles on the topic.
- In 1986, a safety standard entitled "Industrial Robots and Robot Systems: Safety Requirements" was published jointly by the American National Standards Institute and the Robotics Industries Association (RIA) [9].
- In 1987, a journal article [10] entitled "On Robot Reliability and Safety: Bibliography," appeared and it listed most of the publications on the subject.
- In 1991, a book [2] entitled "Robot Reliability and Safety" was published and it presented a comprehensive list of publications on robot safety in addition to the important aspects of robot safety.

Since the early 1990s, there have been many more publications on the topic.

13.2.1 Robot Accidents

Over the years, there have been many accidents involving robots [2, 11–14]. These accidents occurred in many robot user countries: Japan, U.S., Sweden, France, U.K., etc. [11]. According to published sources, there have been at least five fatal accidents; four in Japan and one in the U.S.. Some examples of these fatal accidents are as follows [2]:

- A worker turned on a welding robot during the presence of a person in its work zone. Consequently, the robot pushed that person into the positioning fixture and subsequently the person died.
- During a temporary stoppage cycle of a robot, a repair person climbed over a safety fence without turning off the power to the robot. As the robot commenced its operation, it pushed that person into a grinding machine and, consequently, the person died.
- During the momentary stoppage of a line servicing robot, a worker climbed onto a conveyor belt in motion to recover a faulty part. As the robot started its operation, the person was crushed to death.

TABLE 13.1
National Institute for Occupational Safety and Health Recommendations for Minimizing Robot Related Incidents

Recommendation No.	Recommendation description
1	Allow sufficient clearance distance around all moving parts of the robot system.
2	Provide adequate illumination in control and operational areas of robot system.
3	Provide physical barrier incorporating gates with electrical interlocks, so that when any one gate opens, the robot operation is terminated.
4	Ensure that floors or working surfaces are marked so that the robot work area is clearly visible.
5	Provide back-up to devices such as light curtains, motion sensors, electrical interlocks, or floor sensors.
6	Provide remote "diagnostic" instrumentation so that system maximum troubleshooting can be conducted from outside the robot work area.

Subsequent investigation revealed that in all of the above cases, the humans put themselves in harms way. The robot alone did not kill the human.

Some of the causal factors for robot accidents include robot design, workplace design interfacing, and workplace design guarding. In order to minimize the risk of robot incidents, the National Institute for Occupational Safety and Health proposed various recommendations [15]. In fact, all of these recommendations were directed at three strategic areas: (1) worker training, (2) worker supervision, and (3) robot system design. The robot system design related recommendations are presented in Table 13.1

13.2.2 Robot Hazards and Safety Problems

There are various hazards associated with robots. The three basic types of hazards are shown in Figure. 13.1 [16, 17]. These are impact hazards, the trapping point

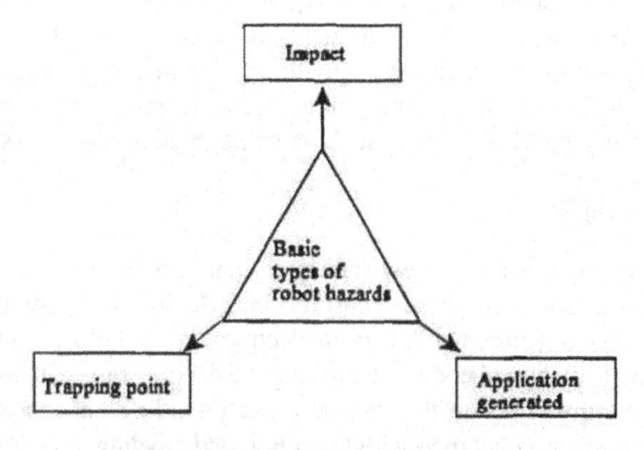

FIGURE 13.1 Basic robot hazards.

hazards, and those that develop from the application itself. The impact hazards are concerned with an individual(s) being struck by a robot's moving parts, by items being carried by the robot, or by flying objects ejected or dropped by the robot. The trapping point hazards usually occur because of robot movements with respect to fixed objects such as machines and posts in the same space. Furthermore, other possible factors include the movements of auxiliary equipment such as work carriages and pallents. The hazards that are generated by the application itself include burns, electric shocks, exposure to toxic substances, and arc flash.

There could be many different causes of the above three basic types of robot hazards but the most prevalent causes include unauthorized access, human error, and mechanical related problems from the robot system itself or from the application.

13.2.3 Safety Considerations in Robot Life Cycle

A robot life cycle could be divided into four distinct phases: (1) design, (2) installation, (3) programming, and (4) operation and maintenance. The overall robot safety problems can only be minimized if careful consideration is given to safety throughout the robot life cycle. Some safety measures that could be taken during each of the four robot life cycle phases are discussed below [2, 7, 18–21]:

Design Phase

The safety measures associated with this phase may be categorized into three distinct groups: electrical, software, and mechanical. The electrical group safety measures include eliminating the risk of an electric shock, minimizing the effects of electromagnetic and radio-frequency interferences, designing wiring circuitry capable of stopping the robot's movement and locking its brakes, and having built-in hose and cable routes using adequate insulation, sectionalization, and panel covers. Some of the safety measures belonging to the software group include having built-in safety commands, periodically examining the built-in self-checking software for safety, having a stand-by power source for robots functioning with programs in random access memory, using a procedure of checks for determining why a failure occurred, and prohibiting a restart by merely resetting a single switch.

The mechanical group related safety measures include providing dynamic brakes for a software crash or power failure, providing drive mechanism covers, providing several emergency stop buttons, providing mechanisms for releasing the stopped energy, and putting guards on items such as gears, belts, and pulleys.

Installation Phase

There are a large number of safety measures that can be taken during the robot installation phase. Some of these measures include installing interlocks, sensing devices, etc.; distancing circuit boards from electromagnetic fields; installing electrical cables according to electrical codes; identifying the danger areas with the aid of codes; placing robot controls outside the hazard zone; providing an appropriate level of illumination to humans concerned with the robot; and labeling stored energy sources.

Programming Phase

During this phase, programmers/setters are the people particularly at risk. For example, according to a study of 131 cases [21], such people were at the highest risk 57% of the time. Some of the safety measures that can be taken during the programming phase include designing the programmer work area so that unnecessary stress with respect to stand accessibility, visibility/lighting, and forced postures, is eliminated; installing hold-to-run buttons; turning off safety-related devices with a key only; planning the programmer's location outside the movement only; and marking the programming position.

Operation and Maintenance Phase

There are a large number of safety related measures that belong to this phase. Some of these are developing appropriate safety operations and maintenance procedures, ensuring functionality of all emergency stops, providing appropriate protective gear to all concerned individuals, minimizing the potential energy of an unexpected motion by having the robot arm extended to its maximum reach position, blocking out all concerned power sources during maintenance, posting the robot's operating weight capacity, observing all government codes and other regulations concerned with robot operation and maintenance, and ensuring that only authorized and trained personnel operate and maintain robots.

13.2.4 Robot Safeguard Approaches

In order to improve robot safety, over the years there have been many safeguard approaches developed [22]. These include flashing lights, warning signs, intelligent systems, electronic devices, infrared light arrays, and physical barriers. Each of these approaches is described below.

Flashing Lights

This approach calls for the installation of flashing lights at the perimeter of the robot working zone or on the robot itself. The primary aim of these lights is to alert concerned individuals that the robot is active or could be in motion at any moment. In the event of pursuing this approach, it is useful to ensure that the flashing lights are energized at all times during the active drive power period.

Warning Signs

This approach is concerned with the use of warning signs. Use of these signs may be considered where robots, because of their speed, size, and inability to impart significant amount of force, cannot injure people. Usually, the warning signs are used for robots used in laboratories and for small-part assembly robots. All in all warning signs are useful for all robot applications, irrespective of the robot's ability to injure people or not.

Intelligent Systems

This approach makes use of intelligent control systems that make decisions through remote sensing, hardware, and software. In order to have an effective intelligent collision-avoidance system, the robot operating environment has to be restricted. Also, special sensors and software should be used. Nonetheless, usually in most industrial settings it is not possible to restrict environment.

Electronic Devices

This approach makes use of ultrasonic for parameter control to seek protection from intrusions. The parameter control electronic barriers use active sensors for intrusion detection. Usually, the use of ultrasonic is considered in those circumstances where unobstructed floor space is a critical issue. An evaluation of robot parameter control devices is presented in Reference 23.

Infrared Light Arrays

This approach makes use of linear arrays of infrared sources known as light curtains. The light curtains are generally reliable and provide an excellent protection to individuals from potential dangers in the robot's operating area. False triggering is the common problem experienced with the use of light curtains. It may occur due to various factors including smoke, flashing lights, or heavy dust in situations where the system components are incorrectly aligned.

Physical Barriers

Even though the use of physical barriers is an important approach to safeguard humans, it is not the absolute solution to a safety problem. The typical examples of physical barriers include safety rails, plastic safety chains, chain-link fences, and tagged-rope barriers. Some of the useful guidelines associated with physical barriers include:

- Consider using safety rails in areas free of projectiles.
- Consider using fences in places where long-range projectiles are considered a hazard.
- Safety rails and chain-link fences are quite effective in places where intrusion is a serious problem.
- When considering the use of a peripheral physical barrier, it is useful to ask questions such as the following:
  - What is being protected?
  - Is it possible that it can be bypassed?
  - How effective is the barrier under consideration?
  - How reliable is the barrier under consideration?

13.2.5 Human Factor Issues in Robotic Safety

One of the single most important components in robotic safety is the issue of human factors. Just like in the case of other automated machines, the problem of human factors in robotics is very apparent and it has to be carefully considered. This section

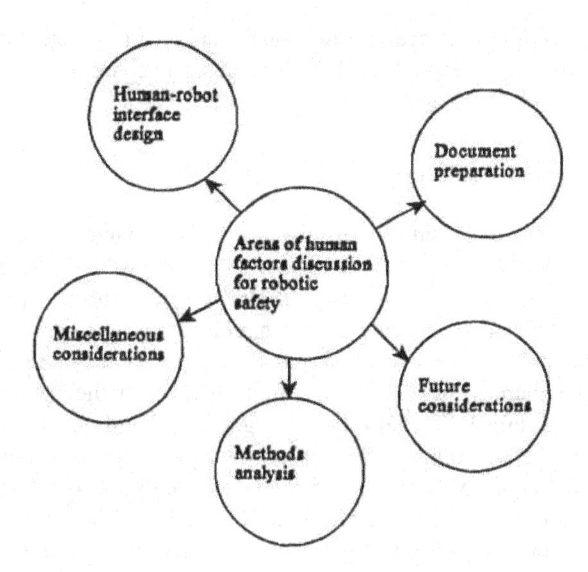

FIGURE 13.2 Areas of human factors discussion for robotic safety.

presents human factor aspects of robotic safety in five distinct areas as shown in Figure 13.2 [24]. These are human-robot interface design, documentation preparation, methods analysis, future considerations, and miscellaneous considerations.

The human-robot interface design is one important area which requires careful consideration of human factors. Here, the primary objective is to develop human-robot interface design so that the probability of the human error occurrence is at minimum. There are various steps that help to prevent the occurrence of human error, including analyzing human actions during robotic processes, designing hardware and software with the intention of reducing human errors, paying careful attention to layouts, and considering factors such as weight, layout of buttons, readability of the buttons' functional descriptions, connectors' flexibility, and hand-held device's shape and size.

The document preparation is concerned with the quality of documentation for use by robot users. These documents must be developed by considering the qualifications and experience of the target group. Some examples of target groups are operators, maintainers, programmers, and designers. Nonetheless, during the preparation of such documents factors, such as easily understandable information, inclusion of practical exercises, completeness of information, and inclusion of pictorial descriptions at appropriate places must be carefully considered.

The classical industrial engineering approach to methods analysis is found to be quite effective in improving robotic safety with respect to human factors. Flow process charts and multiple-activity process charts are two important examples.

Some of the future considerations to improve robotic safety with respect to human factors include the application of artificial intelligence to the worker-robot interface, and considering human factors in factory design with respect to possible robot applications.

The miscellaneous considerations include the analysis of ambient conditions and layout, the preproduction testing of robotic systems, and examination of factors such as noise, lighting, and temperature.

13.3 ROBOT RELIABILITY

A robot not only has to be safe, but it also has to be reliable. An unreliable robot may become the cause of unsafe conditions, high maintenance costs, inconvenience, etc. As robots make use of electrical, mechanical, electronics, pneumatic, and hydraulic parts, their reliability and maintainability problems are very challenging because of many different sources of failures.

Even though there is no clear cut definitive point in the beginning of robot reliability and maintainability field, Engelberger [25] could be regarded as one of the first persons to think in the terms of robot mean time between failures (MTBF) and mean time to repair (MTTR). In 1987, an article entitled "On Robot Reliability and Safety: Bibliography" presented most of the published literature on robot reliability and maintainability [26]. In 1991, a book entitled *Robot Reliability and Safety* covered robot reliability and maintainability in a significant depth along with presenting a comprehensive list of publications on these two areas [27].

13.3.1 Causes and Classifications of Robot Failures

Over the years, there have been various studies performed to determine the causes of robot failures. As a result of such studies, some of the frequently occurring failure causes highlighted were oil pressure valve problems, servo valve malfunctions, noise, printed circuit board troubles, and human errors. In particular, References 28 and 29 pointed out that the robot problems occurred in the following order: control system problems, jig and other tool incompatibility, robot body problems, programming and operation errors, welding gun problems and problems of other tooling parts, deterioration and precision deficiency, runaway, and miscellaneous.

A robot's reliability and its safe operation are primarily affected by four types of failures [30, 31]: (1) random component failures, (2) systematic hardware faults, (3) software faults, and (4) human errors.

The failures that occur unpredictably during the useful life of a robot are known as random failures. The reasons for the occurrence of such failures include undetectable defects, unavoidable failures, low safety factors, and unexplainable causes. The systematic hardware faults occur because of the existence of unrevealed mechanisms in the robot design. One example of the protection against robot hardware faults is to use sensors in the system to detect the loss of pneumatic pressure/line voltage/hydraulic pressure. Reference 32 presents various useful techniques to reduce the occurrence of systematic hardware faults/failures.

The occurrence of software faults is an important factor in the malfunctioning of robots and they may occur due to reasons such as embedded software or the controlling software and application software. Some studies have revealed that over 60% of software errors are made during the requirement and design phases as opposed to less then 40% during the coding phase. The measures such as the performance of failure mode and effects analysis (FMEA), fault tree analysis, and testing help to reduce robot software faults.

During the design, manufacture, test, operation, and maintenance of a robot, various types of human errors may occur. Some studies reveal that the human error represents a significant proportion of total equipment failures. There could be many causes for the occurrence of human error including poor equipment design, poorly written operating and maintenance procedures, poorly trained operation and maintenance manpower, task complexity, inadequate lighting in the work zone, and improper tools used by maintenance personnel.

13.3.2 Robot Reliability Measures

There are various types of reliability-related measures associated with a robot: mean time to robot failure, mean time to robot-related problems, and robot reliability.

Mean Time to Robot Failure (MTRF)

The MTRF can be obtained by using either of the following two formulas:

$$MTRF = \int_0^{\infty} R_b(t)\,dt \tag{13.1}$$

$$MTRF = \left(T_{rp} - T_{drf}\right)/N \tag{13.2}$$

where

$R_b(t)$ is the robot reliability at time t.
T_{rp} is the robot production time expressed in hours.
T_{drf} is the downtime due to robot failures, expressed in hours.
N is the total number of robot failures.

Relationship (13.2) can be used in real life environments to determine mean productive robot time before robot failure. On the other hand, relationship (13.1) can be used to calculate mean time to failure of a single or of redundant robots with defined reliability function.

Example 13.1

Assume that at a robot facility the total robot production hours were 20,000 h with the downtime due to robot failure of 500 h. There have been a total of 15 robot failures. Calculate the robot mean time to failure.

Substituting the above data into Equation (13.2) yields

$$MTRF = (20{,}000 - 500)/15$$
$$= 1300 \text{ h}$$

Thus, the robot mean time to failure is 1300 h.

Example 13.2

Assume that the failure rate of a robot is 0.0004 failure/h and its reliability is expressed by the following equation:

$$R_b(t) = e^{-\lambda_b t} \tag{13.3}$$

where

$R_b(t)$ is the robot reliability at time t.
λ_b is the robot failure rate.

Substituting Equation (13.3) into Equation (13.1) yields

$$\begin{aligned} MTRF &= \int_0^\infty e^{-\lambda_b t}\, dt \\ &= \frac{1}{\lambda_b} \end{aligned} \tag{13.4}$$

Inserting the robot failure rate value into Equation (13.4), we get

$$\begin{aligned} MTRF &= \frac{1}{0.0004} \\ &= 2500 \text{ h} \end{aligned}$$

Thus, the robot mean time to failure is 2500 h.

Mean Time to Robot-Related Problems

The mean time to robot-related problems (MTRP) simply is the average productive robot time prior to the occurrence of a robot-related problem and is expressed by

$$MTRP = \left(T_{rp} - T_{drp}\right)/K \tag{13.5}$$

where

T_{drp} is the downtime due to robot-related problems, expressed in hours.
K is the total number of robot-related problems.

Example 13.3

Assume that at an industrial robot installation the total robot production hours were 25,000 h. Furthermore, downtime due to robot-related problems was 800 h and there have been a total of 30 robot-related problems. Calculate the MTRP.

Substituting the specified data into Equation (13.5), we get

$$MTRP = (25{,}000 - 800)/30$$

$$= 806.67 \text{ h}$$

Thus, the mean time to robot-related problems is 806.67 h.

Robot Reliability

This may be expressed as the probability that a robot will perform its designated mission adequately for the specified time period when used under the designed conditions. Mathematically, the robot reliability may be expressed as follows:

$$R_b(t) = \exp\left[-\int_0^t \lambda_b(t)\,dt\right] \tag{13.6}$$

where

λ_b (t) is the robot time-dependent failure rate or hazard rate.

It is to be noted that Equation (13.6) can be used to obtain robot reliability for any given robot time to failure distribution (e.g., Weibull, gamma, or exponential).

Example 13.4

A robot time to failure is exponentially distributed; thus, its failure rate, λ_b, is 0.0004 failure/h. Calculate the robot reliability for a 15-h operating mission.

Thus, in this case, we have

$$\lambda_b(t) = \lambda \tag{13.7}$$

Inserting Equation (13.7) into Equation (13.6) yields

$$R_b(t) = \exp\left[-\int_0^t \lambda_b\,dt\right] \tag{13.8}$$

$$= e^{-\lambda_b t}$$

Substituting the specified data into Equation (13.8) we get

$$R_b(15) = e^{-(0.0004)(15)}$$

$$= 0.994$$

Thus, there is a 99.4% chance that the robot will operate successfully during the specified mission.

13.3.3 Robot Reliability Analysis and Prediction Methods

In the field of reliability engineering, there are many techniques and methods available that can be used effectively to perform robot reliability and availability analysis and prediction studies. Some of these methods are network reduction technique, fault tree analysis (FTA), failure modes and effect analysis (FMEA), and Markov method [32].

In particular, the two widely used practical methods, i.e., part stress analysis prediction and part count reliability prediction, can also be used to predict robot system reliability during the design phase. Both of these methods are described in Reference 33.

The Markov method is also a widely used approach in reliability engineering and it can equally be applied to perform various kinds of robot reliability studies. Two of its applications are presented in the next section.

13.3.4 Reliability and Availability Analyses of a Robot System Failing with Human Error

This section presents reliability and availability analyses of a mathematical model representing a robot system by using the Markov method [27]. The robot system can fail either due to a human error or other failure. The failed robot system is repaired to its original operating state. The system state space diagram is shown in Figure 13.3. This figure shows three states of the robot system: state 0 (the robot system operating normally), state 1 (the robot system failed due to a human error), and state 2 (the robot system failed due to failures other than human error).

The following assumptions are associated with this model:

- The robot system can fail due to a human error.
- The robot system failures are divisible into two categories: failures resulting from human error, and failures resulting from other than human error.
- The robot system human error and other failures are exponentially distributed or their occurrence rates are constant.
- The failed robot system repair times are exponentially distributed or its repair rates are constant.
- The repaired robot system is as good as new.

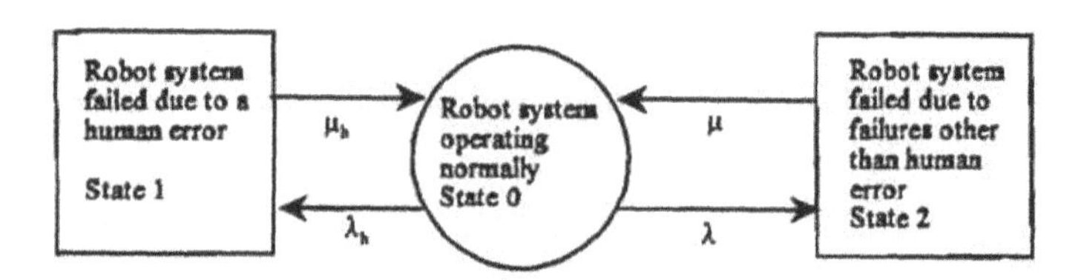

FIGURE 13.3 Robot system state space diagram.

The following symbols were used to develop Equations for the model:

- i is the state of the system: $i = 0$ means the robot system is operating normally; $i = 1$ means the robot system malfunctioned because of a human error; $i = 2$ means the robot system malfunctioned because of a failure other than human error.
- t is time.
- $P_i(t)$ is the probability that the robot system is in state i at time t, for $i = 0, 1, 2$.
- λ_h is the constant robot system human error rate.
- λ is the constant robot system non-human error failure rate.
- μ_h is the constant robot system repair rate from failed state 1.
- μ is the constant robot system repair rate from failed state 2.

Using the Markov technique, from Figure 13.3, we write the following set of differential equations:

$$P_0'(t) + (\lambda + \lambda_h) P_0(t) = \mu_h P_1(t) + \mu P_2(t) \tag{13.9}$$

$$P_1'(t) + \mu_h P_1(t) = \lambda_h P_0(t) \tag{13.10}$$

$$P_2'(t) + \mu P_2(t) = \lambda P_0(t) \tag{13.11}$$

At time $t = 0$, $P_0(0) = 1$ and $P_1(0) = P_2(0) = 0$.

After solving Equations (13.9) through (13.11), we get the following state probability expressions:

$$P_0(t) = \frac{\mu \mu_h}{k_1 k_2} + \left[\frac{(k_1 + \mu)(k_1 + \mu_h)}{k_1 (k_1 - k_2)}\right] e^{k_1 t} - \left[\frac{(k_2 + \mu)(k_2 + \mu_h)}{k_2 (k_1 - k_2)}\right] e^{k_2 t} \tag{13.12}$$

where

$$k_1, k_2 = \frac{-b \pm \sqrt{b^2 - 4(\mu \mu_h + \lambda_h \mu + \lambda \mu_h)}}{2}$$

$$b = \lambda + \lambda_h + \mu + \mu_h$$

$$k_1 k_2 = \mu \mu_h + \lambda \mu_h + \lambda_h \mu$$

$$k_1 + k_2 = -(\mu + \mu_h + \lambda + \lambda_h)$$

$$P_1(t) = \frac{\lambda_h \mu}{k_1 k_2} + \left[\frac{\lambda_h k_1 + \lambda_h \mu}{k_1 (k_1 - k_2)}\right] e^{k_1 t} - \left[\frac{(\mu + k_2) \lambda_h}{k_2 (k_1 - k_2)}\right] e^{k_2 t} \tag{13.13}$$

$$P_2(t) = \frac{\lambda \mu_h}{k_1 k_2} + \left[\frac{\lambda k_1 + \lambda \mu_h}{k_1 (k_1 - k_2)}\right] e^{k_1 t} - \left[\frac{(\mu_h + k_2) \lambda}{k_2 (k_1 - k_2)}\right] e^{k_2 t} \tag{13.14}$$

The robot system availability, AR (t), is expressed by

$$AR(t) = P_0(t) = \frac{\mu \mu_h}{k_1 k_2} + \left[\frac{(k_1 + \mu)(k_1 + \mu_h)}{k_1 (k_1 - k_2)}\right] e^{k_1 t} - \left[\frac{(k_2 + \mu)(k_2 + \mu_h)}{k_2 (k_1 - k_2)}\right] e^{k_2 t} \tag{13.15}$$

As time t becomes very large, the robot system steady-state availability and other state probabilities become

$$AR = \lim_{t \to \infty} AR(t) = \frac{\mu \mu_h}{k_1 k_2} \tag{13.16}$$

$$P_1 = \lim_{t \to \infty} P_1(t) = \frac{\lambda_h \mu}{k_1 k_2} \tag{13.17}$$

$$P_2 = \lim_{t \to \infty} P_2(t) = \frac{\lambda \mu_h}{k_1 k_2} \tag{13.18}$$

where

AR is the robot system steady state availability.
P_1 is the robot system steady state failure probability due to human error.
P_2 is the robot system steady state failure probability due to failures other than human error.

In order to perform robot system reliability analysis, we set $\mu = \mu_h = 0$, in Equations (13.9) through (13.11) and then solve the resulting equations to get

$$P_0(t) = e^{-(\lambda_h + \lambda)t} \tag{13.19}$$

$$P_1(t)=\frac{\lambda_h}{(\lambda+\lambda_h)}\left[1-e^{-(\lambda_h+\lambda)t}\right] \tag{13.20}$$

$$P_2(t)=\frac{\lambda}{(\lambda+\lambda_h)}\left[1-e^{-(\lambda_h+\lambda)t}\right] \tag{13.21}$$

The robot system reliability at time t from Equation (13.19) is given by

$$R_r(t)=P_0(t)=e^{-(\lambda_h+\lambda)t} \tag{13.22}$$

Similarly, the probability of the robot system failure, P_h (t), due to human error from Equation (13.20) is

$$P_h(t)=P_1(t)=\frac{\lambda_h}{(\lambda+\lambda_h)}\left[1-R_r(t)\right] \tag{13.23}$$

Mean time to robot system failure ($MTTR_r$) is expressed by

$$MTTR_r=\int_0^\infty R_r(t)\,dt \tag{13.24}$$

$$=\int_0^\infty e^{-(\lambda_h+\lambda)t}\,dt$$
$$=\frac{1}{\lambda_h+\lambda} \tag{13.25}$$

The robot system overall hazard rate or its total failure rate is expressed by

$$\lambda_r=-\frac{1}{R_r(t)}\frac{d\,R_r(t)}{dt}$$
$$=-\frac{1}{e^{-(\lambda_h+\lambda)t}}\frac{d\left[e^{-(\lambda_h+\lambda)t}\right]}{dt} \tag{13.26}$$
$$=\lambda_h+\lambda$$

Example 13.5

Assume that a robot system can fail either due to human error or other failures and the human error and other failure times are exponentially distributed. Thus, the values of human error and other failure rates are 0.0005 error per hour and 0.002 failure/h. The failed robot system repair times are exponentially distributed with the repair rate value of $\mu = \mu_h = 0.004$ repair per hour. Calculate the robot system steady state availability and its reliability for a 100-h mission.

Thus, we have

$\lambda = 0.002$ failure/h, $\lambda_h = 0.0005$ error/h, $\mu = \mu_h = 0.004$ repair/h, and $t = 100$ h.

Substituting the specified values into Equation (13.11) yields robot system steady state availability

$$AR = \frac{\mu\,\mu_h}{k_1 k_2} = \frac{\mu\,\mu_h}{\mu\,\mu_h + \lambda\,\mu_h + \lambda_h\,\mu}$$

$$= \frac{(0.004)^2}{(0.004)^2 + 0.004(0.002 + 0.0005)}$$

$$= 0.6154$$

Inserting the given data into Equation (13.22) we get

$$R_r(100) = e^{-(0.0005+0.002)(100)}$$

$$= 0.7788$$

Thus, the robot system steady state availability and reliability are 0.6154 and 0.7788, respectively.

13.3.5 Reliability Analysis of a Repairable/Non-Repairable Robot System

This section presents reliability analysis of a mathematical model representing a repairable/non-repairable robot system, by using the Markov technique [34]. The robot system is composed of a robot and an associated safety system/unit. The inclusion of safety systems with robots is often practiced because of robot accidents involving humans. The robot system transition diagram is shown in Figure 13.4.

As shown in Figure 13.4, the normally working robot and its associated safety unit may move to two mutually exclusive states: robot working normally and the safety system has failed, safety system working normally and the robot has failed. From the state robot working normally and its safety system failed, the working robot system may fail safely or with an incident. The following assumptions are associated with the model shown in Figure 13.4:

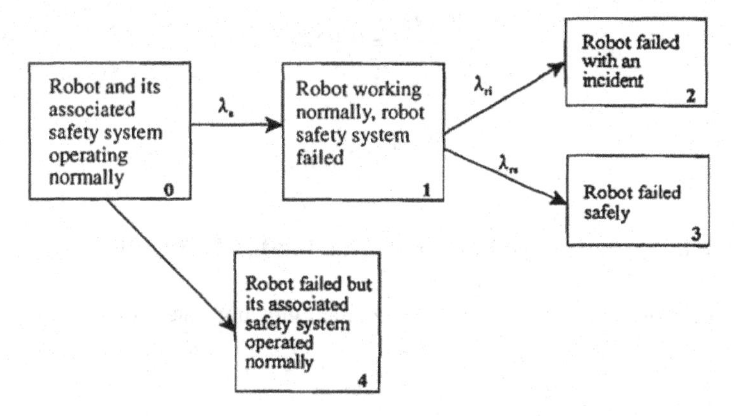

FIGURE 13.4 Robot system transition diagram.

- All failures are statistically independent.
- All failure rates are constant.
- The robot system is composed of a robot and a safety system.
- The robot system fails when the robot fails.

The following notation is associated with the model shown in Figure 13.4:

i ith state of the system: i = 0 (robot and its associated safety system operating normally), i = 1 (robot operating normally, robot safety system failed), i = 2 (robot failed with an incident), i = 3 (robot failed safely), i = 4 (robot failed but its associated safety system operated normally).

$P_i(t)$ The probability that the robot system is in state i at time t; for i = 0, 1, 2, 3, 4.

λ_i ith constant failure rate; i = s (state 0 to state 1), i = ri (state 1 to state 2), i = rs (state 1 to state 3), i = r (state 0 to state 4).

Using the Markov approach, the following system of differential equations is associated with Figure 13.4:

$$\frac{dP_0(t)}{dt} + (\lambda_s + \lambda_r)P_0(t) = 0 \tag{13.27}$$

$$\frac{dP_1(t)}{dt} + (\lambda_{ri} + \lambda_{rs})P_1(t) = P_0(t)\lambda_s \tag{13.28}$$

$$\frac{dP_2(t)}{dt} = P_1(t)\lambda_{ri} \tag{13.29}$$

$$\frac{dP_3(t)}{dt} = P_1(t)\lambda_{rs} \tag{13.30}$$

$$\frac{dP_4(t)}{dt} = P_0(t)\lambda_r \tag{13.31}$$

At time $t = 0$, $P_0(0) = 1$, and $P_1(0) = P_2(0) = P_3(0) = P_4(0) = 0$.

Solving Equations (13.27) to (13.31), we get the following state probability equations:

$$P_0(t) = e^{-At} \tag{13.32}$$

$$P_1(t) = \frac{\lambda_s}{B}\left(e^{-Ct} - e^{-At}\right) \tag{13.33}$$

where

$$A = \lambda_s + \lambda_r$$

$$B = \lambda_s + \lambda_r - \lambda_{rs} - \lambda_{ri}$$

$$C = \lambda_{ri} + \lambda_{rs}$$

$$P_2(t) = \frac{\lambda_{ri}\lambda_s}{AC}\left[1 - \left(Ae^{-Ct} - Ce^{-At}\right)/B\right] \tag{13.34}$$

$$P_3(t) = \frac{\lambda_{rs}\lambda_s}{AC}\left[1 - \left(Ae^{-Ct} - Ce^{-At}\right)/B\right] \tag{13.35}$$

$$P_4(t) = \frac{\lambda_r}{A}\left(1 - e^{-At}\right). \tag{13.36}$$

The reliability of both robot and its safety system working normally is given by

$$R_{rsu}(t) = P_0(t) = e^{-At} \tag{13.37}$$

The reliability of the robot working normally with or without the safety system functioning successfully is:

$$R_{ss}(t) = P_0(t) + P_1(t) = e^{-At} + \frac{\lambda_s}{B}\left(e^{-Ct} - e^{-At}\right). \tag{13.38}$$

The mean time to failure of the robot with the safety system up is expressed by:

$$MTTF_{rsu} = \int_0^{\infty} R_{rsu}(t)\,dt = \frac{1}{A}. \tag{13.39}$$

Similarly, the mean time to failure of the robot with safety system up or down is expressed by:

$$MTTF_{ss} = \int_0^{\infty} R_{ss}(t)\,dt = \frac{1}{A}\left(1 + \lambda_s / C\right). \tag{13.40}$$

If the failed safety system shown in Figure 13.4 (State 1) can be repaired, the robot system reliability will improve considerably. Thus, for constant failed safety system repair rate, μ, from robot system state 1 to 0, the system of differential equations for the new scenario is

$$\frac{dP_0(t)}{dt} + A P_0(t) = P_1(t)\mu \tag{13.41}$$

$$\frac{dP_1(t)}{dt} + D P_1(t) = P_0(t)\lambda_s \tag{13.42}$$

where

$$D = \lambda_{ri} + \lambda_{rs} + \mu$$

$$\frac{dP_2(t)}{dt} = P_1(t)\lambda_{ri} \tag{13.43}$$

$$\frac{dP_3(t)}{dt} = P_1(t)\lambda_{rs} \tag{13.44}$$

$$\frac{dP_4(t)}{dt} = P_0(t)\lambda_r \tag{13.45}$$

At time $t = 0$, $P_0(0) = 1$, and $P_1(0) = P_2(0) = P_3(0) = P_4(0) = 0$.

Solutions to Equations (13.41) through (13.45) are as follows:

$$P_0(t) = e^{-At} + \mu\,\lambda_s \left[\frac{e^{-At}}{(r_1 + A)(r_2 + A)} + \frac{e^{r_1 t}}{(r_1 + A)(r_1 - r_2)} + \frac{e^{r_2 t}}{(r_2 + A)(r_2 - r_1)}\right] \tag{13.46}$$

where

$$r_1, r_2 = \frac{-E \pm \sqrt{E^2 - 4F}}{2}$$

$$E = A + C + \mu$$

$$F = \lambda_{ri}\lambda_s + \lambda_{rs}\lambda_s + \lambda_{ri}\lambda_r + \lambda_{rs}\lambda_r + \mu\lambda_r$$

$$P_1(t) = \lambda_s\left[\left(e^{r_1 t} - e^{r_2 t}\right)/\left(r_1 - r_2\right)\right] \tag{13.47}$$

$$P_2(t) = \frac{\lambda_{ri}\lambda_s}{r_1 r_2}\left[1 + \left(r_1\, e^{r_2 t} - r_2\, e^{r_1 t}\right)/\left(r_2 - r_1\right)\right] \tag{13.48}$$

$$P_3(t) = \frac{\lambda_{rs}\lambda_s}{r_1 r_2}\left[1 + \left(r_1\, e^{r_2 t} - r_2\, e^{r_1 t}\right)/\left(r_2 - r_1\right)\right] \tag{13.49}$$

$$P_4(t) = \frac{\lambda_r}{A}\left(1 - e^{-At}\right) + \mu\,\lambda_s\lambda_r\left[\frac{1}{r_1 r_2 A} - \frac{e^{-At}}{A(r_1 + A)(r_2 + A)} + \frac{e^{r_1 t}}{r_1(r_1 + A)(r_1 - r_2)} + \frac{e^{r_2 t}}{r_2(r_2 + A)(r_2 - r_1)}\right] \tag{13.50}$$

The reliability of both robot and its associated safety system working normally with safety system repair facility from Equation (13.46) is

$$R_{rsr}(t) = e^{-At} + \mu\,\lambda_s\left[\frac{e^{-At}}{(r_1 + A)(r_2 + A)} + \frac{e^{r_1 t}}{(r_1 + A)(r_1 - r_2)} + \frac{e^{r_2 t}}{(r_2 + A)(r_2 - r_1)}\right] \tag{13.51}$$

The reliability of the robot operating normally with or without the safety system operating (but having the safety system repair facility) from Equations (13.46) and (13.47) is

$$R_{ssr}(t) = e^{-At} + \frac{\lambda_s\left(e^{r_1 t} - e^{r_2 t}\right)}{(r_1 - r_2)} + \mu\,\lambda_s\left[\frac{e^{-At}}{(r_1 + A)(r_2 + A)} + \frac{e^{r_1 t}}{(r_1 + A)(r_1 - r_2)} + \frac{e^{r_2 t}}{(r_2 + A)(r_2 - r_1)}\right] \tag{13.52}$$

The mean time to failure of the robot with repair and with safety system operating is expressed by

$$MTTF_{rsr} = \int_0^\infty R_{rsr}(t)\,dt = \frac{1}{A}\left(1+\mu\lambda_s/F\right). \tag{13.53}$$

Similarly, the mean time to failure of the robot with repair and with or without the safety system operating is given by

$$MTTF_{ssr} = \int_0^\infty R_{ssr}(t)\,dt = \frac{1}{A}\left(1+\mu\,\lambda_s/F\right)+\frac{\lambda_s}{A}. \tag{13.54}$$

Example 13.6

Assume that a robot system is composed of a robot and a safety system/unit and the operating robot with failed safety unit can either fail with an incident or safely. The failure rates of robot and safety system/unit are 0.0004 failure per hour and 0.0002 failure per hour, respectively. In addition, the failures rates of the robot failing with an incident or safely are 0.0001 failure per hour and 0.0003 failure per hour, respectively. Calculate the following and comment on the end results:

- The robot system mean time to failure when the safety system is operating. More specifically, use Equation (13.39).
- The robot system mean time to failure with repair. More specifically, use Equation (13.53).

Thus, in this example, we have $\lambda_r = 0.0004$ failure per hour, $\lambda_s = 0.0002$ failure per hour, $\lambda_{ri} = 0.0001$ failure per hour, $\lambda_{rs} = 0.0003$ failure per hour, and $\mu = 0.005$ repair per hour. Substituting the specified data into Equation (13.39) yields

$$MTTF_{rsu} = \frac{1}{A} = \frac{1}{\lambda_s+\lambda_r} = \frac{1}{(0.0002)+(0.0004)} = 1666.67\text{ h}$$

Inserting the given data into Equation (13.53) we get

$$MTTF_{rsr} = \left(1 + \mu\lambda_s/F\right)/A$$

$$= \left(1 + (0.0005)(0.0002)/F\right)/A$$

$$= 2410.7 \text{ h}$$

where

$$F = (0.0001)(0.0002) + (0.0003)(0.0002) + (0.0001)(0.0004) + (0.0003)(0.0004) + (0.005)(0.0004)$$
$$= 2.24 \times 10^{-6}$$
$$A = (0.0002) + (0.0004)$$
$$= 0.0006$$

The above results indicate that with the introduction of repair, the robot system mean time to failure improved from 1666.67 h to 2410.7 h.

13.4 PROBLEMS

1. Discuss the historical developments in robotics with emphasis on robot reliability and safety.
2. Discuss the basic hazards associated with robots.
3. State safety considerations in various phases of the robot life cycle.
4. Describe the following robot safeguard methods:
   - Intelligent systems
   - Infrared light arrays
   - Flashing lights
5. What are the important areas for human factors attention in robotic safety?
6. What are the main causes of robot failures?
7. Define the terms robot reliability and robot mean time to failure.
8. Assume that the times to failure of a robot are Weibull distributed. Obtain an expression for the robot reliability.
9. What is the important difference between the following two terms:
   - Mean time to robot failure
   - Mean time to robot-related problems
10. A robot system can fail either due to human error or other failures. The estimated values of human error and other failure rates are 0.0007 error per hour and 0.004 failure per hour. The repair rate of the failed robot system either due to human error or other failures is 0.01 repair per hour. Calculate the steady state probability of the robot system being in the failed state due to human error.

13.5 REFERENCES

1. Heer, E., Robots in modern industry, in *Recent Advances in Robotics*, Beni, G. and Hackwood, S., Eds., John Wiley & Sons, New York, 1985, pp. 11-36.
2. Dhillon, B.S., *Robot Reliability and Safety*, Springer-Verlag, New York, 1991.
3. Zeldman, M.I., *What Every Engineer Should Known About Robots*, Marcel Dekker, New York, 1984.
4. Worldwide Robotics Survey and Directory, Robot Institute of America, P.O. Box 1366, Dearborn, MI, 1983.
5. Rudall, B.H., Automation and robotics worldwide: reports and surveys, *Robotica*, 14, 243-251, 1996.
6. Parsad, H.P., Safety standards, in *Handbook of Industrial Robots*, Nof, S.Y., Ed., John Wiley & Sons, New York, 1988.
7. An Interpretation of the Technical Guidance on Safety Standards in the Use of Industrial Robots, Japanese Industrial Safety and Health Association, Tokyo, 1985.
8. Bonney, J.F. and Yong, J.F., Eds., *Robot Safety*, Springer-Verlag, New York, 1985.
9. Industrial Robots and Robot Systems: Safety Requirements, American National Standards Institute, New York, 1986.
10. Dhillon, B.S., On robot reliability and safety: Bibliography, *Microelectronics and Reliability*, 27, 105-118, 1987.
11. Dhillon, B.S., Robot accidents, *Proc. First Beijing Int. Conf. Reliability, Maintainability, and Safety*, International Academic Publishers, Beijing, 1992.
12. Altamuro, V.M., Working safely with the iron collar worker, *National Safety News*, 38-40, 1983.
13. Nicolaisen, P., Safety problems related to robots, *Robotics*, 3, 205-211, 1987.
14. Study on Accidents Involving Industrial Robots, prepared by the Japanese Ministry of Labour, Tokyo, 1982. NTIS Report No. PB 83239822, available from the National Technical Information Service (NTIS), Springfield, VA.
15. Request for Assistance in Preventing the Injury of Workers by Robots, prepared by the National Institute for Occupational Safety and Health, Cincinnati, OH, December 1984, NTIS Report No. PB 85236818, available from the National Technical Information Service (NTIS), Springfied, VA.
16. Ziskovsky, J.P., Working safety with industrial robots, *Plant Eng.*, May, 81-85, 1984.
17. Ziskovsky, J.P., Risk analysis and the R^3 factor, *Proc. Robots 8 Conf.*, 2, 15.9–15.21, 1984.
18. Bellino, J.P. and Meagher, J., Design for safeguarding, *Robots East Sem.*, Boston, MA, October 9-11, 1985, pp. 24-37.
19. Russell, J.W., Robot safety considerations: A checklist, *Professional Safety*, December 1983, pp. 36-37.
20. Nicolaisen, P., Ways of improving industrial safety for the programming of industrial robots, *Proc. 3rd Int. Conf. Human Factors in Manufacturing*, November 1986, pp. 263-276.
21. Jiang, B.C., *Robot Safety: User's Guidelines in Trends in Ergonomics/Human Factors III*, Karwowski, W., Ed., Elsevier, Amsterdam, 1986, pp. 1041-1049.
22. Addison, J.H., Robotic Safety Systems and Methods: Savannah River Site, Report No. DPST-84-907 (DE 35-008261), December 1984, issued by E.I. du Pont de Nemours & Co., Savannah River Laboratory, Aiken, SC.
23. Lembke, J.R., An Evaluation of Robot Perimeter Control Devices, Topical Report No. 705349, Report No. DBX-613-3031, Bendix (Kansas City Division), January 1984.

24. Zimmers, E.W., Human factors aspects of robotic safety, *Proc. Robotic Industries Assoc. (RIA) Robot Safety Sem.*, Chicago, April 24, 1986, pp. 1-8.
25. Engelberger, J.F., *Robotics in Practice*, Kogan Page, London, 1980.
26. Dhillon, B.S., On robot reliability and safety, *Microelectronics and Reliability*, 27, 105-118, 1987.
27. Dhillon, B.S., *Robot Reliability and Safety*, Springer-Verlag, New York, 1991.
28. Sato, K., Case study of maintenance of spot-welding robots, *Plant Maintenance*, 14, 28, 1982.
29. Sugimoto, N. and Kawaguchi, K., Fault tree analysis of hazards created by robots, *Proc. 13th Int. Symp. Industrial Robots*, 1983, pp. 9.13–9.28.
30. Khodabandehloo, K., Duggan, F., and Husband, T.F., Reliability of industrial robots: A safety viewpoint, *Proc. 7th British Robot Assoc. Annu. Conf.*, 233-242, 1984.
31. Khodabandehloo, K., Duggan, F., and Husband, T.F., Reliability assessment of industrial robots, *Proc. 14th Int. Symp. Industrial Robots*, 209-220, 1984.
32. Dhillon, B.S., *Reliability Engineering in Systems Design and Operation*, Van Nostrand Reinhold Company, New York, 1983.
33. MIL-HDBK-217F, Reliability Prediction of Electronic Equipment, U.S. Department of Defense, Washington, D.C.
34. Dhillon, B.S. and Yang, N., Reliability analysis of a repairable robot system, *J. Quality Maintenance Eng.*, 2, 30-37, 1996.

14 Medical Device Reliability

14.1 INTRODUCTION

Even though the history of Reliability Engineering goes back to World War II, the serious thinking of the applications of reliability engineering concepts to medical devices or equipment is not that old. Nowadays, there are encouraging signs that both the engineers and manufacturers in the medical field are breaking away from their conventional outlook. There could be various reasons for this new phenomenon; government requirements, cost effectiveness, public demand, advances made in other areas, etc. For example, because of the dependance of human life on the performance and reliability of the myriad of medical devices, in the U.S. the two basic important sources of regulation and control are the government (e.g., Food and Drug Administration (FDA)) and the legal system (e.g., product liability) [1]. Basically, both of these factors have played an instrumental role in driving the medical device industry to implement more and stringent controls on its products.

Therefore, today an engineer involved in the design of various medical products must address the issue of the degree of reliability desired as a function of the specific application. This is very crucial because a consequence of a failure may vary from one application area to another. For example, harm to the patient could result from a failure in a defibrillator; however, on the other hand, failure of a monitoring device may be tolerated without directly endangering the patient.

The latter part of the 1960s may be regarded as the real beginning of the medical device reliability field as this period witnessed the start of various publications on the subject [2–6]. In 1980 an article entitled "Bibliography of Literature on Medical Equipment Reliability" provided a comprehensive list of publications on the subject [7]. In 1983, a book on reliability engineering devoted a chapter to medical equipment reliability [8]. Since the early 1980s, many more publications on the field have appeared and this chapter discusses various aspects related to the medical device reliability.

14.2 FACTS AND FIGURES, GOVERNMENT CONTROL AND LIABILITY

The modern health care system has been in existence for over a century; for example, the American Hospital Association (AHA) was founded in 1896 and various types of medical devices have been designed and manufactured for a long time. With advances in modern technology, the sophistication of medical devices has also increased at a significant rate. For example, Greatbatch invented and patented the implantable cardiac pacemaker somewhere between 1958 and 1960 and the first successful human implant was performed in Buffalo in 1960 [9]. Today, total artificial heart and assist devices are under development and they are expected to play an important role in the improvement of health care throughout the world. Nonetheless, some of the facts and figures directly or indirectly associated with the medical devices are as follows:

- There are approximately 5000 institutional, 600 associate, and 40,000 personal members of the AHA. The institutional members include short and long term hospitals, headquarters of health care systems, hospital affiliated educational programs, etc. and the associate members are made up of organizations such as commercial firms, consultants, and suppliers. The individuals working in the health care field, health care executive assistants, full-time students studying hospital administration, etc. are classified under the personal member category.
- In 1974 dollars, the total assets of the AHA registered hospitals were in the order of $52 billion [10].
- Today, modern hospitals use over 5000 different medical devices ranging from tongue depressors to sophisticated devices such as pacemakers.
- Medical devices account for approximately $120 billion worldwide market [11] and in 1984, medical devices using electronics accounted for over $11 billion annual worldwide sales [12–14].
- In 1997, there were 10,420 registered medical device manufacturers in the U.S. [15].
- In the middle of the 1990s, 93% of medical devices had markets worth less than $150 million [16].
- The committee on hospitals (NFPA) stated 1200 deaths per year due to faulty instrumentation [17, 18].
- In 1990, an FDA study reported that approximately 44% of the quality related problems that resulted in voluntary medical device recall during the period from October 1983 to September 1989 were attributable to deficiencies/errors that could have been eradicated by effective design controls [19].
- During the early half of the 1990s, over a period of five years, the top 10 orthopedic companies in the U.S. increased their regulatory affairs staff by 39% [20].
- Operator errors account for well over 50% of all technical medical equipment problems [21].
- According to the findings of the Philadelphia Emergency Care Research Institute, from 4 to 6% of hospital products were sufficiently dangerous to warrant immediate correction. These findings were based on testing of a sample of 15,000 different hospital products [21].
- In 1990, the U.S. Congress passed the Safe Medical Device Act (SMDA) that empowered the FDA to implement the Preproduction Quality Assurance program. This program requires/encourages medical device manufacturers to address deficiencies in product design contributing to failures.

As medical devices can range from a complexity of a modern magnetic resonance imaging (MRI) scanner to the simplicity of dental floss, many countries around the world regulate them and impose restrictions on their manufacture and import including quality and reliability. As the competition is a good policeman of economy, the market imposes significant expectations of performance and reliability of the medical devices. In fact, in many times such expectations surpass the regulatory requirements quite significantly.

In the U.S., the FDA regulates medical devices to ensure their safety and effectiveness [22]. For example, two years after the enactation of the Medical Device Amendments of 1976, the FDA issued the medical device good manufacturing practices (GMPs) regulation. This document included requirements prescribing the facilities, techniques/methods, and controls to be employed in the medical device manufacture, packaging, and storage [23]. However, the U.S. Congress Subcommittee on Oversight and Investigations held hearings on March 13, 1984 concerning medical device failures, particularly deaths resulting from cardiac pacemaker leads that malfunctioned due to design defects [19]. It concluded that the FDA was unable to address such problems under the 1976 amendments.

Nowadays, the FDA requires reports concerning incidents of death, serious injury or serious illness, or failure of a certain device likely to cause or contribute to a death or serious injury if the failure were to recur [24]. Manufacturers, distributors, and user facilities of medical devices have certain reporting requirements. Noncompliance to such requirements can lead to FDA regulatory actions including warning letters, product seizure, injunction, or civil penalties. Also, it is mandatory that any organization (including non-U.S. manufacturers) that manufactures medical devices for the U.S. market must establish a quality system that is in line with the GMPs [22].

A different strategy to regulate medical devices is followed by the European Union (EU). The Medical Device Directive (MDD) of the European Parliament defines the requirements for medical devices in the EU countries. After May 1998, a firm (including non-EU companies) manufacturing medical devices to market in the EU countries must establish a quality system in conformance to the MDD. The MDD outlines a set of requirements relating to the medical device safety covering a broad range, including electromagnetic compatibility, biological compatibility, and electrical safety [22].

Currently, product liability is another factor for manufacturing safe and reliable medical devices. For example on June 26, 1996, the U.S. Supreme Court's ruling on the *Medtronic, Inc. vs. Lohr* case handed out a sweeping victory to American consumers by rejecting the Medtronic's claim that federal law provides immunity to medical device manufacturers from all civil liability, including when individuals are injured or killed due to defective products [25]. This case was filed by a Florida woman named Lora Lohr who was implanted with a cardiac pacemaker to rectify her abnormal heart rhythm. The pacemaker manufactured by Medtronic failed and resulted in requiring emergency surgery plus additional medical procedures. The lawsuit filed by Lohr alleged that the failure of the pacemaker was the result of defective design, manufacturing, and labeling.

In delivering the judgment on the case, Justice John Paul Stevens wrote "Medtronic argument is not only unpersuasive, it is implausible" [25]. The Supreme Court's decision has put additional pressure on manufacturers to produce safe and reliable medical devices.

14.3 MEDICAL ELECTRONIC EQUIPMENT CLASSIFICATION

There is a large variety of electronic equipment used in the health care system. Basically, such equipment may be grouped under three broad categories [3, 21].

- Category A
- Category B
- Category C

Category A includes equipment requiring high reliability because it is directly and immediately responsible for the patient's life or which may become so in emergencies. When such equipment fails, there is seldom much time for even emergency repairs; thus, it must always function at the moment of need. Some examples of such equipment are respirators, cardiac defibrillator, electro-cardiographic monitors, and cardiac pacemakers.

A vast majority of medical electronic equipment fall under Category B. Such equipment is used for routine or semi-emergency diagnostic or therapeutic purposes. Failure of this equipment does not create the same emergency as in the case of Category A equipment; in addition, there is time for repair. Some examples of this category of equipment are electronyography equipment, electrocardiograph and electroencephalograph recorders and monitors, ultrasound equipment, gas analyzers, colorimeters, diathermy equipment, and spectrophotometers.

Equipment belonging to Category C is not essential to a patient's life or welfare but serves as a convenience equipment. Some of the examples are wheel chairs, electric beds, and bedside television sets. One should note here that there is considerable overlap, especially in Categories A and B. More specifically, say some equipment of Category B may be used as Category A equipment. One example of such equipment is electrocardiograph recorder or monitor.

14.4 MEDICAL DEVICE RECALLS

In the U.S., defective medical devices are subject to recall and repair. The passage of the Medical Device Amendments of 1976 has helped to increase the public's awareness of these activities and their corresponding number. The FDA Enforcement Report publishes data on medical device recalls. For example, during the period of 1980–1982, there were a total of 230 medical device related recalls [26]. These recalls were classified into nine problem areas as follows [21]:

- Faulty product design
- Defective components

- Defects in material selection and manufacturing
- Missassembly of parts
- Electrical problems
- Contamination
- Mislabeling
- Radiation (X-ray) violations
- No premarket approval and failure to comply with GMPs.

The number of recalls associated with each of these categories was 40, 13, 54, 10, 17, 40, 27, 25, and 4, respectively. It is interesting to note that faulty product design, product contamination, mislabeling, and defects in material selection and manufacturing accounted for 70% of the medical device recalls. The components of the faulty product design category were premature failure, alarm defects, potential for malfunction, electrical interference, failure to perform as required, and potential for leakage of fluids into electrical components. Four subcategories of the product contamination were defective package seals, non-sterility, other package defects, and other faults. The problem areas of the mislabeling category were incomplete labeling, inadequate labeling, disparity between label and product, and misleading or inaccurate labeling. There were five specific problems in the material selection and manufacturing category: inappropriate materials, manufacturing defects, separation of bonded components, actual or potential breakage/cracking, and material deterioration.

14.5 MEDICAL DEVICE DESIGN QUALITY ASSURANCE

Just like in the life cycle of other engineering devices, the design phase is the most important phase in the life cycle of a medical device. The inherent safety and effectiveness of the medical devices can only be improved through design enhancement. Careful planning and management of the design process can help in establishing effectiveness, reliability, and safety in a device.

It is emphasized that a deficiency in a device design could be very costly once the device is manufactured and released for field use. The costs may not only be made up of replacement and redesign but also of liability and loss of customer faith in the device. Nonetheless, according to FDA studies, one of the major causes of device failure is defective design. Thus, the design quality assurance is very important to produce reliable and safe medical devices.

In 1989, for the first time the FDA published a notice of availability for design control recommendations entitled "Preproduction Quality Assurance Planning: Recommendations for Medical Device Manufacturers" [23]. It encouraged all medical device manufacturers to follow these recommendations by establishing a program for assessing the reliability, safety, and effectiveness of device design before releasing the design to manufacturing. The FDA preproduction or design quality assurance program is composed of 12 elements:

1. organization
2. specifications

3. design review
4. reliability assessment
5. parts/materials quality control
6. software quality control
7. design transfer
8. labeling
9. certification
10. test instrumentation
11. manpower
12. quality monitoring subsequent to the design phase.

Each of these elements is described below.

14.5.1 Organization

The formal documentation should contain the organizational elements and authorities desirable to develop the preproduction quality assurance (PQA) program, to execute program requirements, and to achieve program goals. Also, it is essential to formally assign and document the responsibility for implementing the entire program and each program element. Audit the PQA program periodically and update it as experience is gained.

14.5.2 Specifications

After establishing desired device design characteristics such as physical, chemical, performance, etc., translate them into written design specifications. Specifications should address performance characteristics, as applicable, such as reliability, safety, stability, and precision, important to meet the device's end objective. It is important to remember that these preliminary specifications provide the vehicle with which to develop, control, and evaluate the device design.

Evaluate and document changes made to the specifications during research and development, in particular, ensure that the product's safety or effectiveness is not compromised by these changes. All in all, for the sake of effectiveness of specifications, the qualified individuals from areas such as reliability, quality assurance, manufacturing, and marketing should be asked to review the specification document.

14.5.3 Design Review

The purpose of design review is to detect and correct deficiencies during the design phase of a medical device. As the cornerstone of the PQA program, the design reviews should be conducted at various stages of device design to assure conformance to design criteria and to detect weak design spots. The sooner a design review is initiated, the better, as the design weaknesses can be detected earlier and in turn it will cost less to rectify the problem. The extent and frequency of design reviews depend on the significance and complexity of the device under consideration. The items to be assessed during these reviews include subsystems (including software, if applicable), components, drawings, test results and specifications, packaging, analysis results, and labeling.

Usually, the members of the design review team belong to areas such as engineering, quality assurance, manufacturing, purchasing, servicing, and marketing.

In particular, the design reviews should include, as applicable, failure mode and effect analysis. In addition, such analysis should be performed at the initial stage of the design effort and as part of each design review to detect design flaws, especially with respect to device safety, reliability, and performance. Usually, the following two failure mode analysis techniques are employed:

- Failure Mode Effects and Criticality Analysis (FMECA)
- Fault Tree Analysis (FTA)

FMECA is an inductive "bottom-up" approach and assumes the existence of basic defects at the component level and then evaluates the effects on assembly levels. Furthermore, it is a structured approach for proceeding part-by-part through the system to determine effects of failures and is described in detail in Reference 26.

FTA is another approach to failure mode analysis, but it is a deductive and "top-down" method. The approach is especially useful for application to medical devices because human/device interfaces can be taken into consideration. The approach is described in detail in the chapter devoted to the method.

14.5.4 Reliability Assessment

This is the prediction and demonstration process for estimating the basic reliability of a device, and it can be used for both new and modified designs. More specifically, reliability assessment may be described as a continuous process involving reliability prediction, demonstration, data analysis, then reliability re-prediction, re-demonstration, and re-data analysis continually.

Reliability assessment can be started by theoretical/statistical approaches by first determining each part's reliability, then progressing upward, establishing each subassembly and assembly reliability, until the establishment of the entire device's reliability. Even though a good estimate of reliability can be made by testing complete devices under simulated operating environments, the most meaningful reliability data can only be obtained from field use. By making the reliability assessment an essential component of the PQA program, it will certainly help to produce reliable medical devices.

14.5.5 Parts/Materials Quality Control

In order to assure that all parts/materials (P/M) used in device designs have the desirable reliability, it is essential to have an effective P/M quality assurance program. The essential elements of the program are P/M selection, specification, qualification, and ongoing quality verification, i.e., irrespective of whether P/M was fabricated in-house or supplied by vendors.

It is useful to develop a preferred P/M list during the preliminary design stage and refine it as the design progresses.

14.5.6 Software Quality Control

The establishment of a software quality assurance program (SQAP) is essential when a medical device design incorporates software developed in-house. The basic purpose of this program is to outline a systematic approach to software development, particularly including a protocol for the device software formal review and validation to meet overall functional reliability. Nonetheless, the primary goals of the software quality assurance are reliability, maintainability, correctness, and testability. Reference 27 is a useful document for writing software quality assurance plans.

14.5.7 Design Transfer

Once the medical device design is transformed into a physical entity, its technical adequacy, reliability, and safety verifications are sought through testing under the real life operating or simulated use conditions. It is to be noted that in going from laboratory to scaled-up production, it is quite possible that standards or techniques and procedures may not be accurately transferred or additional manufacturing processes are added. Thus, this factor requires careful attention. All in all, the clinical trials of a device under consideration can only begin after the verification of its safety and reliability under the simulated operating environments, specifically at the expected performance limits.

14.5.8 Labeling

Labeling includes items such as display labels, manuals, inserts, software for cathode ray tube (CRT) displays, panels, charts, and recommended test and calibration protocols. Design reviews are a useful vehicle to ensure that the labeling is in compliance with laws and regulations as applicable and the directions for the device usage can be understood easily by the user community.

14.5.9 Certification

After passing the preproduction qualification testing by the initial production units, it is essential to perform a formal technical review to assure adequacy of the design, production, and quality assurance approaches/procedures. In addition, the review should incorporate a determination of items such as those listed below.

- Overall adequacy of quality assurance plan.
- Resolution of any discrepancy between the standards/procedures employed to construct the design during the research and development phase and those identified for the production phase.
- Adequacy of specifications.
- Suitability of test approaches employed for evaluating compliance with the approved specifications.
- Adequacy of specification change control program.
- Resolution of any discrepancy between the approved specifications for the device and the end produced device.

14.5.10 Test Instrumentation

This is concerned with effectively calibrating and maintaining, as per the formal calibration and maintenance program, the equipment used in qualification of the design.

14.5.11 Manpower

Only qualified and competent individuals must be allowed to perform device associated design activities such as design review, analysis, and testing.

14.5.12 Quality Monitoring Subsequent to the Design Phase

The quality assurance effort does not end at the design phase of a medical device, but it continues in manufacturing and field use phases. In order to meet this challenge, it is expected that manufacturers of medical devices establish an effective program for activities such as follows:

- Identifying trends or patterns associated with device failures.
- Performing analysis of quality related problems.
- Initiating appropriate corrective measures to stop recurrence of the above failures or problems.
- Reporting in a timely manner the problems found either in-house or in the field use.

The device manufacturers should also establish an appropriate mechanism to assure that failure data resulting from user complaints and service records relating to design are reviewed by the individuals involved with the design of the device.

14.6 HUMAN ERROR OCCURRENCE AND RELATED HUMAN FACTORS

Human errors are universal and committed each day; some are trivial and others are serious or fatal. In the medical field, human errors can result in serious injuries and even deaths [28, 29]. Some examples and facts concerning human error occurrence in the health care system are as follows [29, 30]:

- Up to 90% of accidents both generally and in medical devices are due to human error [31, 32].
- A human error caused a fatal radiation overdose accident involving the Therac radiation therapy device [33].
- A study of anesthesiology errors revealed that 24% of the anesthesiologists responded positively to committing an error with fatal results [34].
- A patient was seriously injured by over-infusion because a nurse mistakenly read the number 7 as a 1 [30].

- During the treatment of an infant patient with oxygen, a physician set the flow control knob between 1 and 2 liters per minute without realizing the fact that the scale numbers represented discrete, instead of continuous, settings. As the result of the physician's action, the patient became hypoxic [30].

14.6.1 Control/Display Related Human Factors Guidelines

As many medical devices usually contain large consoles with rows of controls and displays, they could be the important cause for many human errors if designed without seriously considering human factors. The guidelines such as those listed below have proven quite useful for displays and controls [30]:

- Ensure that controls, displays, and workstations are designed considering user capabilities such as reach, memory, strength, hearing, vision, and dexterity.
- Ensure that the design of switches and control knobs correspond to the user conventions outlined in current medical device standards and so on.
- Ensure that all design facets are consistent with user expectations as much as possible.
- Ensure that all displays and labels are made so that they can be read with ease from normal distances and viewing angles. In particular, pay attention to factors such as color, symbol size, display depth, and controls.
- Ensure that the design of control and display arrangements is well organized and uncluttered.
- Ensure that tactile feedback is provided by controls.
- Ensure that knobs, switches, and keys are arranged and designed so that the likelihood of inadvertent activation is reduced to a minimum.
- Ensure that color and shape coding, as appropriate, are used to facilitate the efficient identification of controls and displays.
- Ensure that the intensity and pitch of auditory signals are such that they can be easily heard by the device users.
- Ensure that visual signal's brightness is such that it can be easily perceived by users performing their tasks under various ambient illumination conditions.
- Ensure that items such as acronyms, abbreviations, text, and symbols placed on or displayed by the device are consistent with the instructional manual.

14.6.2 Medical Device Maintainability Related Human Factor Problems

As poor maintenance can affect safe and reliable operation of medical devices, proper emphasis is required during design to simplify their maintenance. Some of the problems frequently encountered by the maintainers include poor self-diagnostic capability, ambiguous part arrangements, poor design for easy cleaning, difficulty in reaching or manipulating screws/parts, inadequate labeling, coding, or numbering of parts, difficulty in locating parts visually or by touch, and requirements for difficult-to-find tools [30].

14.6.3 Human Factor Pointers for Already Being Used/to be Purchased Medical Devices

The recognition of human factor design related problems with already being used or to be purchased medical devices is important for their safe, reliable, and effective operation. When the design of a medical device has serious deficiencies with respect to human factors, patterns may emerge. Experienced users, maintenance personnel, and others can be a valuable source of information. Specifically, the symptoms of problems with already being used devices may be categorized in four groups [30]:

1. Complaints
2. Observations
3. Installation Problems
4. Incidents

Some examples of the problems under the complaint group are poorly located or labeled controls, illogical and confusing device operation, difficult to read or understand displays, annoying device alarms, and difficult to hear or distinguish alarms. The items belonging to the observation group include slow and arduous training, refusal of the staff to use the device, and only a few individuals seem to be able to use the device correctly. Some of the installation related problems are accessories installed incorrectly, frequently parts become detached, involved individuals find installation of accessories difficult, confusing, or overly time-consuming, and alarms and batteries fail often. One important element of the incident category is that highly competent people are involved in accidents or "near-misses". More specifically, this could be a warning.

It is important to consider the means of assessing usability of medical devices prior to their procurement, particularly if they are to be used for life-sustaining or life-supporting purposes. Some of the associated steps include the following:

- Determine if the manufacturer has performed human factors/usability testing of the device under consideration.
- Review published evaluation data concerning the device under consideration.
- Determine from staff and other sources about the human factor related performance concerning predecessor device models produced by the same manufacturer.
- Negotiate a trial period prior to the actual procurement of the device under consideration.
- Make contact with facilities that have already used the device under consideration.

14.6.4 Issues in Considering the Need for, and Performance of, Human Factors Analysis and Testing of Medical Devices

There are many issues that could be raised in considering the need for, and performance of, human factors analysis and testing of medical devices. Some of these issues are consequences of error for the patient/user, fulfillment of user requirements, requirement of user interaction with the device with respect to operation, maintenance, cleaning, or parts installation, reviewing of the literature and company files for useful human factors information, likelihood of the combination of user interface, user population, and operating conditions leading to errors, types of studies, analyses, and tests being conducted, testing in simulated and/or actual environments by the project team, user involvement in actual testing, individuals belonging to the design team focusing on the user-related issues during actual testing, existence of a test plan, development of user requirements and their update, and coordination of hardware and software designers, technical writers, and the efforts of others with respect to human factors [30].

14.7 MEDICAL DEVICE SOFTWARE

Nowadays many medical devices use software and its reliability is as important as the hardware reliability of such devices. For example, as per References 35 and 36, "software failures have lead to patient fatalities, and the potential for more such hazards is increasing as software becomes complex and increasingly crucial in the manufacture of medical devices". There are primarily two reasons for medical device software failures: inadequacies in the functionality of the software and errors in its implementation [37]. Before we examine the subject of software quality assurance/reliability in detail, let us first explore the definition of a medical device because the FDA medical device statutory and regulatory provisions application is subject to items falling under the Federal Food, Drug, and Cosmetic Act (FD&C Act) definition.

The definition of a medical device under this Act basically is any instrument/apparatus/machine/etc. or other similar or related article, including any part, component, or accessory intended for application in the diagnosis of disease/prevention of disease/etc. or intended to affect any function or the structure of body [38, 39]. Any software satisfying this definition is subject to concerned FDA rules and regulations. FDA classifies software into two categories: stand-alone software (i.e., it can be a device itself) and software that is a part, component, element, or accessory to a device. Some examples of the stand-alone software are as follows [39]:

- Blood bank software systems controlling donor deferrals and the release of blood products.
- Software that helps diagnose osteoporosis.

- Software that analyzes potential therapeutic interventions for a certain patient.
- Information systems used in hospitals.

The typical examples of component or part software are the software used in pacemakers, diagnostic X-ray systems, ventilators, infusion pumps, magnetic resonance imaging devices, etc. Similarly, examples of the accessory software are the softwares used for the conversion of pacemaker telemetry data, computation of bone fracture risk from bone densitometry data, and calculation of rate response for a cardiac pacemaker.

14.7.1 Software Testing for Improving Medical Device Safety

Medical device safety with respect to software failures could be significantly improved through software testing. The basic purpose of software testing is to ensure that a device performs its assigned task and its potential for hazard is minimized. The testing accounts for a very large proportion of the total software development process. Nonetheless, software testing may be grouped into two areas: manual software testing and automated software testing. Each of these areas is described below [36].

Manual Software Testing

There are many traditional approaches to manual software testing including functional testing, safety testing, free-form testing, white box testing, and error testing as shown in Figure 14.1.

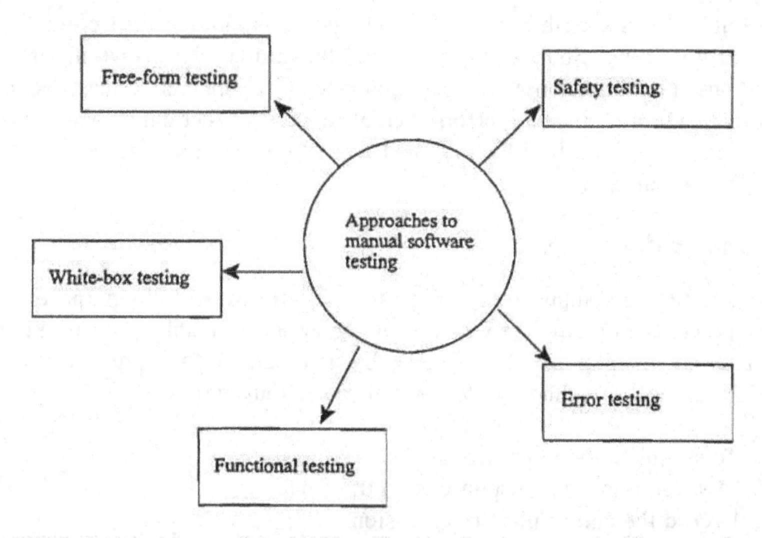

FIGURE 14.1 Some of the traditional approaches to manual software testing.

Functional testing is the heart of traditional validation and verification procedures and focuses test efforts on the performance expected from the device or, more specifically, the device must perform as per its design specification. This type of testing usually receives the most attention and the greatest test cases because each function must undergo exhaustive testing. Even though functional testing is very time-consuming, it is an orderly process. However, it is extremely boring, and boredom frequently leads to human error in the tests and omissions in the documentation, irrespective of the testers' effort in accuracy.

The purpose of the safety testing is that a device must operate safely. Thus, this type of testing is focused on areas of potential harm to people such as patients or users. Safety testing is useful for verifying that software-associated hazards have been mitigated [40].

As no formal testing can find each and every important software bug, it is advisable to conduct free-form testing on top of the formal approaches such as described earlier. Free-form testing is not performed according to device specifications; thus, it defies formalization and subjects the device to additional stresses. This form of testing may simply be described as a process in which testers with varying knowledge levels operate the device in unexpected and unconventional ways to provoke failures. Even though free-form testing can discover a large percentage of the errors or bugs, there is no guarantee that it will cover all potential problem areas.

The white-box testing approaches are employed in situations where the tester is required to verify the internal operations of the device software in order to test it satisfactorily. This type of testing allows the tester to look into the interior of the device and develop tests to discover weak spots in the software program's internal logic. The white-box testing techniques can be incorporated into other manual test approaches such as functional testing, free-form testing, and safety testing.

As a device is expected to handle error conditions effectively, the error testing is a useful tool to assure that the device under consideration operates correctly when things do go wrong. Some examples of the things that can go wrong are power fluctuations or outages, nonsense values generated by peripheral devices, and internal components failures. Tests for errors such as these are rather cumbersome to create and execute because of the difficulty involved in recreating the errors in a systematic and repeatable manner.

Automated Software Testing

This makes use of computer technology to test software with the purpose to make the test process more efficient, accurate, and complete, in addition to backing it up with better documentation. More specifically, automated software test systems make use of the computer technology for the following four purposes:

1. To stimulate the test target.
2. Monitor associated response from the device.
3. Record the end results or conclusions.
4. Control the total process.

The approaches such as functional testing, error testing, free-form testing, safety testing, and white-box testing can be improved through automation.

14.7.2 Cardioplegia Delivery System Software Reliability Program Core Elements and Software Safety Improvement with Redundancy

The cardioplegia delivery system is used to provide for multiple combinations of blood, arresting agent, crystalloids, and additive solutions under a perfusionist's programmable control. During the development of the operating software for a cardioplegia delivery system, the core elements of the software reliability program that were implemented included design simplification, documented performance requirements, code inspections, periodic audits and audit trails, safety hazard analysis, design and coding standards, requirement tracing, a clearly defined user interface, quality practices and standards, reliability modeling to predict operational reliability, extensive testing beginning at the module level, and documentation and resolution of all defects [41].

It is considered that knowledge about these core elements will provide a useful input in developing software for other medical devices.

The past experience indicates that errors in software will still occur even with the best of development methodologies and intentions. However, it is possible to improve software safety quite significantly by designing with redundant safety mechanisms [37]. The process starts by conducting hazard analysis [42]. After the identification of important hazards and their associated severity and potential causes, it is feasible to devise electromechanical systems for detecting such conditions and overriding the device's processor. If such systems can be developed orthogonal to the software (i.e., the unique point of common functionality is in the discovery of the error), then it is possible to have true redundancy and improvement in safe response to hazardous situations.

14.8 SOURCES FOR ADVERSE MEDICAL DEVICE REPORTABLE EVENTS AND FAILURE INVESTIGATION DOCUMENTATION

In order to ensure the safety of patients and others, directly or indirectly, the medical device reporting (MDR) rules and regulations in both the U.S. and EU countries state that adverse events concerning medical devices must be tracked and reported in an effective manner. The FDA requires medical device manufacturer/user facilities such as hospitals, nursing homes, distributors/importers, and ambulatory surgical facilities to report MDR-type incidents effectively. Failure to comply with its requirements may lead to product seizure, injunction, warning letters, etc.

In particular, it is necessary that the manufacturers of medical devices must examine all possible sources for MDR-type reportable incidents. Some of those sources are as follows [43]:

- Servicing/repair reports
- User/distributor records
- Published/unpublished literature sources
- The manufacturer's product complaint-handling mechanism
- In-house research/testing evaluation/etc. records
- Legal records
- Sales representative contacts
- Technical service customer contacts.

The documentation of an investigation of a device, process, or quality system failure is very important and requires careful consideration. Thus, the main objective of a medical device failure investigation documentation should be to provide all necessary information about the problem as well as to serve as an information repository that can be utilized meaningfully by other individuals including external auditors and assessors. Nonetheless, an effective documentation of a failure investigation should include, at a minimum, proper information on items such as unique identification of the failure investigation, problem description, level of investigation, investigation method/procedures/tests/inspections, equipment used in the investigation and their unique idenfication, reviewed documentation/components/parts/labeling/packaging/processes descriptions, the results of all inspections/testing examinations, analysis methods and analysis of the results, cross-references to other failure investigations/complaints/etc., investigators' signature and dates of test and examinations performed, actual results of the testing and examinations, report of the conclusions to demonstrate adequate closure, recommendations for corrective actions to prevent recurrence of the problem, and formal acceptance and closure of the failure investigation [44].

14.9 A SMALL INSTRUMENT MANUFACTURER'S APPROACH TO PRODUCE RELIABLE AND SAFE MEDICAL DEVICES

Over the years many companies have developed their own approaches to manufacture safe and reliable medical devices. One such approach was developed by Bio-Optronics. The Bio-Optronics's product development plan is composed of 13 steps listed below [45].

1. Analyze existing medical problems.
2. Develop a product concept to find a solution to a specific medical problem.
3. Evaluate environments under which the medical device is operating.
4. Evaluate the people expected to operate the product under consideration.
5. Build a prototype.
6. Test the prototype under laboratory environment.
7. Test the prototype under the field operating environment.
8. Make changes to the device design as appropriate to satisfy the actual field requirements.

9. Perform laboratory and field tests on the modified device.
10. Build pilot units to conduct appropriate tests.
11. Invite impartial experts to test pilot units under the field environment.
12. Release the product/device design for production.
13. Study the field performance of the device and support with device maintenance.

14.10 AEROSPACE AND MEDICAL EQUIPMENT RELIABILITY AND RELIABILITY APPROACH COMPARISONS

The development of the reliability discipline was basically associated with the manufacture of aerospace equipment. In the manufacture of medical devices, the application of the reliability concepts is relatively new. Nonetheless, comparisons of a number of factors associated with aerospace and medical product reliability are presented below [8, 46].

- **Aerospace Product Reliability**
  - Costly equipment.
  - A well-developed field for reliability.
  - Human well being directly or indirectly is a factor.
  - Large manufacturing concerns with well-developed reliability engineering departments with many years of experience.
  - Individuals with previous reliability related experience with aerospace equipment use sophisticated reliability methods.

- **Medical Product Reliability**
  - Relatively less expensive products.
  - A relatively new area for reliability concept applications.
  - Patients' well being or lives are involved.
  - Relatively smaller manufacturing concerns with relatively recent interest with the application of reliability principles in product development.
  - Individuals with less previous reliability related experience with medical devices use relatively simple reliability approaches.

The driving force in the aerospace industry to consider and analyze reliability by imposing a complex set of standards and specifications that control all aspects of equipment design has been the U.S. Department of Defense. In contrast, the medical device industry driven by factors such as market forces, limited government regulation, and safety considerations, has been lagging in imposing reliability requirements in the device design and development process. A rough comparison of the approaches followed and controls imposed by both aerospace and medical device industries with respect to some important factors affecting the product reliability is given in Table 14.1 [47].

TABLE 14.1
Aerospace and Medical Industries Approach to Selected Reliability Factors

No.	Reliability factor	Aerospace industry's approach	Medical industry's approach
1	Stress screening	MIL-HDBK-2164 [48] is used for control.	None.
2	Part quality	MIL-STD-883 [49] is used for control.	Left to individual firm.
3	Predictions	MIL-HDBK-217 [50] is used for control.	Bell Communication Research (BELLCORE) [51] is used for control.
4	Design margins	MIL-HDBK-251 [52] is used for control.	Safety standards produced by bodies such as International Electro-Technical Commission (IEC) and American Society of Mechanical Engineers (ASME) are used for control.
5	Complexity	MIL-HDBK-338 [53] is used for control.	Left to individual firm.
6	Testing and verification	MIL-STD-781 [54] and MIL-HDBK-781 [55] are used for control.	None for reliability.

14.11 GUIDELINES FOR RELIABILITY AND OTHER PROFESSIONALS ASSOCIATED WITH HEALTH CARE

There are various types of professionals involved in the manufacture and use of medical devices. Reliability engineers are one of them. References 8 and 46 present some suggestions for all of these professionals. Thus, suggestions for reliability and other professionals associated with health care are given below.

- **Reliability Professionals**
  - Focus your attention on critical device failures as not all device failures are equally important.
  - Pay proper attention to cost effectiveness and remember that some reliability improvement decisions need very little or no additional expenditure.
  - In order to obtain immediate results employ approaches such as design review, parts review, qualitative FTA, and FMEA.
  - Use simple reliability techniques instead of some sophisticated methods used by the aerospace industry. Examples of the techniques that can be quite useful in medical device industry are FMEA and qualitative FTA.
  - Remember that the device reliability responsibility rests with the manufacturer during the design and manufacturing phase, but in the operational phase it is basically the responsibility of the user.

- **Other Professionals**
  - Keep in the back of your mind that probably the largest single expense in a business organization is the cost of failures. Such failures could be associated with people, equipment, business systems, and so on. The cost of business can be decreased quite significantly by reducing such failures.
  - The manufacturer is responsible for device reliability during the design and manufacturing phases, but during field use the user must use the device according to design conditions. More specifically, both parties are expected to accept their responsibilities for the total success.
  - Make comparison of the human body and medical device failures. Both of them need positive actions from reliability engineers and doctors to improve device reliability and extend human life, respectively.
  - Recognize that failures are the cause of poor medical device reliability and positive thinking and actions can improve device reliability.
  - Remember that the applications of reliability principles have successfully improved the reliability of aerospace systems and such applications to medical devices will generate dividends of a similar nature.

14.12 PROBLEMS

1. Make a comparison of medical device reliability with aerospace system reliability.
2. Write an essay on medical device failures.
3. Discuss at least five typical important reasons for the medical device recalls in the U.S.
4. State measures proposed by the FDA to produce reliable devices.
5. Give at least five examples of human error in the health care system.
6. Discuss human factors with respect to medical devices.
7. What are the two software classifications used by the FDA? Describe both of them in detail.
8. Discuss the following types of manual software testing:
   - Functional testing
   - Safety testing
   - Error testing
9. Compare manual software testing with automated software testing.
10. Discuss the possible sources of MDR-type reportable incidents or events.

14.13 REFERENCES

1. Bell, D.D., Contrasting the medical-device and aerospace-industries approach to reliability, *Proc. Annu. Reliability Maintainability Symp.*, 125–127, 1995.
2. Johnson, J.P., Reliability of ECG instrumentation in a hospital, *Proc. Annu. Symp. Reliability*, 314-318, 1967.
3. Crump, J.F., Safety and reliability in medical electronics, *Proc. Annu. Symp. Reliability*, 320-322, 1969.

4. Gechman, R., Tiny flaws in medical design can kill, *Hosp. Top.*, 46, 23-24, 1968.
5. Meyer, J.L., Some instrument induced errors in the electrocardiogram, *J. Am. Med. Assoc.*, 201, 351-358, 1967.
6. Taylor, E.F., The effect of medical test instrument reliability on patient risks, *Proc. Annu. Symp. Reliability*, 328-330, 1969.
7. Dhillon, B.S., Bibliography of literature on medical equipment reliability, *Microelectronics and Reliability*, 20, 737-742, 1980.
8. Dhillon, B.S., *Reliability Engineering in Systems Design and Operation*, Van Nostrand Reinhold Company, New York, 1983.
9. Eberhard, D.P., Qualification of high reliability medical grade batteries, *Proc. Annu. Reliability Maintainability Symp.*, 356–362, 1989.
10. Fairhurst, G.F. and Murphy, K.L., Help wanted, *Proc. Annu. Reliability Maintainability Symp.*, 103-106, 1976.
11. Murray, K., Canada's medical device industry faces cost pressures, regulatory reform, *Medical Device and Diagnostic Industry Magazine*, 19(8), 30-39, 1997.
12. Bassen, H., Sillberberg, J., Houston, F., and Knight, W., Computerized medical devices: Trends, problems, and safety, *IEEE Aerospace and Electronic Systems (AES) Magazine*, September 1986, pp. 20-24.
13. Dickson, C., World medical electronics market: An overview, *Medical Devices and Diagnostic Products*, May 1984, pp. 53-58.
14. *World Medical Electronics Markets 1985 Yearbook*, Market Intelligence Research Company, Palo Alto, CA, March 1985.
15. Allen, D., California home to almost one-fifth of U.S. medical device industry, *Medical Device and Diagnostic Industry Magazine*, 19(10), 64-67, 1997.
16. Bethune, J., The cost effectiveness bugaboo, *Medical Device and Diagnostic Industry Magazine*, 19(4), 12-15, 1997.
17. Walter, C.W., *Electronic News*, January 27, 1969.
18. Micco, L.A., Motivation for the biomedical instrument manufacturer, *Proc. Annu. Reliability Maintainability Symp.*, 242-244, 1972.
19. Schwartz, A.P., A call for real added value, *Medical Industry Executive*, February/March, 1994, pp. 5-9.
20. Allen, R.C., FDA and the cost of health care, *Medical Device and Diagnostic Industry Magazine*, 18(7), 28-35, 1996.
21. Dhillon, B.S., Reliability technology in health care systems, *Proc. IASTED Int. Symp. Comp. Adv. Technol. Med. Health Care Bioeng.*, 84-87, 1990.
22. O'Leary, D.J., International standards: Their new role in a global economy, *Proc. Annu. Reliability Maintainability Symp.*, 17-23, 1996.
23. Hooten, W.F., A brief history of FDA good manufacturing practices, *Medical Device and Diagnostic Industry Magazine*, 18(5), 96, 1996.
24. Cady, W.W. and Iampietro, D., Medical Device Reporting, *Medical Device and Diagnostic Industry Magazine*, 18(5), 58-67, 1996.
25. Bethune, J., Ed., On product liability: Stupidity and waste abounding, *Medical Device and Diagnostic Industry Magazine*, 18(8), 8-11, 1996.
26. MIL-STD-1629A, Procedures for Performing a Failure Mode, Effects, and Criticality Analysis, Department of Defense, Washington, D.C.
27. ANSI/IEEE-STD-730-1984, IEEE Standard for Software Quality Assurance Plans, American National Standards Institute (ANSI), New York, 1984.
28. Bogner, M.S., Ed., *Human Error in Medicine*, Lawrence Erlbaum Associates, Hillsdale, NJ, 1994.

29. Maddox, M.E., Designing medical devices to minimize human error, *Medical Device and Diagnostic Industry Magazine,* 19(5), 166-180, 1997.
30. Sawyer, D., Do It By Design: Introduction to Human Factors in Medical Devices, Center for Devices and Radiological Health (CDRH), Food and Drug Administration (FDA), Washington, D.C., 1996.
31. Nobel, J.L., Medical devices failures and adverse effects, *Pediat-Emerg. Care,* 7, 120-123, 1991.
32. Bogner, M.S., Medical devices and human error, in *Human Performance in Automated Systems: Current Research and Trends,* Mouloua, M. and Parasuraman, R., Eds., Lawrence Erlbaum Associates, Hillsdale, NJ, 1994, pp. 64-67.
33. Casey, S., *Set Phasers on Stun: and Other True Tales of Design Technology and Human Error,* Aegean, Inc., Santa Barbara, CA, 1993.
34. McDonald, J.S. and Peterson, S., Lethal errors in anaesthesiology, *Anesthesiol,* 63, A 497, 1985.
35. Wood, B.J. and Ermes, J.W., Applying hazard analysis to medical devices, Part II: Detailed hazard analysis, *Medical Device and Diagnostic Industry Magazine,* 15(3), 58-64, 1993.
36. Weide, P., Improving medical device safety with automated software testing, *Medical Device and Diagnostic Industry Magazine,* 16(8), 66-79, 1994.
37. Levkoff, B., Increasing safety in medical device software, *Medical Device and Diagnostic Industry Magazine,* 18(9), 92-101, 1996.
38. Federal Food, Drug, and Cosmetic Act, as Amended, Sec. 201 (h), U.S. Government Printing Office, Washington, D.C., 1993.
39. Onel, S., Draft revision of FDA's medical device software policy raises warning flags, *Medical Device and Diagnostic Industry Magazine,* 19(10), 82-92, 1997.
40. Mojdehbakhsh, R., Tsai, W., and Kirani, S., Retrofitting software safety in an implantable medical device, *Trans. Software,* 11, 41-50, 1994.
41. Heydrick, L. and Jones, K.A., Applying reliability engineering during product development, *Medical Device and Diagnostic Industry Magazine,* 18(4), 80-84, 1996.
42. IEEE-STD-1228, Standard for Software Safety Plans, Institute of Electrical and Electronic Engineers (IEEE), New York, 1994.
43. Cady, W.W. and Iampietro, D., Medical device reporting, *Medical Device and Diagnostic Industry Magazine,* 18(5), 58-67, 1996.
44. Thibeault, A., Documenting a failure investigation, *Medical Device and Diagnostic Industry Magazine,* 19(10), 68-74, 1997.
45. Rose, H.B., A small instrument manufacturer's experience with medical instrument reliability, *Proc. Annu. Reliability Maintainability Symp.,* 251-254, 1972.
46. Taylor, E.F., The reliability engineer in the health care system, *Proc. Reliability and Maintainability Symp.,* 245-248, 1972.
47. Bell, D.D., Contrasting the medical-device and aerospace-industries approach to reliability, *Proc. Annu. Reliability Maintainability Symp.,* 125-127, 1995.
48. MIL-STD-2164, Environment Stress Screening Process for Electronic Equipment, Department of Defense, Washington, D.C.
49. MIL-STD-883, Test Methods and Procedures for Microelectronics, Department of Defense, Washington, D.C.
50. MIL-HDBK-217, Reliability Prediction of Electronic Equipment, Department of Defense, Washington, D.C.
51. Reliability Prediction Procedure for Electronic Equipment, Report No. TR-NWT-000332, issue 4, Bell Communication Research, New Jersey, 1992.

52. MIL-HDBK-251, Reliability/Design Thermal Applications, Department of Defense, Washington, D.C.
53. MIL-HDBK-338, Electronic Reliability Design Handbook, Department of Defense, Washington, D.C.
54. MIL-STD-781, Reliability Design Qualification and Production Acceptance Tests: Exponential Distribution, Department of Defense, Washington, D.C.
55. MIL-HDBK-781, Reliability Test Methods, Plans and Environments for Engineering Development, Qualification and Production, Department of Defense, Washington, D.C.

15 Design Maintainability and Reliability Centered Maintenance

15.1 INTRODUCTION

Maintainability is a design and installation characteristic that imparts to an item a greater inherent ability to be maintained, and consequently results in factors such as a decrease in required maintainability, tools, skill levels, facilities, and manhours.

The history of maintainability can be traced back to 1901 in the Army Signal Corps Contract with the Wright brothers to develop an airplane. This document contained a clause that the aircraft under consideration be "simple to operate and maintain" [1]. Various studies conducted during the period between World War II and the early 1950s by the U.S. Department of Defense indicated startling results concerning the state of reliability and maintainability of equipment used by the three services [2, 3]. For example, an Army study revealed that approximately between two-thirds and three-fourths of its equipment were either under repair or out of service.

In 1956, a series of articles appeared in *Machine Design* covering areas such as design electronic equipment for maintainability, recommendations for designing maintenance access in electronic equipment, design recommendations for test points, and factors to consider in designing displays [4]. In 1959, the U.S. Department of Defense released a document (i.e., MIL-M-26512) on maintainability specification. In fact, an appendix to this document contained an approach for planning a maintainability demonstration test.

In 1960, the first commercially available book entitled *Electronic Maintainability* appeared [5]. Since the 1960s a large number of publications on various aspects of maintainability has appeared [6, 7].

Reliability centered maintenance (RCM) is a systematic methodology employed to highlight the preventive maintenance tasks necessary to realize the inherent reliability of an item at the lowest resource expenditure. As per the published sources, the original development of the RCM concept could be traced back to the late 1960s in the commercial aircraft industry. For example, in 1968 a document (MSG 1) entitled "Maintenance Evaluation and Program Development" was prepared by the U.S. Air Transport Association (ATA) for use with the Boeing 747 aircraft [8]. At a later date, MSG 1 was revised to handle two other wide-body aircrafts: L-1011 and DC-10. The revised version became known as MSG 2 [9].

In 1974, the U.S. Department of Defense commissioned United Airlines to prepare a document on processes used by the civil aviation industrial sector to develop maintenance programs for aircraft [10]. The resulting document was entitled "Reliability Centered Maintenance" by United Airlines.

MSG 2 was revised by the ATA in 1980 to include maintenance programs for two more aircraft: Boeing 756 and 767. The revised version was named MSG 3 [11]. In Europe, MSG 3 became known as "European MSG 3" because it included maintenance programs for two more aircraft: A-300 and Concorde [12]. In the 1970s, U.S. forces were attracted by the RCM methodology and in 1985 the military published two documents entitled "Guide to Reliability Centered Maintenance" [13] and "Reliability Centered Maintenance for Aircraft, Engines, and Equipment" [14]. Today, RCM methodology is actively being practiced in many different parts of the world.

This chapter describes design maintainability and RCM.

15.2 DESIGN MAINTAINABILITY: MAINTAINABILITY NEED AND RELATIONSHIP BETWEEN RELIABILITY AND MAINTAINABILITY

There are several factors responsible for the need of product maintainability. In particular, two important factors are the alarmingly high operating and support costs because of failures and subsequent maintenance. For example, one study conducted by the U.S. Air Force in the 1950s reported that one third of all Air Force operating cost was due to maintenance, and one third of Air Force manpower was directly or indirectly associated with maintenance [15]. Another study performed by the Army reported that 60 to 75% of the electronic equipment total life cycle cost was due to its operation and maintenance [16, 17].

The main objectives of both reliability and maintainability include: to assure that the system/equipment manufactured will be in the readiness state for operation when required, capable of performing effectively its designated functions, and meeting all the required maintenance characteristics during its life time.

In order to understand the relationship between reliability and maintainability, some of the important aspects of both reliability and maintainability are described below.

Reliability

This is a design characteristic that results in durability of the system/equipment to carry out its designated function under a stated condition and time period. It is accomplished through various actions including controlling processes, selecting optimum engineering principles, testing, and adequate component sizing. Nonetheless, there are many specific general principles of reliability: design for simplicity, use less number of parts to perform multiple functions, design to minimize the occurrence of failures, maximize the use of standard parts, provide fail safe designs, use parts with proven reliability, minimize stress on parts, provide satisfactory safety factors between strength and peak stress values, provide redundancy when required, and provide for simple periodic adjustment of parts subject to wear [18].

Maintainability

This is a built-in design and installation characteristic that provides the end product/equipment an inherent ability to be maintained, thus ultimately resulting in lower maintenance cost, required skill levels, man-hours required, required tools and equipment, and improved mission availability.

Some of the specific general principles of maintainability are reduce mean time to repair (MTTR), lower life cycle maintenance costs, provide for maximum interchangeability, consider benefits of modular replacement vs. part repair or throwaway design, reduce or eliminate altogether the need for maintenance, establish the extent of preventive maintenance to be performed, reduce the amount, frequency, and complexity of required maintenance tasks, and lower amount of supply supports required [18].

15.2.1 Maintainability and Availability Analyses During Design and Development Phase and Maintainability Design Characteristics

There are various types of maintainability and availability analyses performed during a product's design and development phase. Such analyses serve a number of purposes including evaluating the design performance with respect to both qualitative and quantitative requirements, developing most appropriate ways and means to reduce the need for maintenance, generating maintainability related data for application in maintenance planning and in conducting support analysis, and quantifying maintainability requirements at the design level [19].

There are a number of maintainability-related basic characteristics that must be carefully considered during the product design phase. The maintainability design characteristics include those features/factors that help to lower product downtime and unavailability (e.g., maintenance ease). Nonetheless, some of the particular features that affect the maintainability aspects of engineering design include standardization, safety, interchangeability, equipment packaging, and human factors. Each of these items is described below [1, 20].

Standardization

This is basically concerned with imposing limitations on the variety of different items that can be used to satisfy the product requirements. Some of the specific goals of the standardization are to minimize the use of different types of parts, use the maximum number of common parts in different products, and maximize the use of interchangeable parts. The benefits of standardization include reduction in item life-cycle acquisition and support costs and improvement in item reliability and maintainability.

Safety

This is an important maintainability design factor because individuals performing various types of maintenance tasks may be exposed to hazards and accident situations.

Hazards and accidents could be the result of unsatisfactory attention given to safety during the design phase. Nonetheless, the human safety guidelines include installing appropriate fail-safe devices, carefully studying the potential sources of injury by electric shock, fitting all access openings with appropriate fillets, providing appropriate emergency doors, installing items requiring maintenance such that hazard in accessing them is minimized, and providing adequate amount of tracks, guides, and stops to facilitate equipment handling.

Interchangeability

It means that a given component can be replaced by any similar part, and the replacing part can carry out the required functions of the replaced part effectively. There are various factors that must be considered carefully in determining needs for interchangeability including cost effectiveness of manufacture and inspection, and field conditions. Maximum interchangeability can only be assured if the design professionals consider factors such as providing adequate level of information in the task instructions, physical similarities (include shape, size, mounting, etc.), and non-existence of physical interchangeability in places where functional interchangeability is not expected.

Equipment Packaging

This is an important maintainability factor and is basically concerned with the manner in which equipment is packaged: ease of parts removal, item layout, access, and so on. In order to achieve the effectiveness of equipment packaging, careful attention must be given to factors such as accessibility requirements, modularization needs, manufacturing requirements, reliability and safety factors, environmental factors, and standardization needs.

In particular, modularization and accessibility are very important and are described in more detail. Modularization is the division of a product into distinct physical and functional units to assist removal and replacement. The advantages of modular construction include reduction in maintenance time and cost, usually easily maintainable product, shorter design times because of simplified design, and requires relatively low skill levels and fewer maintenance tools.

On the other hand, accessibility is the relative ease with which an item can be reached for actions such as repair, inspection, or replacement. The accessibility is affected by factors such as types of tools required, access usage frequency, type of task to be performed, hazards involved with respect to access usage, and distance to be reached.

Human Factors

As maintainability depends on both the maintainer and the operator, human factors are very important. For example, one aircraft maintenance study reported that over a period of 15 months, human error resulted in 475 accidents and incidents in flight and ground operations [1, 20]. Furthermore, most of the accidents happened shortly after periodic inspections and 95 aircraft were damaged or destroyed with a loss of 14 lives. In human factors, environment plays an important role since it can vary quite considerably from one application to another. Environment can be classified into three categories: physical, operational, and human. The physical category includes factors such as temperature, noise, vibration, radiation, pressure, and dust. The components of the operational category are work duration, illumination, maintenance workspace arrangement, acoustics, ventilation, etc. The factors belonging to the human category include physical, psychological, human limitations, and physiological.

15.2.2 Design Reviews

Design reviews are an important aspect of equipment development and they allow parties involved in the development process to assess factors such as the adequacy of program planning and the maturity of the design effort. There are various types of design reviews but, in particular, two important reviews are preliminary design review and critical design review [19]. The preliminary design review is performed after the availability of hardware development specifications and is a formal review of the basic design approach for a system or a functionally related group of lower-level components. The issues addressed during the preliminary design reviews from the maintainability aspect include quantitative maintainability requirements, repair time data sources, repair rate prediction methods, system/product maintainability characteristics, conformance to the maintainability ground rules and design criteria, plans for maintainability demonstrations, preventive maintenance schedules, and comparison between quantitative maintainability requirements and available preliminary predictions.

The critical design review is a formal review for a system/functionally related group of lower-level elements or parts, performed before releasing fabrication or production blue prints to ensure that detailed design solutions effectively meet the requirements established by the hardware development specification. The maintainability related issues addressed during this review include quantitative maintainability requirements and their comparison with available preliminary predictions, details of plans for maintainability demonstrations, methods for automatic, semiautomatic, and manual recovery from failure, conformance to the maintainability ground rules and design criteria, preventive-maintenance schedules for compatibility with system requirements, and system conformance to the planned maintenance concept and unique maintenance procedures (if any) [19].

On all of the above issues, various questions are raised during design reviews. One typical example of such questions is as follows:

- What are the major failure modes of the item under consideration and how will they affect the system operation? Two examples of the item are a replaceable unit and a module.

In addition, for each failure mode, questions on areas such as those listed below could be asked.

- Mechanism for operator or user to know if the item has failed.
- Mechanism employed by the built-in test system to identify failure.
- Remove and replace sequence for the failure.
- Training or experience necessary for performing the maintenance action.
- Required tools or equipment.
- Hazards involved in performing the maintenance action.
- Actions for reducing or eliminating failure mode frequency.
- Measures to reduce the maintenance time needed to accomplish the maintenance action.

15.2.3 Maintainability Measures

As quantitative maintainability specifications are based on desired limiting conditions imposed on product/equipment down time, maintenance labor-hours, etc., the knowledge of various maintainability parameters during the design phase is important for an effective end product maintainability. Some of the typical maintainability measures are mean time to repair (MTTR), medium corrective maintenance time, maximum corrective maintenance time, mean preventive maintenance time, and the probability of completing repair in a given time interval (i.e., the maintainability function). Each of these measures is presented below.

Mean Time to Repair

In maintainability analysis, this is probably the most widely used maintainability measure or parameter. MTTR is also referred to as mean corrective maintenance time and a typical corrective maintenance or repair cycle is made up of steps such as fault/failure detection, fault/failure isolation, disassemble to gain access, repair, and reassemble. Usually probability distributions such as exponential, log-normal, and normal are used to represent corrective maintenance times. In particular, the exponential distribution is assumed for electronic equipment having an effective built-in test capability along with a rapid remove and replace maintenance concept. However, it must be remembered that the exponential assumption for corrective maintenance times may result in incorrect conclusions as most repair actions consume some repair time.

The lognormal distribution is frequently used to represent corrective maintenance times associated with electronic equipment without having a built-in test capability. In addition, it can also be used for electromechanical systems possessing widely individual repair times.

The normal distribution is often assumed for mechanical/electro-mechanical equipment with a remove and replace maintenance concept.

The system MTTR is expressed by

$$MTTR = \left[\sum_{j=1}^{k} \lambda_j T_j\right] \Big/ \sum_{j=1}^{k} \lambda_j \tag{15.1}$$

where

T_j is the corrective maintenance or repair time required to repair unit j; for j = 1, 2, 3, ..., k.
k is the total number of units.
λ_j is the constant failure rate of unit j; for j = 1, 2, 3, ..., k.

Example 15.1

An engineering system is made up of six replaceable subsystems 1, 2, 3, 4, 5, and 6 with constant failure rates $\lambda_1 = 0.0002$ failures/h, $\lambda_2 = 0.0002$ failures/h, $\lambda_3 = 0.0001$ failures/h, $\lambda_4 = 0.0004$ failures/h, $\lambda_5 = 0.0005$ failures/h, and $\lambda_6 = 0.0006$ failures/h, respectively. Corrective maintenance times associated with subsystems 1, 2, 3, 4, 5, and 6 are $T_1 = 1$ h, $T_2 = 1$ h, $T_3 = 0.5$ h, $T_4 = 2$ h, $T_5 = 3$ h, and $T_6 = 3.5$ h, respectively. Estimate the system MTTR.

Inserting the given values into Equation (15.1) yields

$$MTTR = \frac{(0.0002)(1)+(0.0002)(1)+(0.0001)(0.5)+(0.0004)(2)}{(0.0002)+(0.0002)+(0.0001)+(0.0004)}$$

$$\frac{+(0.0005)(3)+(0.0006)(3.5)}{+(0.0005)+(0.0006)}$$

$$= 2.425 \text{ h}$$

Thus, the system mean time to repair is 2.425 h.

Median Corrective Maintenance Time

It is the time within which 50% of all corrective maintenance actions can be performed. The measure is dependent upon the probability density function describing the times to repair (e.g., exponential, lognormal). Thus, the median corrective maintenance time (MCMT) for the exponential distribution is expressed by

$$MCMT = 0.69/\mu \quad (15.2)$$

where

μ is the constant repair rate, thus the reciprocal of the MTTR.

Similarly, the MCMT for the lognormal distribution is given by

$$MCMT = MTTR/\exp\left(\sigma^2/2\right) \quad (15.3)$$

where

σ^2 is the variance around the mean value of the natural logarithm of repair times.

Maximum Corrective Maintenance Time

This is another maintainability measure and is used to estimate the time to accomplish a specified percentage of all potential repair actions. The most commonly specified percentiles are 90th and 95th. For example, the 90th percentile means that no greater than 10% of the maintenance activities may exceed the maximum corrective maintenance time.

The estimation of maximum corrective maintenance time depends on the probability distribution representing the times to repair. The maximum corrective maintenance times for exponential, normal, and lognormal distributions are given below.

Exponential

The maximum corrective maintenance time is defined by

$$T_{max} = C\,(MTTR) \tag{15.4}$$

where

T_{max} is the maximum corrective maintenance time.
C is equal to 2.312 or 3 for the 90th and 95th percentiles, respectively.

Normal

The maximum corrective maintenance time is expressed by

$$T_{max} = MTTR + m\,\sigma_n \tag{15.5}$$

where

m is equal to 1.28 or 1.65 for the 90th and 95th percentiles, respectively.
σ_n is the standard deviation of the normally distributed maintenance time.

Lognormal

The maximum corrective maintenance time is given by

$$T_{max} = \text{antilog}\left(t_a + m\,\sigma_\ell\right) \tag{15.6}$$

where

t_a is the average of the logarithms of the repair times.
σ_ℓ is the standard deviation of the logarithms of the repair times.

Mean Preventive Maintenance Time

Various preventive maintenance related activities such as inspections, calibrations, and tuning are performed to keep the item or equipment at a specified performance level. A well-planned and well-executed preventive maintenance program can be quite useful to reduce equipment/item downtime and improve its performance.

The mean preventive maintenance time is expressed by

$$MPMT = \left[\sum_{j=1}^{n} T_{pj} f_{pj}\right] \Big/ \sum_{j=1}^{n} f_{pj} \quad (15.7)$$

where

MPMT is the mean preventive maintenance time.
n is the total number of preventive maintenance tasks.
f_{pj} is the frequency of preventive maintenance task j; for j = 1, 2, 3, …, n.
T_{pj} is the elapsed time for preventive maintenance task j; for j = 1, 2, 3, …, n.

In computing the value of MPMT, it is to be remembered that if the frequencies, f_{pj}, are given in maintenance tasks per hour, then the T_{pj} must be expressed in hours.

Maintainability Function

This is used to calculate probability of completing repair in a specified time interval. After the identification of the repair time distributions, the corresponding maintainability functions can be obtained by using the following relationship [1]:

$$M(t) = \int_0^t f_{rp}(t)\,dt \quad (15.8)$$

where

t is the time.
$M(t)$ is the maintainability function.
$f_{rp}(t)$ is the probability density function of the repair times.

Maintainability functions for exponential, lognormal, and normal repair time distribution are obtained below.

Exponential

The probability density function representing corrective maintenance times is defined by

$$f_{rp}(t) = \mu \exp(-\mu t) \quad (15.9)$$

where

t is the variable repair time.
μ is the constant repair rate or the reciprocal of the MTTR.

Inserting Equation (15.9) into relationship (15.8) yields

$$M(t) = \int_0^t \mu \exp(-\mu t) dt = 1 - \exp(-\mu t) \tag{15.10}$$

Since μ = 1/MTTR, Equation (15.10) becomes

$$M(t) = 1 - \exp(-t/MTTR) \tag{15.11}$$

Example 15.2

After performing analysis of repair actions associated with an electric generator, it is established that the generator's repair times are exponentially distributed with a mean value of 4 h. Calculate the probability of accomplishing a repair in 5 h.

By inserting the given data into Equation (15.11), we get

$$M(5) = 1 - \exp(-5/4) = 0.7135$$

It means that there is an approximately 72% chance that the repair will be completed within 5 h.

Lognormal

This is a widely used probability distribution in maintainability work and its probability density function representing corrective maintenance times is given by

$$f_{rp}(t) = \frac{1}{(t-\gamma)\sigma\sqrt{2\pi}} \exp\left[-\frac{1}{2}\{\ln(t-\gamma)-\theta\}^2\right] \tag{15.12}$$

where

- γ is a constant denoting the shortest time below which no maintenance activity can be performed.
- θ is the mean of the natural logarithms of the maintenance times.
- σ is the standard deviation with which the natural logarithms of the maintenance times are spread around the mean θ.

The following relationship defines the mean:

$$\sigma = [\ln t_1 + \ln t_2 + \ln t_3 + \ldots + \ln t_k]/k \tag{15.13}$$

where

t_j is the maintenance time j; for j = 1, 2, 3, ..., k.
k is the number of maintenance times.

The standard deviation, σ, is expressed by the following relationship:

$$\sigma = \left[\sum_{j=1}^{k} \left(\ln t_j - \theta \right)^2 / (k-1) \right]^{1/2} \tag{15.14}$$

The maintainability function, M(t), for this distribution is expressed by

$$\begin{aligned} M(t) &= \int_0^\infty t f_{rp}(t) dt \\ &= \frac{1}{\sigma\sqrt{2\pi}} \int_0^\infty \exp\left[-\frac{1}{2} \left(\frac{\ln t - \theta}{\sigma} \right)^2 \right] dt \end{aligned} \tag{15.15}$$

Normal

Corrective maintenance times can also be represented by the normal distribution. The distribution repair time probability density function is defined by

$$f_{rp}(t) = \frac{1}{\sigma\sqrt{2\pi}} \exp\left[-\frac{1}{2} \left(\frac{t-\mu}{\sigma} \right)^2 \right] \tag{15.16}$$

where

μ is the mean of maintenance times
σ is the standard deviation of the variable maintenance time t around the mean μ.

By substituting Equation (15.16) into Equation (15.8), we get

$$M(t) = \frac{1}{\sigma\sqrt{2\pi}} \int_{-\infty}^{t} \exp\left[-\frac{1}{2} \left(\frac{t-\mu}{\sigma} \right)^2 \right] dt \tag{15.17}$$

The mean of the maintenance times is expressed by

$$\mu = \sum_{j=1}^{k} t_j / k \tag{15.18}$$

where

t_j is the maintenance time j; for j = 1, 2, 3, ..., k.
k is the total number of maintenance times.

The standard deviation is

$$\sigma = \left[\sum_{j=1}^{k} \left(t_j - \mu\right)^2 \Big/ (k-1) \right]^{1/2} \tag{15.19}$$

15.2.4 Maintainability Design Guidelines and Common Maintainability Design Errors

Over the years professionals involved with maintainability have developed many guidelines that can be used to produce effective design with respect to maintainability. These guidelines include design for minimum maintenance skills, provide for visual inspection, use standard interchangeable parts, design for minimum tools and adjustments, avoid the use of large cable connectors, design for safety, provide test points, use color-coding, label units, provide handles on heavy parts for the ease of handling, use captive-type chassis fasteners, provide troubleshooting techniques, group subsystems, and use plug-in rather than solder-in modules [21].

During the design phase of an equipment, there are many errors committed that adversely affect its maintainability. Some of the common design errors are placing an adjustment out of arm's reach, omitting handles, using access doors with numerous small screws, placing removable parts such that they cannot be dismantled without taking the entire unit from its case, locating adjustable screws close to a hot component/an exposed power supply terminal, placing adjustable screws in locations cumbersome for maintenance personnel to discover, providing unreliable built-in test equipment, locating fragile parts just inside the bottom edge of the chassis where the maintenance personnel are expected to place their hands, using chassis and cover plates that drop when the last screw is taken out, placing low-reliability parts beneath other parts, placing screwdriver adjustments beneath modules, and providing insufficient space for maintenance personnel to get their gloved hands into the unit to perform required adjustment [21].

15.3 RELIABILITY CENTERED MAINTENANCE

The commercial airline industry developed the reliability centered maintenance (RCM) concept after becoming fully aware that the traditional approach to maintenance with its heavy emphasis on scheduled inspection, servicing, and removal was not impacting the equipment reliability as per expectations. In fact, it became very apparent after the introduction of more complex aircraft systems that the relationship between reliability and preventive maintenance tended to disappear [12]. RCM is based on the premise that the inherent reliability or safety of a system cannot be improved through maintenance and good maintenance practices can only preserve such characteristics. The RCM philosophy calls for the performance of scheduled maintenance on critical components only under the following circumstances:

- It will stop a decrease in reliability and/or degradation of safety to unacceptable levels or
- It will decrease equipment life cycle cost.

In contrast, the RCM philosophy also calls for not performing scheduled maintenance on noncritical items unless it will lower equipment life cycle cost.

15.3.1 Traditional Maintenance, RCM, and Questions for Selecting Assets for RCM

In order to provide a better understanding of RCM, here we briefly describe both traditional maintenance and RCM. *Webster's Encyclopaedic Dictionary* defines the words "maintain" and "maintenance", respectively, as follows [22]:

- To cause to remain unaltered or unimpaired.
- A maintaining or being maintained.

In contrast, RCM may be described as a process employed to determine the maintenance needs of any physical facility in its operating context [10].

Prior to the start of performing analysis of maintenance needs of the facilities, it is essential to fully know about these facilities and to select the ones to be subjected to the RCM review process. Nonetheless, the RCM process entails asking various questions concerning the chosen facilities on areas such as functions and associated performance standards of the facility in its current operating context, the ways the facility fails to meet its required functions, each functional failure's causes, consequences of each failure, ways each failure matter, preventive measures to stop each failure, and measures to be taken if an appropriate preventive task cannot be found.

15.3.2 RCM Process Related Maintenance Task Classification Areas and Steps

The RCM process calls for maintenance tasks to be categorized into three distinct areas: hard-time maintenance, on-condition maintenance, and condition monitoring [23].

The hard-time maintenance is for those failure modes that need scheduled maintenance at pre-determined time or usage intervals.

The on-condition maintenance is for those failure modes that need scheduled inspections or tests for measuring the level of item deterioration so that on the basis of this deterioration level either corrective maintenance can be carried out or the item can still remain in service.

The condition monitoring is for those failure modes that need unscheduled tests or inspection on parts where it is possible to tolerate a failure during system operation or where it is possible to detect an impending failure through the means of regular monitoring during normal operations.

The RCM process is made up of a number of steps and each such step is an integral part of RCM engineering planning and analysis. Nonetheless, the application of this process as well as the resulting preventive maintenance program ensure that systems or equipment are kept in good working condition and, consequently, are quite effective to support their operational requirements. However, to attain such results, it is absolutely necessary that the RCM process be initially practiced during equipment design and development phase to develop maintenance plans and procedures, and continued during the operational phase utilizing actual experience data to make changes, as appropriate, to the maintenance plans.

The seven basic steps of the RCM process are as follows [23]:

1. Determine maintenance-important items.
2. Collect necessary failure data.
3. Conduct fault tree analysis.
4. Use decision logic to important failure modes.
5. Record maintenance categories.
6. Implement RCM decisions.
7. Apply sustaining engineering based on real-life experience information.

Each of the above steps is described below.

Determine Maintenance-Important Items

Traditionally to identify maintenance-important items the failure mode, effects, and criticality analysis (FMECA) approach has been used.

Also, the approach provided the basis for developing detailed preventive/corrective maintenance requirements to be used during the system/equipment operation phase. However, for more effective results, a better method is required that uses the results of data collection programs and field operating experience to develop new and upgrade existing maintenance programs. Fault tree analysis (FTA) is considered to be a quite effective approach for this purpose because it can use field failure data to determine maintenance-importance items.

Collect Necessary Failure Data

This step is concerned with obtaining required data for each basic fault associated with the fault tree. For assessing criticality and determining occurrence probabilities, the essential inputs include part failure rate, operator error probability, and inspection efficiency data. Some of the sources of obtaining part failure rate data are the actual real life experience, MIL-HDBK-217 [24], and generic failure data banks. Many of the sources for obtaining human error related data are listed in Reference 25.

Conduct Fault Tree Analysis

This step is concerned with calculating the occurrence probabilities of all fault events (i.e., basic, intermediate, and top) on the basis of combinatorial properties of the logic elements in the fault tree. The sensitivities (sensitivity or conditional probability may be defined as the probability that an occurrence of a basic fault event will lead to a safety incident) are calculated by assigning a probability of unity to an elementary or basic fault event and then determining a safety incident's resultant probability. In turn, the criticality of each basic fault event is calculated by using these sensitivity results.

Criticality may simply be described as a measure of the relative seriousness or impact of each and every fault event on the undesired or top fault event. It (i.e., criticality) involves both qualitative and quantitative fault tree analyses and provides a base mechanism to rank the fault events in their order of severity.

In quantitative terms, the criticality is expressed below [26].

$$C = P(y)P(F|y) \tag{15.20}$$

where

$P(y)$ is the probability of occurrence of the basic fault event, y.
$P(F|y)$ is the sensitivity or conditional probability that an occurrence of a basic fault event will lead to a safety incident.
C is the criticality.

Use Decision Logic to Important Failure Modes

This step is concerned with the RCM decision logic designed to result in, by asking standard assessment questions, the most effective preventive maintenance task combinations. Each and every critical failure mode highlighted by the FTA is reviewed by applying the decision logic for each maintenance-important item. Judgments are passed as to the need for various maintenance tasks as each and every failure mode is processed. The tasks considered necessary along with the intervals established to be appropriate, form the overall scheduled preventive maintenance program.

The decision logic is composed of the following two levels:

- **Level 1.** This is concerned with the evaluation of each failure mode for determining consequence category: hidden safety, evident safety, operational economic, non-operational economic, or non-safety/economic. Four questions associated with this level are as follows:
  1. Does the failure or fault cause a direct adverse effect on operating performance?
  2. Does the failure/secondary failure lead to a safety incident?
  3. Can operator(s) detect failures or faults when they occur?
  4. Does the hidden failure itself, or the combination of a hidden failure plus an additional failure of a system-related or back-up function, lead to a safety incident?
- **Level 2.** This is concerned with taking into consideration the failure causes for each failure mode for choosing the specific type of task. Twenty-one

questions are asked in this level [23]. Some examples of those questions are as follows:

- Is an operator monitoring task effective and applicable?
- Is a check for verifying operation/detecting impending failure effective and applicable?
- Is a servicing task effective and applicable?

Record Maintenance Categories

This step is concerned with applying the decision logic of the previous step to group maintenance requirements into three categories (i.e., hard-time maintenance requirements, on-condition maintenance requirements, and condition monitoring maintenance requirements) and defining a maintenance task profile. This task profile basically highlights part number and failure mode, and the preventive-maintenance task-selection for RCM logic questions with answer "yes". Subsequently, the maintenance-task profile is used for determining applicable preventive maintenance tasks to each part under consideration.

Implement RCM Decisions

Subsequent to the establishment of the maintenance-task profile, the task frequencies/intervals are established and enacted as part of the overall maintenance plan.

Apply Sustaining Engineering Based on Real-life Experience Information

As the RCM process has a life-cycle perspective, the main objective in this step is to reduce scheduled maintenance burden and support cost while keeping the equipment/systems in a desirable operational-readiness state. After the system is operational and field data start to accumulate, one of the most pressing steps is to reassess all RCM default decisions.

All in all, it must be remembered that the key objective is to eliminate all excessive maintenance related costs while maintaining established and required reliability and safety levels.

15.3.3 RCM Advantages

Over the years many companies have applied the RCM methodology and have experienced many advantages to a varying degree. Some of these advantages are as follows [10, 23]:

- **Better safety and environmental integrity.** Prior to considering the impact of each failure mode on operations, the RCM methodology first considers its effects on safety and environments. It means appropriate steps are taken to minimize or eliminate all identifiable equipment-associated safety and environmental hazards.

- **Improved motivation of individuals.** The RCM effort is especially useful to improve the motivation of the individuals involved in the review process. This occurs because of factors such as clearer understanding of the functions of the facility under consideration, clear general understanding of the issues involved, and the knowledge that each group member played a role in formulating goals.
- **Better operating performance.** As the RCM philosophy dictates that all types of maintenance have value to a certain degree, thus it (RCM) outlines rules for deciding which is the most effective in every circumstance. Consequently, the most effective form of maintenance is chosen for each machine and appropriate actions are pursued in cases where maintenance can be effective.
- **Improved maintenance cost-effectiveness.** In most of the industrial sector, the maintenance activity is the third biggest component of operating costs. Thus, controlling maintenance costs is very important. The application of RCM helps to lower these costs in several ways including less routine maintenance, less need to acquire services of costly experts, clearer guidelines for acquiring new maintenance technology, and better buying of maintenance services.
- **Improved team effort.** The application of the RCM methodology helps to provide a common and easily comprehensible language for all individuals involved with the maintenance effort. This, in turn, not only fosters teamwork within the review groups themselves but also improves communication and cooperation among production and maintenance departments, management, repair persons, operators, designers, vendors, and so on.
- **Better maintenance database.** RCM information and decision worksheets are a comprehensive maintenance database, which in turn has advantages such as reducing the effects of staff turnover, more accurate drawings and manuals, and adapting to changing circumstances.
- **Improved useful life of costly items.**

15.3.4 Reasons for the RCM Methodology Application Failures

When the RCM methodology is applied correctly and effectively, it can provide fast results and associated benefits. However, its every application may not yield its full potential. In fact, some of the applications may only achieve very little or nothing at all. Some of the major possible reasons for the RCM methodology application failures are too hurried or too superficial application, too much emphasis on failure data (e.g., on parameters such as mean time between failures (MTBF) and mean time between to repair (MTTR), and the analysis carried out at too low a level [10].

15.4 PROBLEMS

1. Define the term "Reliability Centered Maintenance" and discuss its historical developments.
2. Discuss the following terms:
   - Maintainability
   - Interchangeability
   - Standardization
3. Write an essay on design maintainability.
4. What are the important factors that affect the maintainability aspects of an equipment design?
5. What are the issues addressed during the preliminary design review of an equipment from the stand point of maintainability?
6. An electronic system is composed of four subsystems 1, 2, 3, and 4. The time to failure of each subsystem is exponentially distributed. The failure rates of subsystems 1, 2, 3, and 4 are $\lambda_1 = 0.01$ failure/h, $\lambda_2 = 0.03$ failure/h, $\lambda_3 = 0.04$ failure/h, and $\lambda_4 = 0.05$ failure/h, respectively. In addition, the corrective maintenance times associated with subsystems 1, 2, 3, and 4 are $T_1 = 2$ h, $T_2 = 4$ h, $T_3 = 1$ h, and $T_4 = 3$ h, respectively. Calculate the electronic system mean time to repair.
7. Obtain a maintainability function when the probability density function of the repair times is described by the Weibull distribution.
8. Compare traditional maintenance with RCM.
9. Describe the RCM methodology process.
10. What are the benefits of applying the RCM methodology?

15.5 REFERENCES

1. AMCP 706-133, 1976, Engineering Design Handbook: Maintainability Engineering Theory and Practice, prepared by the Army Material Command, Department of the Army, Washington, D.C.
2. Shooman, M.L., *Probabilistic Reliability: An Engineering Approach,* McGraw-Hill, New York, 1968.
3. Moss, M.A., *Minimal Maintenance Expense,* Marcel Dekker, New York, 1985.
4. Retterer, B.L. and Kowalski, R.A., Maintainability: A historical perspective, *IEEE Trans. Reliability,* 33, 56-61, 1984.
5. Akenbrandt, F.L., Ed., *Electronic Maintainability,* Engineering Publishers, Elizabeth, NJ, 1960.
6. Dhillon, B.S., *Reliability Engineering in Systems Design and Operation,* Van Nostrand Reinhold Company, New York, 1983.
7. Dhillon, B.S., *Reliability and Quality Control: Bibliography on General and Specialized Areas,* Beta Publishers, Gloucester, Canada, 1993.
8. MSG 1, Maintenance Evaluation and Program Development, 747 Maintenance Steering Group Handbook, Air Transport Association, Washington, D.C., July 1968.
9. MSG 2, Airline/Manufacturer Maintenance Program Planning Document, Air Transport Association, Washington, D.C., September 1980.
10. Moubray, J., *Reliability Centered Maintenance,* Industrial Press, New York, 1992.

11. MSG 3, Airline/Manufacturer Maintenance Program Planning Document, Air Transport Association, Washington, D.C., September 1980.
12. Anderson, R.T., *Reliability Centered Maintenance: Management and Engineering Methods,* Elsevier Applied Science Publishers, London, 1990.
13. U.S. AMC Pamphlet 750-2, Guide to Reliability Centered Maintenance, U.S. Army, Department of Defense, Washington, D.C., 1985.
14. MIL-STD-1843, Reliability Centered Maintenance for Aircraft, Engines, and Equipment, Department of Defense, Washington, D.C., 1985.
15. Hall, A.C., Relationship between operational effectiveness of electronic systems and maintenance minimization and training, *Proc. Symp. Electron. Mainten.,* Washington, D.C., 1955.
16. AMC Memo AMCRD-ES, Subject: Economy vs. Life Cycle Cost, signed by Maj. Gen. Guthrie, J.R., Director, Research, Development, and Engineering, Headquarters, Washington, D.C., January 21, 1971.
17. Ankenbrandt, F.L., Ed., *Electronic Maintainability,* Vol. 3, Engineering Publishers, Elizabeth, NJ, 1960.
18. AMCP-706-134, Maintainability Guide for Design, prepared by the Department of the Army, Department of Defense, Washington, D.C., 1972.
19. Grant Ireson, W., Coombs, C.F., and Moss, R.Y., *Handbook of Reliability Engineering and Management,* McGraw-Hill Companies, New York, 1996.
20. Dhillon, B.S., *Engineering Design: A Modern Approach,* Richard D. Irwin, Chicago, IL, 1996.
21. Pecht, M., Ed., *Product Reliability, Maintainability, and Supportability Handbook,* CRC Press, Boca Raton, FL, 1995.
22. *Webster's Encyclopaedic Dictionary,* Lexicon Publication, New York, 1988.
23. Brauer, D.C. and Brauer, G.D., Reliability centered maintenance, *IEEE Trans. Reliability,* 36, 17-24, 1987.
24. MIL-HDBK-217, Reliability Prediction of Electronic Equipment, Department of Defense, Washington, D.C.
25. Dhillon, B.S., *Human Reliability: With Human Factors,* Pergamon Press, New York, 1986.
26. NURG-0492, Fault Tree Handbook, prepared by the U.S. Nuclear Regulatory Commission, Washington, D.C., 1981.

16 Total Quality Management and Risk Assessment

16.1 INTRODUCTION

Total quality management (TQM) is an enhancement to the conventional approach of conducting business and it may simply be stated as a philosophy of pursuing continuous improvement in each and every process through the integrated efforts of all concerned persons associated with the organization.

The roots of the total quality movement may be traced back to the early 1900s in the works of Frederick W. Taylor, the father of scientific management, concerning the time and motion studies [1, 2]. In 1924, to control product variables, Walter A. Shewhart, working as a quality control inspector for Bell Telephone Laboratories, developed a statistical chart. This development is probably considered as the beginning of the statistical quality control.

In the late 1940s, the efforts of quality gurus such as W.E. Denning, J. Juran, and A.V. Feigenbaum played an instrumental role in the strengthening of the TQM movement [3]. In 1950, Deming lectured on the principles of statistical quality control to 230 Japanese engineers and scientists at the request of the Japanese Union of Scientists and Engineers (JUSE) [4]. In turn, in 1951, JUSE established the Deming prize to be awarded to a company demonstrating the most effective implementation of quality measures and policies [5]. However, the term "Total Quality Management" was not coined until 1985 and, in fact, it is credited to an American behavioral scientist, Nancy Warren [5].

After witnessing the success of the Deming prize in Japan, the U.S. government established the Malcolm Baldrige Award in 1987 for companies demonstrating effective implementation of quality assurance policies and measures. In 1988, the Cellular Telephone Division of Motorola was the first recipient of the Malcolm Baldrige Award for its achievements in reducing defects from $1060/10^6$ to $100/10^6$ during the time frame from 1985 to 1988 [6, 7].

Since the late 1980s many other important events related to TQM have occurred.

Risk is present in all human activity and it can either be health and safety related or economic. An example of the economic related risk is loss of equipment/production due to accidents involving fires, explosions, and so on. Nonetheless, risk may simply be described as a measure of the probability and security of a negative effect to health, equipment/property, or the environment [8]. Insurance is one of the oldest strategies for dealing with risks and its history can be traced back to about 4000 years in Mesopotamia (modern Iraq) when the Code of Hamurabi, in 1950 BC, formalized bottomry contracts containing a risk premium for the chance of losing ships and their cargo [9]. In the same area (i.e., Tigris-Euphrates Valley) around 3200 BC a group of people known as the Asipu served as risk analysis consultants to other people involved in making risky, uncertain, or difficult decisions.

In the fourth century BC a Greek doctor, Hippocrates, correlated occurrence of diseases with environmental exposures and in the sixteen century AD Agricola identified the correlation between occupational exposure to mining and health [9].

The basis of modern quantitative risk analysis was developed by Pierre Laplace in 1792 by calculating the probability of death with and without smallpox vaccination. In the twentieth century, the conceptual development of risk analysis is due to factors such as the development of nuclear power plants and concerns about their safety and the establishment of such U.S. bodies as the Environmental Protection Agency (EPA), Occupational Safety and Health (OSHA), and National Institute for Occupational Safety and Health (NIOSH). A more detailed history of risk analysis and risk management is provided in Reference 10.

Both TQM and risk assessment are described below.

16.2 TQM

A key to success in the corporate world has been the ability to improve on business processes and operational tasks. Today, the playing field for company products has changed forever because competition is no longer based solely on price, but rather on a combination of price and quality. It means the combination must surpass the competition offered by the competitive products for the ultimate success. Consequently, it may be said that companies must continually improve and upgrade their capabilities in order to satisfy changing customer needs. Thus, quality goals and measuring attainment of such goals is a dynamic process. In other words, achieving quality is a journey without end.

A survey of 100 U.S. executives conducted in 1989 revealed that only 22% of them stated that their company had done all it could to create a quality-fostering environment [11, 12]. This was probably an important factor for picking up steam by the TQM movement in the U.S. Nonetheless, according to Reference 11, the rationale for embarking upon TQM varied among the U.S. companies and the principal reasons included increased competition, customer demands, employees, and internal crisis.

It is important to remember that many times there could be good scientific innovations but due to an effective practice of a TQM methodology, the organizations/nations may lose the competitive advantage. For example, during the period between the mid-1950s to the mid-1980s, British researchers won 26 Nobel Prizes for new discoveries in comparison to only 4 by the Japanese [12]. But during the same time period, the Japanese industry successfully took a major share of the world market.

16.2.1 Traditional Quality Assurance Vs. TQM

In order to effectively practice TQM, it is important to understand basic differences between the traditional quality assurance approach and TQM. The main areas of differences are shown in Figure 16.1. Table 16.1 presents differences in these areas [5, 7].

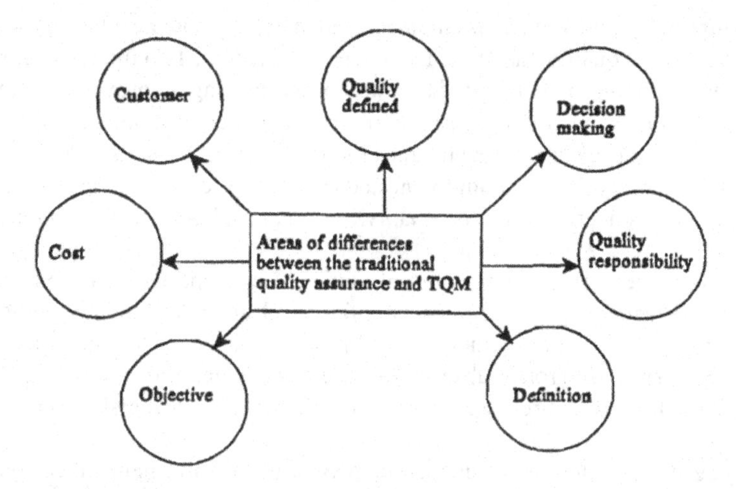

FIGURE 16.1 The main areas of differences between the traditional quality assurance program and TQM.

TABLE 16.1
Differences Between the Traditional Quality Assurance Approach (TQAA) and TQM

Area	TQAA	TQM
Objective	Find errors	Prevent errors
Definition	Product driven	Customer driven
Cost	Improvements in quality increase cost	Improvements in quality lower cost and increase productivity
Customer	Ambiguous understanding of customer needs	An effective approach defined to comprehend and meet customer requirements
Quality defined	Products meet specifications	Products suitable for consumer use
Quality responsibility	Inspection center/quality control department	All employees in the organization involved
Decision making	Practiced top-down approach	Practiced team approach with team of employees

16.2.2 TQM Principles and Elements

Two fundamental principles of TQM are customer satisfaction and continuous improvement. An organization involved in manufacturing or providing goods or services could have either internal or external customers. The internal customers belong to the organization itself, whereas the external customers are not part of the organization. The practice of the "market in" concept allows a healthy customer orientation and recognizes that every work process is made up of various stages [14]. Thus, at each stage, inputs from customers are sought for the purpose of determining the changes to be made to better satisfy their needs.

Continuous improvement is another fundamental principle and an important factor to meet the quality challenge facing an organization. Two important functions performed by management include continuously making improvements to techniques and procedures currently used, through process control and directing efforts to make important breakthroughs in concerned processes. Seven important elements of TQM are management commitment and leadership, customer service, supplier participation, team effort, cost of quality, training, and statistical techniques and methods [14].

For the success of the application of the TQM concept, the management commitment and leadership role is very essential. A clear understanding of the TQM concept by the senior management is very important in order to count on its commitment and leadership role. Subsequently, the management can establish new goals and policies for the entire organization and then play a leadership role in their realization.

As the demand for better quality increases, the involvement of customers is becoming increasingly important. The main purpose of such involvement is to represent customer interests and ultimately reflect them in goals, plans, and controls developed by the quality team.

A company's ability to produce a good quality product to a certain degree depends upon the type of relationship among the parties involved in the process: the processor, the customer, and the supplier. Even though mutual trust and partnership are the essential ingredients for supplier responsiveness, some organizations require their suppliers to have formal TQM programs [15].

The purpose of the team effort or approach is to involve all concerned parties in the TQM program including customers, vendors, and subcontractors. The quality team membership is on a voluntary basis and 3 to 15 individuals form the team. However, the team members possess skills in areas such as cost-benefit analysis, brainstorming approaches, statistics, planning and controlling projects, public relations, flow charting, and presentation methods.

The cost of quality is a primary quality measurement tool and is applied to select quality improvement projects, monitor the effectiveness of the TQM process, and justify the cost to doubters. The two important advantages of the cost of quality are to communicate to management the benefit of applying the TQM concept in terms of dollars and to raise quality awareness.

A Japanese axiom states that quality begins with training and ends with training [16]. As under the TQM program, quality is the responsibility of each employee in the organization; thus, it is essential that the training effort must be targeted for each hierarchy level of the firm.

Statistical techniques and methods are used to solve various TQM related problems. Some examples of the statistical techniques and methods are control charts, scatter diagrams, graphs and histograms, cause and effect diagrams, and Pareto diagrams [5, 17-20]. Such statistical approaches may be used for purposes such as verifying, repeating, and reproducing measurements based on data, identifying and separating quality problem causes, making decisions on facts based on data rather than the opinions of individuals, and communicating the information in a language that can easily be understood by all quality team members.

16.2.3 TQM Contributors

Over the years, there have been many people who directly or indirectly contributed to the development of the TQM concept. Three of these people who played important roles were W.E. Deming, J.M. Juran, and P.B. Crosby. Each of these individuals lay down series of points/steps to improve quality. Such points/steps are presented below [1].

Deming Points

W.E. Deming, a graduate in engineering, mathematics, and physics, outlined his philosophy in 14 points to improve quality. His 14 points are as follows [1, 5, 21, 22]:

- Create constancy of purpose for improving products/services. This means develop a mission statement by considering issues such as quality philosophy, investors, long term corporate objectives, employees, and profit distribution.
- Adopt the new philosophy or lead to promote change. More specifically, the current acceptable levels of defects/mistakes/delays are unacceptable and every concerned body is alerted to determine reasons for their (i.e., defects/mistakes/delays) existence. Ultimately, the team effort is practiced to rectify problems under consideration.
- Stop depending on mass inspection to achieve quality and build quality into the product.
- Stop awarding business or contracts on the basis of low bids and develop long-term relationships on the basis of performance.
- Continuously improve product, quality, service.
- Institute training that includes modern methods, techniques, and approaches.
- Take advantage of modern supervision techniques and approaches.
- Eliminate fear so that everyone may perform effectively.
- Break down barriers between departments/units and demand team effort.
- Eradicate slogans, numerical goals, etc. for the workforce because they generate adversarial relationships.
- Remove numerical quotas and management by objectives (MBO) and instead substitute leadership.
- Remove obstacles to employee pride in workmanship.
- Encourage a vigorous education and self-improvement program.
- Make the transformation every individual's task and make every individual work on it.

Juran Steps

J.M. Juran, another guru of quality, stated his quality philosophy in a number of steps. His steps to quality improvement are the following:

- build awareness of both the need for improvement and opportunities for improvement
- develop goals for acquired improvements

- organize to achieve the set goals
- impart training as appropriate
- implement projects directed at finding solutions to problems
- report progress periodically
- give proper recognition
- communicate results to all concerned bodies
- keep scores
- sustain momentum by making improvements to the organization's regular systems.

Juran's quality philosophy is described in detail in References 1 and 23.

Crosby Steps

P.B. Crosby, another major contributor to quality, outlined his quality philosophy in 14 steps as follows [1, 23]:

- Ensure that management is committed to quality for the long period.
- Establish cross-departmental teams to improve quality.
- Highlight areas with current and potential problems.
- Estimate the cost of quality and also describe how it is being utilized by management.
- Enhance the quality awareness and all employees' personal commitment to quality as much as possible.
- Take corrective measures as soon as possible to rectify highlighted problems.
- Develop program aim at zero defects.
- Provide appropriate training to supervisors so that they can perform their responsibilities in the quality program effectively.
- Hold a zero defects day for the purpose of ensuring that everyone in the organization is aware of the new direction.
- Encourage both individuals and teams to develop personnel and team improvement goals with respect to quality.
- Encourage all employees to make management aware of difficulties encountered by them in achieving set quality goals.
- Effectively recognize participating employees.
- Establish quality councils to promote communication on an on-going basis.
- Repeat everything to demonstrate that quality improvement is a journey without an end.

16.2.4 TQM Methods

Various appropriate methods or tools are needed to implement TQM. The success of such implementation will depend on the degree of understanding of these methods by the management as well as adopting the most suitable of such methods. Nonetheless,

TABLE 16.2
Selective TQM Methods Belonging to the Analytical Classification

Method No.	Method name
1	Taguchi methods
2	Cause and effect analysis
3	Force field analysis
4	Domainal mapping
5	Solution effect analysis
6	Failure mode and effect analysis
7	Tolerance design
8	Fault tree analysis
9	Minute analysis
10	Paired comparisons

the primary role of the TQM methods in problem-solving for improving quality is to be effective in satisfying customer needs. Furthermore, the methods can also be useful to identify possible root causes and their potential solutions, in addition to utilizing data/information to choose the most suitable alternatives for managing quality.

There are a large number of methods that can be used in TQM work. In fact, 100 such methods are described in Reference 20. This reference classifies the TQM methods into four distinct categories:

- Analytical
- Management
- Idea generation
- Data collection, analysis, and display

Tables 16.2 through 16.5 present selective methods belonging to each of the above four categories, respectively. Some of these TQM methods are described below [20-22].

Pareto Analysis

This is performed to separate the most important causes of a problem from the trivial ones. It is named after an Italian economist called Vilfredo Pareto (1848–1923) and its popular use in quality work is due to J.M. Juran who emphasized that 80% of quality problems are the result of only 20% of the causes. Thus, Pareto analysis is an extremely useful TQM method to highlight areas for a concerted effort.

Sometime Pareto analysis is also called the 80/20 rule. Nonetheless, the following steps are used to perform such analysis [20]:

TABLE 16.3
Selective TQM Methods Belonging to the Management Classification

Method No.	Method name
1	Kaizen
2	Pareto analysis
3	Quality circles
4	Quality function deployment (QFD)
5	Potential problem analysis
6	Deming wheel (PDCA)
7	Error proofing (pokayoke)
8	Benchmarking
9	Arrow diagram
10	Mystery shopping

TABLE 16.4
Selective TQM Methods Belonging to the Idea Generation Classification

Method No.	Method name
1	Brainstorming
2	Morphological forced connections
3	Opportunity analysis
4	Mind mapping
5	Nominal group technique
6	Imagineering
7	Snowballing
8	Buzz groups
9	List reduction
10	Multi-voting

- List activities/causes in a tabular form and count their occurrences.
- Arrange them in descending order.
- Compute the total for the entire list.
- Determine for each cause/activity the percentage of total.
- Develop a Pareto diagram by showing percentages vertically and their corresponding causes horizontally.
- Conclude from the end results.

TABLE 16.5
Selective TQM Methods Belonging to the Data Collection, Analysis, and Display Classification

Method No.	Method name
1	Histograms
2	Process analysis
3	Box and whisker plots
4	Checksheets
5	Scatter diagrams
6	Hoshin Kanri (quality policy deployment)
7	Flowcharts
8	Spider web diagrams
9	Statistical process control (SPC)
10	Dot plots

Hoshin Kanri (Quality Policy Deployment)

This method is used to delight customers through the manufacturing and servicing process by implementing the company goals. The method is applied when through top-down and bottom-up consultation the objectives are identified at each organizational level and the overall organizational goals have been established as particular targets. The following factors are associated with Hoshin Kanri [20, 24]:

- Establish long and short term organizational goals.
- Identify the goals that can be measured.
- Identify the important processes involved in achieving the above goals.
- Direct the team members to reach consensus on performance indicators at appropriate process stages.
- Challenge each process level to force organization to make changes to its quality culture.
- Use organizational goals as measurable goals for the purpose of making employees comprehend the criticality of the quality improvement process.

Cause-and-Effect Diagram

This is also known as the Ishikawa diagram, named after its Japanese originator K. Ishikawa, who developed it in the early 1950s. Another name used to describe the cause-and-effect diagram is "Fishbone" diagram because of its resemblance to the skeleton of a fish as shown in Figure 16.2.

The cause-and-effect diagram method is an extremely useful approach to determine the root causes of a given problem and to generate relevant ideas. Pictorially, the right-hand side of the diagram or the fish head represents effect and the left-hand

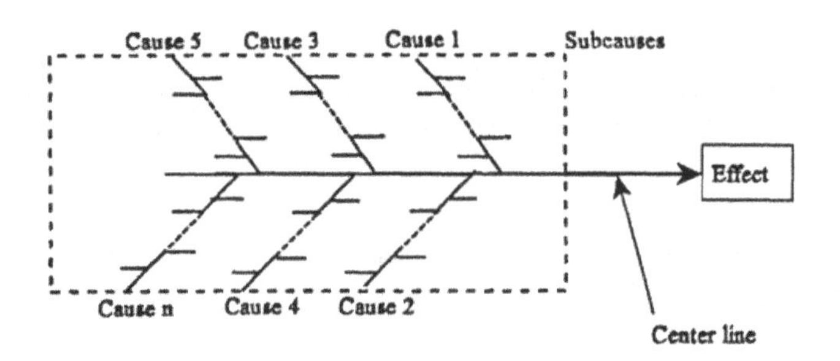

FIGURE 16.2 A general cause-and-effect diagram.

side represents all the possible causes which are connected to the central line known as the "Fish spine".

In particular, the cause-and-effect diagram with respect to TQM can be described as follows: the customer satisfaction could be the effect, and manpower, materials, methods, and machinery are the major causes. The following steps are useful in developing a cause-and-effect diagram:

- Develop problem statement.
- Brainstorm to highlight all possible causes.
- Establish main cause classifications by stratifying into natural groupings and process steps.
- Develop the diagram by connecting the causes by following essential process steps and fill in the problem or the effect in the diagram box on the extreme right.
- Refine cause classifications through questions such as follows:
  - What causes this?
  - Why does this condition exist?

There are many advantages of the cause-and-effect diagram, including useful to highlight root causes, a useful tool to generate ideas, an effective mechanism to present an orderly arrangement of theories, and a useful vehicle to guide further inquiry.

Quality Function Deployment (QFD)

This is a method used for optimizing the process of developing and manufacturing new products as per customer need. The method was developed by the Japanese in 1972 [22, 25] to translate customer needs into appropriate technical requirements. The approach can be applied in areas such as research, product development, engineering, manufacturing, and marketing. The technique makes use of a set of matrices to relate customer needs to counterpart characteristics expressed as technical specifications and process control requirements. The important QFD planning documents include customer needs planning matrix, process plan and quality control charts, operating instructions, and product characteristic deployments matrix.

The customer need planning matrix is used to translate the consumer requirements into product counterpart characteristics and the purpose of the process plan and quality control charts is to identify important process and product parameters along with control points. The operating instructions are useful to identify operations that must be accomplished to achieve critical parameters and on the other hand the product characteristic deployment matrix is used to translate final product counterpart characteristics into critical component characteristics.

A QFD matrix is frequently called the "House of Quality" because of its resemblance to a house. The following steps are necessary to build the house of quality [22, 25]:

- Identify the needs of customers.
- Identify the essential product/process characteristics that will meet the customer requirements.
- Establish the relationship between the customer needs and the counterpart characteristics.
- Conduct evaluation analysis of competing products.
- Develop counterpart characteristics of competing products and formulate appropriate goals.
- Identify counterpart characteristics to be utilized in the remaining process.

One clear cut advantage of the QFD is that it encourages companies to focus on the process itself rather than focusing on the product or service. Furthermore, the development of correlations between what is required and how it is to be acquired, the important areas become more apparent, help in making decisions.

One important limitation associated with the application of QFD is that the exact needs must be identified in complete detail.

16.2.5 Designing for Quality and Goals for TQM Process Success

It may be said that quality starts with the product design specification writing phase and continues to its operational phase. Nonetheless, there are a number of product design elements that can adversely affect quality. Some of these elements are poor part tolerances, lack of design refinement resulting in the use of more parts than required to carry out the desired functions, difficulty in fabricating parts because of poor design features, and the use of fragile parts.

There are various measures that can be useful to improve quality during the product design phase including eliminating the need for adjustments, using repeatable and well-understood processes, using parts that can easily withstand process operations, reducing parts and part numbers, eliminating engineering changes on released products, simplifying assembly and making it foolproof, designing for robustness using Taguchi methods, and designing for efficient and satisfactory testing [26].

There are a number of goals that must be satisfied in order to achieve the effectiveness of the TQM process. These goals include:

- A clear understanding of internal/external customer needs by all organization employees.
- Meeting of control guidelines as per customer requirements by all significant process/systems.
- Use of a system to continuously improve processes that better fulfill the current and future needs of customers.
- Establishment of real rewards and incentives for employees when process control and customer satisfaction results are achieved.

16.2.6 Deming's "Deadly Diseases" of American Management and Common Errors During the Starting Quality Initiative Phase

W.E. Deming not only put forward 14 points to improve quality, but he also pointed out seven "deadly diseases" of American Management that are counter-productive to quality. These "deadly diseases" are excessive medical costs (e.g., Blue Cross is the highest paid supplier of General Motors), excessive cost of liabilities (e.g., the U.S. is the leading country in the world in lawsuits), lack of constancy of purpose, job hopping by management personnel (this works against understanding and long-term efforts), performance reviews (they are detrimental to team effort and build fear), running an organization on visible data and information alone (effects of a dissatisfied customer cannot be measured), and emphasis on short-term profits [1, 27].

Beside putting in half-hearted implementation efforts and setting unrealistic expectation goals, many companies also make various common errors when starting quality initiatives. These common errors are as follows [1, 28]:

- **Delegation of authority and poor management leadership.** In some organizations, the management delegate responsibility to a hired expert when starting quality initiatives rather than taking the appropriate leadership role to get everyone on board.
- **Team mania.** Ultimately, all employees should be involved with teams. Also, supervisors and employees must learn how to be effective coaches and team players, respectively. The team effort can only succeed after cultural change in the organization.
- **Practicing a narrow, dogmatic approach.** Some companies follow Deming philosophy, Juran philosophy, or Crosby philosophy without realizing that none of these philosophies is truly a one-size-fits-all proposition. More specifically, the quality programs must be tailored to satisfy their individual specific needs.
- **Deployment process.** Some companies develop quality initiatives without simultaneously developing effective mechanisms to integrate them into all different aspects of the organization; for example, marketing, operations, and budgeting.
- **Fuzziness about the differences among factors such as inspiration, awareness, skill building, and education.**

16.3 RISK ASSESSMENT

Risk assessment may simply be described as the process of risk analysis and risk evaluation [8, 29]. In turn, risk analysis is concerned with the utilization of available data for determining the risk to human, environment, or property/equipment from hazards and usually is made up of steps such as scope definition, hazard identification, and risk determination. The stage at which values and judgments make entry to the decision process is known as risk evaluation.

The complete process of risk assessment and risk control is called risk management. The term "risk control" is simply the decision making process concerned with managing risk as well as the implementations, enforcement, and re-evaluation of its effectiveness periodically, using risk assessment end results as one of the input factors.

16.3.1 Risk Analysis Role and Required Expertise Areas

Risk can only be managed effectively after its comprehension analysis. Risk analysis serves as a useful tool in identifying health and safety problems and approaches to uncover their solutions, satisfying regulatory requirements, and facilitating objective decisions on the risk acceptability.

A multi-disciplinary approach is often required to conduct risk analysis and it may require adequately sufficient knowledge in areas such as probability and statistics, engineering (electrical, mechanical, chemical, or nuclear), systems analysis, health sciences, social sciences, and physical, chemical, or biological sciences.

16.3.2 Risk Analysis Objectives in Hazardous System Life Cycle

The life cycle of hazardous systems may be divided into three major phases as shown in Figure 16.3.

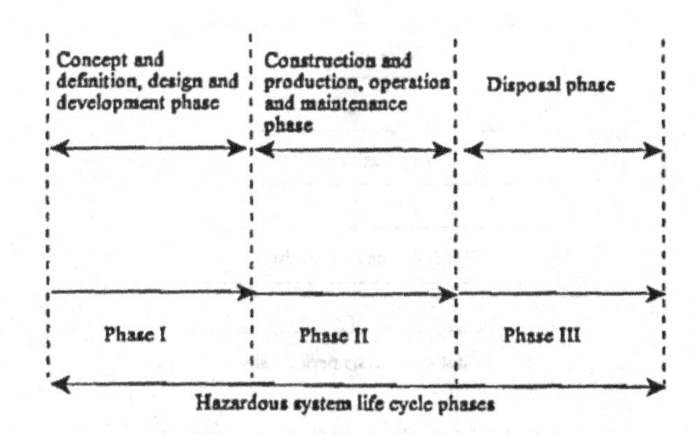

FIGURE 16.3 Major phases of hazardous systems.

The major phases shown in the figure are as follows [8]:

- Phase I: Concept and definition, design, and development phase.
- Phase II: Construction and production, operation, and maintenance phase.
- Phase III: Disposal phase.

The risk analysis objectives concerning Phase I include: identify major contributors to risk, provide input to design process and assess overall design adequacy, provide input to help in establishing procedures for normal and emergency conditions, and provide appropriate level of input to the evaluation of the acceptability of proposed potentially hazardous facilities or activities.

Some of the specific risk analysis objectives belonging to Phase II are to gauge and assess experience for the purpose of making comparisons between actual performance and relevant requirements, to update information on major risk contributors, to provide input on plant risk status in operational decision making, and to provide input to the optimization of normal and emergency procedures.

The two important risk analysis objectives of Phase III are to provide input to disposal policies and procedures, and to assess the risk associated with process disposal activities so that appropriate requirements can be satisfied effectively.

16.3.3 Risk Analysis Process Steps

The risk analysis process is composed of six steps as shown in Figure 16.4.

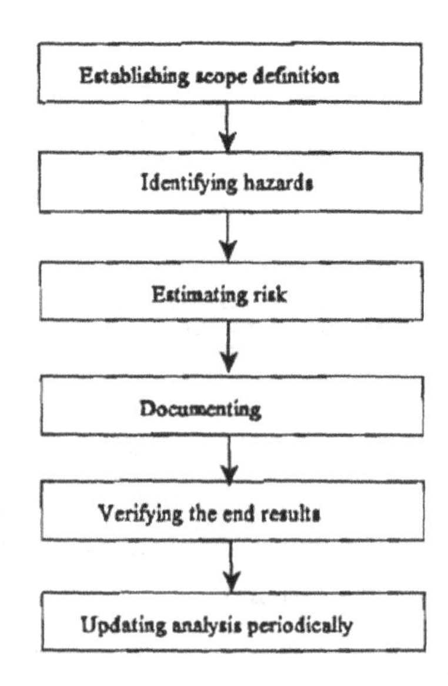

FIGURE 16.4 Risk analysis process steps.

The establishing scope definition is the first step of the risk analysis process and the risk analysis scope is defined and documented at the beginning of the project after a thorough understanding of the system under consideration. Nonetheless, the following five basic steps are involved in defining the risk analysis scope:

1. Describe the problems leading to risk analysis and then formulate risk analysis objectives on the basis of major highlighted concerns.
2. Define the system under consideration by including factors such as system general description, environment definition, and physical and functional boundaries' definition.
3. Describe the risk analysis associated assumptions and constraints.
4. Highlight the decisions to be made.
5. Document the total plan.

Identifying hazards is the second step of the risk analysis process and is basically concerned with the identification of hazards that will lead to risk in the system. This step also calls for the preliminary evaluation of the significance of the identified hazardous sources. The main purpose of this evaluation is to determine the appropriate course of action.

Estimating risk is the third step of the risk analysis process and risk estimation is conducted in the following steps:

- Investigate sources of hazards to determine the probability/likelihood of occurrence of the originating hazard and associated consequences.
- Conduct pathway analysis to determine the mechanisms and likelihood through which the receptor under consideration is influenced.
- Choose risk estimation methods/approaches to be used.
- Identify data needs.
- Discuss assumptions/rationales associated with methods/approaches/data being utilized.
- Estimate risk to evaluate the degree of influence on the receptor under consideration.
- Document the risk estimation study.

Documenting is the fourth step of the risk analysis process and is basically concerned with effectively documenting the risk analysis plan, the preliminary evaluation, and the risk estimation. The documentation report should contain sections such as title, abstract, conclusions, table of contents, objectives/scope, assumptions/limitations, system description, analysis methodology description, results of hazard identification, model description and associated assumptions, quantitative data and associated assumptions, results of risk estimation, references, appendices, discussion of results, and sensitivity analysis.

The fifth step of the risk analysis process is basically concerned with verifying the end results. More specifically, it may be stated that verification is a review process used to determine the integrity and accuracy of the risk analysis process. Verification is conducted at appropriate times by person(s) other than the involved analyst(s).

The final step of risk analysis is concerned with periodically updating the analysis as more up-to-date information becomes available.

16.3.4 Risk Analysis Techniques for Engineering Systems

Over the years many different methods to conduct risk analysis have been developed [8, 30-32]. It is important to carefully consider the relevance and suitability of these techniques prior to their proposed applications. Some of the factors to be considered specifically in this regard are appropriateness to the system under study, scientific defensibility, format of the results with respect to improvement in understanding of the risk occurrence and risk controllability, and simplicity.

The additional factors that should be used as the basis for selecting risk analysis technique(s) by the analyst include study objectives, development phase, level of risk, system type and hazard being analyzed, information and data needs, manpower requirement, level of expertise required, updating flexibility, and resource requirement.

The risk analysis methods used for engineering systems may be grouped into two categories: (1) hazard identification and (2) risk estimation. Examples of the techniques belonging to the hazard identification category are hazard and operability study (HAZOP), event tree analysis (ETA), and failure modes and effect analysis (FMEA). Similarly, the examples of the techniques belonging to the risk estimation category include frequency analysis and consequence analysis.

Four of these techniques are discussed below.

Consequence Analysis

Consequence analysis is concerned with estimating the impact of the undesired event on adjacent people, property, or the environment. Usually, for risk estimations concerning safety, it consists of calculating the probability that people at different distances (and in different environments) from the undesired event source will suffer injury/illness. Some examples of the undesired event are explosions, projection of debris, fires, and release of toxic materials. It means there is a definite need to use consequence models for predicting the extent and probability of casualties due to such undesired events. Nonetheless, it is appropriate that consequence analysis takes into consideration the factors such as analysis based on chosen undesirable events, measures to eradicate consequences, explanation of any series of consequences resulting from the undesirable events, measures to eradicate consequences, explanation of any series of consequences resulting from the undesirable events, outlining of the criteria used for accomplishing the identification of consequences, and immediate and aftermath consequences.

Hazard and Operability Study (HAZOP)

This is a form of FMEA and was developed for applications in chemical industries. HAZOP is a systematic approach used to identify hazards and operational problems throughout a facility. Over the years, it is proven to be an effective tool for identifying unforeseen hazards designed into facilities due to various reasons, or introduced into existing facilities due to factors such as changes made to process conditions or operating procedures.

The approach has primarily threefold objectives: develop a full facility/process description, systematically review each and every facility/process part to identify how deviations from the design intentions can occur, and pass judgment on whether such deviations can result in hazards/operating problems.

HAZOP can be used to analyze design at various stages as well as to perform analysis of process plants in operation. However, the application of HAZOP in the early design phase often leads to a safer detailed design. The following basic steps are associated with HAZOP [8]:

- Develop study objectives and scope.
- Form HAZOP team by ensuring that its membership is comprised of persons from design and operation.
- Collect appropriate drawings, process description, and other relevant documentation; for example, layout drawings, process control logic diagrams, operation and maintenance procedures, and process flowsheets.
- Conduct analysis of all major pieces of equipment, supporting equipment, etc.
- Effectively document the study.

Frequency Analysis

This is basically concerned with estimating the occurrence frequency of each undesired event or accident scenario. Two commonly used approaches for performing frequency analysis are as follows:

- Making use of the frequency data of past relevant undesired events to predict frequency of their future occurrences.
- Using methods such as ETA and fault tree analysis (FTA) to calculate the occurrence frequencies of undesired events.

All in all, each of the above two techniques has strengths where the other has weaknesses; thus, each such approach should be used to serve as a check for the other wherever it is feasible.

Event Tree Analysis (ETA)

This is a "bottom-up" approach used to identify the possible outcomes when an initiating event's occurrence is known. The approach is useful for analyzing facilities having engineered accident-mitigating characteristics to identify the events' sequence that follows the initiating event and generate given sequences. Usually, it is assumed that each sequence event is either a success or a failure. It is important to note that often the ETA approach is used to perform analysis of more complex systems then the ones handled by the FMEA approach [30, 33, 34].

Because of the inductive nature of ETA, the fundamental question asked is "What happens if ...?" ETA highlights the relationship between the success or failure of various mitigating systems as well as the hazardous event that follows the initiating event. Additional factors associated with ETA include

- A comprehensive risk assessment requires identification of all potential initiating events.
- An excellent tool to identify events that require further investigation using FTA.
- It is rather difficult to incorporate delayed success or recovery events when performing ETA.
- ETA application always leaves room for missing some important initiating events.

16.3.5 Advantages of Risk Analysis

There are many benefits to performing risk analysis. Some of those are as follows [8]:

- It identifies potential hazards and failure modes.
- It provides in-depth understanding of the system.
- It helps to make comparisons of risks to those of other similar systems.
- It is useful to suggest modifications to reduce risks.
- It provides quantitative risk statements.
- It helps to identify major contributors to risk.
- It is useful to develop priorities for expenditures on improved safety.

16.4 PROBLEMS

1. Write an essay on TQM.
2. Define the following terms:
    - Risk
    - Risk management
    - Risk assessment
    - Risk control
3. Compare traditional quality assurance program with TQM.
4. Discuss Deming's 14 points concerning TQM.
5. List 10 most important methods of TQM. Provide a short discussion on each of these methods.
6. What are the Deming's "deadly diseases" of American management with respect to quality?
7. What are the common errors made when starting quality initiatives?
8. Describe the risk analysis process.
9. What are the benefits of risk analysis?
10. Discuss the following two risk analysis methods:
    - HAZOP
    - ETA

16.5 REFERENCES

1. Goetsch, D.L. and Davis, S., *Implementing Total Quality*, Prentice-Hall, Englewood Cliffs, NJ, 1995.
2. Rao, A., Carr, L.P., Dambolena, I., Kopp, R.J., Martin, J., Raffi, F., and Schlesinger, P.F., *Total Quality Management: A Cross Functional Perspective*, John Wiley & Sons, New York, 1996.
3. Gevirtz, C.D., *Developing New Products with TQM*, McGraw-Hill, New York, 1994.
4. Dobyns, L. and Crawford-Mason, C., *Quality or Else*, Houghton Mifflin, Boston, 1991.
5. Schmidt, W.H. and Finnigan, J.P., *The Race without a Finish Line: America's Quest for Total Quality*, Jossey-Bass Publishers, San Francisco, CA, 1992.
6. Van Ham, K., Setting a total quality management strategy, in *Global Perspectives on Total Quality*, The Conference Board, New York, 1991.
7. Muadu, C.N. and Chu-hua, K., Strategic total quality management (STQM), in *Management of New Technologies for Global Competitiveness*, Madu, C.N., Ed., Quorum Books, Westport, CT, 1993, pp. 3-25.
8. Risk Analysis Requirements and Guidelines, CAN/CSA-Q6340-91, prepared by the Canadian Standards Association, 1991. Available from Canadian Standards Association, 178 Rexdale Boulevard, Rexdale, Ontario, Canada.
9. Molak, V., Ed., *Fundamentals of Risk Analysis and Risk Management*, CRC Press, Boca Raton, FL, 1997.
10. Covello, V.T. and Manpower, J., Risk analysis and risk management: A historical perspective, *Risk Analysis*, 5, 103-120, 1985.
11. Farquhar, C.R. and Johnston, C.G., *Total Quality Management: A Competitive Imperative Report* No. 60-90-E, 1990, The Conference Board of Canada, Ottawa, Ontario, Canada.
12. Caropreso, F., *Making Total Quality Happen*, The Conference Board, New York, 1990.
13. Spenley, P., *World Class Performance Through Total Quality*, Chapman and Hall, London, 1992.
14. Burati, J.L., Matthews, M.F., and Kalidindi, S.N., Quality management organizations and techniques, *J. Construction Eng. Manage.*, 118, 112-128, 1992.
15. Matthews, M.F. and Burati, J.L., Quality Management Organizations and Techniques, Source Document 51, The Construction Industry Institute, Austin, Texas, 1989.
16. Imai, M., *Kaizen: The Key to Japan's Competitive Success*, Random House, New York, 1986.
17. Ishikawa, K., *Guide to Quality Control*, Asian Productivity Organization, Tokyo, 1982.
18. Kume, H., *Statistical Methods for Quality Improvement*, The Association for Overseas Technology Scholarship, Tokyo, 1985.
19. Perisco, J., Team up for quality improvement, *Quality Progress*, 22, 33-37, 1989.
20. Kanji, G.K. and Asher, M., *100 Methods for Total Quality Management*, Sage Publications Ltd., London, 1996.
21. Heizer, J. and Render, B., *Production and Operations Management*, Prentice-Hall, Upper Saddle River, New Jersey, 1995.
22. Mears, P., *Quality Improvement Tools and Techniques*, McGraw-Hill, New York, 1995.
23. Uselac, S., *Zen Leadership: The Human Side of Total Quality Team Management*, Mohican Publishing Company, Londonville, OH, 1993.
24. Akao, Y., *Hoshin Kanri: Policy Deployment for Successful TQM*, Productivity Press, Cambridge, MA, 1991.
25. Yoji, K., *Quality Function Deployment*, Productivity Press, Cambridge, MA, 1991.

26. Daetz, D., The effect of product design on product quality and product cost, *Quality Progress*, June 1987, pp. 63-67.
27. Coppola, A., Total quality management, in *Tutorial Notes, Annu. Reliability Maintainability Symp.*, 1992, pp. 1-44.
28. Clemmer, J., Five common errors companies make starting quality initiatives, *Total Quality*, 3, 4-7, 1992.
29. Kunreuther, H. and Slovic, P., Eds., Challenges in risk assessment and risk management, in *Annu. Am. Acad. Political Soc. Sci.*, 545, 1-220, 1996.
30. Wesely, W.E., Engineering risk analysis, in *Technological Risk Assessment*, Rice, P.F., Sagan, L.A., and Whipple, C.G., Eds., Martinus Nijhoff Publishers, The Hague, 1984, pp. 49-84.
31. Covello, V. and Merkhofer, M., *Risk Assessment and Risk Assessment Methods: The State-of- the-Art*, NSF report, 1984, National Science Foundation (NSF), Washington, D.C.
32. Dhillon, B.S. and Rayapati, S.N., Chemical systems reliability: A survey, *IEEE Trans. Reliability*, 37, 199-208, 1988.
33. Cox, S.J. and Tait, N.R.S., *Reliability, Safety, Risk Management*, Butterworth-Heinemann Ltd., Oxford, 1991.
34. Ramakumar, R., *Engineering Reliability: Fundamentals and Applications*, Prentice-Hall, Englewood Cliffs, NJ, 1993.

17 Life Cycle Costing

17.1 INTRODUCTION

Today in the global economy and due to various other market pressures, the procurement decisions of many products are not entirely made on initial acquisition costs but on their total life cycle costs; in particular is the case of expensive products. Many studies performed over the years indicate that the product ownership costs often exceed procurement costs. In fact, according to Reference 1, the product ownership cost (i.e., logistics and operating cost) can vary from 10 to 100 times the procurement cost. Even the assertion of the ownership cost being high could be detected from the overall annual budgets of various organizations. For example, in 1974, the operation and maintenance cost accounted for 27% of the total U.S. Department of Defense budget as opposed to 20% for procurement [2].

Life cycle cost of a product may simply be described as the sum of all costs incurred during its life span, i.e., the total of procurement and ownership costs. The history of life cycle costing goes back to the mid-1960s when Logistics Management Institute prepared a document [3] entitled "Life Cycle Costing in Equipment Procurement" for the U.S. Assistant Secretary of Defense, Installations and Logistics. Consequently, the Department of Defense released a series of three guidelines for life cycle costing procurement [4-6].

In 1974, Florida became the first U.S. state to formally adopt the concept of life cycle costing and in 1978, the U.S. Congress passed the National Energy Conservation Policy Act. The passage of this Act made it mandatory that every new federal government building be life cycle cost effective. In 1981, Reference 7 presented a list of publications on life cycle costing. Over the years many people have contributed to the subject of life cycle costing and Reference 8 presents a comprehensive list of publications on the subject. This chapter discusses different aspects of life cycle costing.

17.2 REASONS AND USES OF LIFE CYCLE COSTING AND REQUIRED INPUTS

Life cycle costing is increasingly being used in the industry to make various types of decisions. Some of the reasons for this upward trend could be increasing maintenance cost, competition, increasing cost effectiveness awareness among product users, budget limitations, costly products (e.g., aircrafts, military systems), and greater ownership costs in comparison to procurement costs.

The life cycle costing concept can be used for various different purposes including selecting among options, determining cost drivers, formulating contractor incentives, choosing the most beneficial procurement strategy, assessing new technology application, forecasting future budget needs, making strategic decisions and design trade-offs, providing objectives for program control, deciding the replacement of aging equipment, improving comprehension of basic design associated parameters in product design and development, optimizing training needs, and comparing logistics concepts [9, 10].

A life cycle costing project can only be accomplished effectively if the required information is available. Nonetheless, the professionals involved in conducting life cycle cost studies should seek information on general items such as those listed below prior to the start of the project under consideration [8, 9, 11]:

- Estimate goal
- Time schedule
- Required data
- Involved individuals
- Required analysis format
- Required analysis detail
- Treatment of uncertainties
- Ground rules and assumptions
- Analysis constraints
- Analysis users
- Limitations of funds
- Required accuracy of the analysis

Nonetheless, the specific data required to conduct life cycle cost studies for an item include useful life, acquisition cost, periodic maintenance cost, transportation and installation costs, discount and escalation rates, salvage value/cost, taxes (i.e., investment tax credit, tax benefits from depreciation, etc.), periodic operating costs (e.g., energy cost, labor cost, insurance, cost of materials, and cost of supplies) [12].

17.3 LIFE CYCLE COSTING STEPS AND ACTIVITIES

A life cycle costing study can be conducted in seven steps as shown in Figure 17.1 [13]. There are many activities associated with life cycle costing. Some of those activities are as follows [14]:

- Develop an accounting breakdown structure.
- Define activities that generate ownership costs of products.
- Establish cause-and-effect relationships.
- Identify cost drivers.
- Determine or define the life cycle of items/products.
- Develop cost estimate relationships for every element in the life cycle cost breakdown structure.
- Conduct sensitivity analysis.
- Develop escalated and discounted life cycle costs.

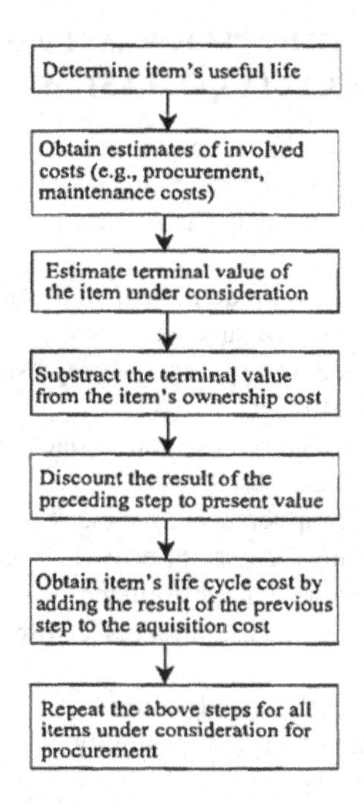

FIGURE 17.1 Life cycle costing steps.

17.4 SKILL REQUIREMENT AREAS OF A LIFE CYCLE COSTING ANALYST AND HIS/HER ASSOCIATED PROFESSIONALS

In order to perform an effective life cycle cost analysis, the life cycle cost analyst must possess skills in many areas including engineering, finance and accounting, estimating, statistical analysis, reliability and maintainability engineering, logistics, and contracting. Nonetheless, the professionals such as those listed below provide input to life cycle cost studies:

- Design engineers
- Maintenance engineers
- Tooling engineers
- Reliability and maintainability engineers
- Planning engineers
- Test engineers
- Quality control engineers
- Manufacturing engineers

17.5 LIFE CYCLE COSTING PROGRAM EVALUATION AREAS AND LIFE CYCLE COST ESTIMATE REPORT

In order to keep a life cycle costing program functioning effectively, it is evaluated periodically. There are many areas in which questions could be raised by the equipment manufacturer/user management. These areas include [9, 15]:

- Inclusion of life cycle cost requirements into subcontracted designs.
- Cost performance review of subcontractors.
- Validity of cost-estimating approaches being used.
- Identification of cost drivers.
- Effectiveness of inflation and discounting factors being used.
- Cost estimating database's broadness.
- Buyer's awareness of top ten cost drivers.
- Management's defining of cost priority in relation to factors such as product delivery schedule and performance.
- Planning and design groups' awareness of life cycle cost requirements concerning them.
- Coordination of life cycle costing effort.
- Life cycle cost requirements' compatibility with reliability, maintainability, and safety programs.
- Qualifications of professionals managing the life cycle costing program.
- Cost estimates' validation by an independent appraisal.
- Cost model construction.
- Performance of trade-off studies.

The documentation of the results of the life cycle cost analysis is as important as the analysis itself. Poor documentation can dramatically decrease the effectiveness of actual analysis. Needless to say, the writing of life cycle cost estimate reports requires careful consideration. Even though such reports may vary from one project/organization to another, they should include information on items such as those listed below [8].

- Data sources and their validity/accuracy
- The life cycle cost model
- Ground rules and assumptions
- Sensitivity analyses
- Cost-estimating relationships used
- Important cost breakdown estimates
- Contractual cost incentives/requirements (e.g., reliability improvement warranty and design to cost)
- Comparisons of life cycle cost estimate with other similar estimates (if applicable)
- Significance and conclusions of end estimates.

17.6 TIME DEPENDENT FORMULAS FOR LIFE CYCLE COST ANALYSIS

As the value of money changes with the passage of time in estimating the life cycle cost of a product, in particular its ownership cost has to be discounted by using various time dependent formulas before it is added to the procurement cost. In modern society, interest and inflation rates are used to take into consideration the time dependent value of money. Nonetheless, the history of the interest may be traced back to 2000 B.C. in Babylon where interest on borrowed commodities (e.g., grain) was paid in the form of grain or through other means. Around 575 B.C., the concept of interest reached to a such level that it led to the establishment of a firm of international bankers with home offices in Babylon [16].

This section presents selective formulas to take into consideration the time value of money for use in life cycle cost analysis [8, 17, 18].

17.6.1 Single Payment Compound Amount Formula

In this case, P amount of money is invested for m interest periods (i.e., normally years) at compound interest rate i (i.e., interest rate per interest period). Thus, the future amount, FA, of the invested money is

$$FA = P(1+i)^m \tag{17.1}$$

Example 17.1

A person deposited $700 dollars in a bank at a compound interest rate of 5% per year. Calculate its future amount after 7 years.

By inserting the given data into Equation (17.1) we get

$$FA = 700(1+0.05)^7$$
$$\simeq \$985$$

Thus, the future amount after 7 years will be $985.

17.6.2 Single Payment Present Value Formula

This formula is concerned with determining the present value of a single amount of money to be received or spent after m interest periods or years at compound interest rate i.

Thus, from Equation (17.1), we get

$$P = FA(1+i)^{-m} \tag{17.2}$$

where

P is the present value of FA.

Example 7.2

A company sold a personal computer and the buyer agreed to pay \$5000 after 5 years. The estimated interest rate is 6% per year. Calculate the present value of the amount to be paid by the buyer.

Inserting the specified data into Equation (17.2) yields

$$P = 5{,}000(1+0.06)^{-5}$$
$$\simeq \$3736$$

Thus, the present value is \$3736.

17.6.3 Uniform Periodic Payment Present Value Formula

In this case X amount of equal payments are made at the end of m interest periods and all these payments are deposited at interest rate i compounded after each interest period. Thus, the present value, PV, of all these payments is expressed by

$$\begin{aligned} PV &= PV_1 + PV_2 + PV_3 + ----+ PV_m \\ &= \frac{X}{(1+i)} + \frac{X}{(1+i)^2} + \frac{X}{(1+i)^3} + ----+ \frac{X}{(1+i)^m} \end{aligned} \quad (17.3)$$

where

PV_i is the present value of payment X made at the end of interest period i; for i = 1, 2, - - - -, m.

In order to find the sum of the geometric series represented by Equation (17.3), we multiply both sides of this equation by 1/(1 + i) to get

$$\frac{PV}{(1+i)} = \frac{X}{(1+i)^2} + \frac{X}{(1+i)^3} + \frac{X}{(1+i)^4} + ----+ \frac{X}{(1+i)^{m+1}} \quad (17.4)$$

Subtracting Equation (17.3) from Equation (17.4) yields

$$\frac{PV}{(1+i)} - PV = \frac{X}{(1+i)^{m+1}} - \frac{X}{(1+i)} \quad (17.5)$$

By rearranging Equation (17.5) we get

$$PV = X\left[\frac{1-(1+i)^{-m}}{i}\right] \quad (17.6)$$

Example 17.3

A machine's expected useful life is 10 years and its use will generate an estimated revenue of $20,000 at the end of each year. Calculate the present value of the total revenue to be generated by the machine, if the annual interest rate is expected to be 5%.

Substituting the given data into Equation (17.6) yields

$$PV = (20{,}000)\left[\frac{1-(1+0.05)^{-10}}{0.05}\right]$$

$$\approx \$154{,}435$$

Thus, the present value of the total machine revenue is $154,435.

17.6.4 Uniform Periodic Payment Future Amount Formula

This represents the same situation as for Section 17.6.3 except that instead of finding the total present value of the periodic payments, the total future amount is sought. Thus, the total future amount, FA_t, of all the periodic payments is given by

$$FA_t = X + X(1+i) + ---- + X(1+i)^{m-2} + X(1+i)^{m-1} \quad (17.7)$$

As Equation (17.7) is a geometric series, its sum can be found. Thus, we multiply both sides of Equation (17.7) by factor (1 + i) to get

$$(1+i)FA_t = X(1+i) + X(1+i)^2 + --- + X(1+i)^{m-1} + X(1+i)^m \quad (17.8)$$

Subtracting Equation (17.7) from Equation (17.8) yields

$$(1+i)FA_t - FA_t = X(1+i)^m - X \quad (17.9)$$

By rearranging Equation (17.9) we get

$$FA_t = X\left[(1+i)^m - 1\right]/i \quad (17.10)$$

Example 17.4

In Example 17.3, instead of calculating the present value of the total income, find the total future amount.

Substituting the given data into Equation (17.10) yields

$$FA_t = (20,000)\left[(1+0.05)^{10} - 1\right]/0.05$$

$$\simeq \$251,558$$

Thus, the total future amount is \$251,558.

17.7 LIFE CYCLE COST ESTIMATION MODELS

Over the years there have been many mathematical models developed to estimate life cycle cost of various items. A life cycle cost estimation model of an item usually requires inputs such as mean time between failures, mean time to repair, acquisition cost, training cost, warranty coverage period, labor cost per corrective maintenance action, average material cost of a failure, cost of carrying spares, cost of installation, and time spent for travel [19]. Nonetheless, the life cycle cost estimation models may be divided into two major classifications as follows:

- General models
- Specific models

Both these classifications are discussed below.

17.7.1 General Life Cycle Cost Estimation Models

These models are not specifically tied to any particular system or equipment but they may vary from one organization to another or within an organization. For example, the life cycle cost estimation model used by the U.S. Navy is different than the one used by the Army [14]. Nonetheless, the basic idea is to sum up all the costs associated with an item over its life span. Three general life cycle cost estimation models are presented below [8, 11, 17].

General Life Cycle Cost Model I

In this case, the item life cycle cost is divided into two categories, i.e., recurring cost (RC) and nonrecurring cost (NRC). Thus, the item life cycle cost (LCC) can be expressed by

$$LCC = RC + NRC \tag{17.11}$$

In turn, RC has five major components: maintenance cost, operating cost, support cost, labor cost, and inventory cost.

Similarly, the elements of the NRC are procurement cost, training cost, LCC management cost, support cost, transportation cost, research and development cost, test equipment cost, equipment qualification approval cost, installation cost, and reliability and maintainability improvement cost.

General Life Cycle Cost Model II

In this case, the equipment life cycle cost is categorized into four major classifications: research and development cost (RDC), production and construction cost (PCC), operation and support cost (OSC), and disposal cost (DC). Thus, mathematically, the equipment LCC is given by

$$LCC = RDC + PCC + OSC + DC \tag{17.12}$$

The major elements of the RDC are engineering design cost, software cost, life cycle management cost, product planning cost, test and evaluation cost, research cost, and design documentation cost.

PCC is made up of five important elements: construction cost, manufacturing cost, initial logistics support cost, industrial engineering and operations analysis cost, and quality control cost.

The major components of the OSC are product distribution cost, product operation cost, and sustaining logistic support cost.

Finally, DC is expressed by

$$DC = URC + \left[CF(\theta)\,(IDC - RV)\right] \tag{17.13}$$

where

θ is the number of unscheduled maintenance actions.
CF is the condemnation factor.
URC is the ultimate retirement cost of the equipment under consideration.
IDC is the equipment disposal cost.
RV is the reclamation value.

General Life Cycle Cost Model III

In this case, the equipment life cycle cost is made up of three major components: initial cost (IC), operating cost (OC), and failure cost (FC). Thus, mathematically, the equipment LCC can be expressed as follows [20]:

$$LCC = IC + OC + FC \tag{17.14}$$

The average cost of a failure is given by

$$AC = \left[HC(MTTR + RT) + ALC\right] + \left[LSC(MTTR + RT) + MAC\right] + RPC \tag{17.15}$$

where

RPC is the cost of replaced parts.
ALC is the administrative loss cost due to a failure.
MTTR is the mean time to repair.
RT is the repaired item response time.
HC is the cost of the repair organization per hour.
MAC is the mission-abort loss cost which occurs with each interruption of service, and is also called the event-related, one-time cost.
LSC is the loss of service cost per hour.

Thus, the failure cost is expressed by

$$FC = (AC)(LT)/MTBF \tag{17.16}$$

where

MTBF is the mean time between failures in hours.
LT is the life time of the item in hours.

The failure loss cost rate per hour, FCR, is defined by

$$FCR = AC/MTBF \tag{17.17}$$

Substituting Equations (17.16) and (17.17) into Equation (17.14) yields

$$LCC = IC + OC + (FCR)(LT) \tag{17.18}$$

17.7.2 Specific Life Cycle Cost Estimation Models

These models are developed to estimate life cycle cost of specific items or equipment. Three such models are presented below.

Switching Power Supply Life Cycle Cost Model

This was developed to estimate life cycle cost of switching power supplies [21]. The life cycle cost is expressed by

$$LCC = IC + FC \tag{17.19}$$

In turn, FC is given by

$$FC + \lambda \alpha (RC + SC) \tag{17.20}$$

where

λ is the constant unit failure rate.
α is the product's expected life.
RC is the repair cost.
SC is the cost of spares.

The cost of spares is given by

$$SC = (USC)Q \tag{17.21}$$

where

USC is the unit spare cost.
Q is the fractional number of spares for each active unit.

Inserting Equations (17.20) and (17.21) into Equation (17.19) yields

$$LCC = IC + \lambda\alpha\left[RC + (USC)Q\right] \tag{17.22}$$

Early Warning Radar Life Cycle Cost Model

This model was developed to estimate life cycle cost of an Early Warning Radar [8]. The life cycle cost of this equipment was divided into three parts: acquisition cost (AC), operation cost (OC), and logistic support cost (LSC). Mathematically, this is expressed as follows:

$$LCC = AC + OC + LSC \tag{17.23}$$

The predicted breakdown percentages of the LCC for AC, OC, LSC were 28%, 12%, and 60%, respectively.

The four major components of the acquisition cost were fabrication cost, installation and checkout cost, design cost, and documentation cost. Their predicted breakdown percentages were 20.16%, 3.92%, 3.36%, and 0.56%, respectively.

OC was made up of three principal elements (each element's predicted breakdown percentage is given in parentheses): cost of personnel (8.04%), cost of power (3.84%), and cost of fuel (0.048%).

The logistic support cost was expressed by

$$LSC = RLC + ISC + AC + RMC + ITC + RSC \tag{17.24}$$

where

RLC is the repair labor cost.
ISC is the initial spares' cost.

AC is the age cost.
RMC is the repair material cost.
ITC is the initial training cost.
RSC is the replacement spares' cost.

The predicted breakdown percentages of the LCC for each of these six cost elements were 38.64%, 3.25%, 1.186%, 5.52%, 0.365%, and 11.04%, respectively.

Appliance Life Cycle Cost Model

This model is proposed to estimate life cycle cost of major appliances [22]. In this case, the life cycle cost of an appliance is expressed by

$$LCC = APC + \sum_{k=1}^{n} EC_k \left[\frac{FC(1+FER)^k}{(1+i)^k} \right] \tag{17.25}$$

where

APC is the appliance acquisition cost in dollars.
EC_k is the appliance energy consumption in year k, expressed in million BTUs.
n is the appliance useful life in years.
FC is the fuel cost in year one, expressed in constant dollars per million BTUs.
FER is the annual fuel escalation rate (%) in constant dollars.
i is the discount rate (%) in constant dollars.

If EC_k and FER are constant over the useful life span of the appliance, Equation (17.25) becomes

$$LCC = APC + (EC)(FC) \sum_{k=1}^{n} \left[\frac{FC(1+FER)}{(1+i)} \right]^K \tag{17.26}$$

The typical useful life of appliances such as refrigerators, freezers, ranges and ovens, electric dryers, and room air conditioners are 15, 20, 14, 14, and 10 years, respectively.

Example 17.5

A company using a machine to manufacture certain engineering parts is contemplating replacing it with a better one. Three different machines are being considered for its replacement and their data are given in Table 17.1. Determine which of the three machines should be purchased to replace the existing one with respect to their life cycle costs.

TABLE 17.1
Data for Three Machines Under Consideration

No.	Description	Machine A	Machine B	Machine C
1	Procurement price	$175,000	$160,000	$190,000
2	Expected useful life in years	12	12	12
3	Annual failure rate	0.02	0.03	0.01
4	Annual interest rate	5%	5%	5%
5	Annual operating cost	$5,000	$7,000	$3,000
6	Cost of a failure	$4,000	$3,000	$5,000

The expected failure costs of Machines A, B, and C, respectively, are given by

$$FC_A = (0.02)(4{,}000)$$
$$= \$80$$

$$FC_B = (0.03)(3{,}000)$$
$$= \$90$$

and

$$FC_C = (0.01)(5{,}000)$$
$$= \$50$$

Using the given and calculated data in Equation (17.6), we obtain the following present value of Machine A failure cost over its useful life:

$$PVFC_A = 80\left[\frac{1-(1+0.05)^{-12}}{0.05}\right]$$
$$\simeq \$709$$

where

$PVFC_A$ is the present value of Machine A failure cost over its life span.

Similarly, the present values of Machines B and C failure costs over their useful lives are as follows:

$$PVFC_B = 90\left[\frac{1-(1+0.05)^{-12}}{0.05}\right]$$

$$\simeq \$798$$

and

$$PVFC_C = 50\left[\frac{1-(1+0.05)^{-12}}{0.05}\right]$$

$$\simeq \$443$$

The present values of Machines A, B, and C operating costs over their useful life span, using Equation (17.6) and the given data, are

$$PVOC_A = 5{,}000\left[\frac{1-(1+0.05)^{-12}}{0.05}\right]$$

$$\simeq \$44{,}316$$

$$PVOC_B = 7{,}000\left[\frac{1-(1+0.05)^{-12}}{0.05}\right]$$

$$\simeq \$62{,}043$$

and

$$PVOC_C = 3{,}000\left[\frac{1-(1+0.05)^{-12}}{0.05}\right]$$

$$\simeq \$26{,}590$$

where

$PVOC_A$ is the present value of Machine A operating cost over its useful life.
$PVOC_B$ is the present value of Machine B operating cost over its useful life.
$PVOC_C$ is the present value of Machine C operating cost over its useful life.

The life cycle costs of Machines A, B, and C are

$$LCC_A = 175{,}000 + 709 + 44{,}316$$

$$= \$220{,}025$$

$$LCC_B = 160{,}000 + 798 + 62{,}043$$

$$= \$222{,}841$$

and

$$LCC_C = 190,000 + 443 + 26,590$$

$$= \$217,033$$

As the life cycle cost of Machine C is the lowest, thus it should be purchased.

17.8 COST ESTIMATION MODELS

Over the years, various types of mathematical models have been developed to estimate costs for use in LCC models. Many such models are presented in Reference 8. This section presents three cost estimation models taken from published literature.

17.8.1 Cost-Capacity Model

This is a useful model to obtain quick approximate cost estimates for similar new equipment, plants, or projects of different capacities. Mathematically, the cost-capacity relationship is expressed as follows:

$$C_n = C_{od}\left(KP_n / KP_{od}\right)^{\theta} \tag{17.27}$$

where

C_n is the cost of the new item under consideration.
KP_n is the capacity of the new item.
C_{od} is the cost of the old but similar item.
KP_{od} is the capacity of the old but similar item.
θ is the cost-capacity factor and its frequently used value is 0.6. For items such as pumps, heat exchanger, tanks, and heaters, the proposed values of this factor are 0.6, 0.6, 0.7, and 0.8, respectively [8, 23-25].

Example 17.6

An electric utility spent $1.5 billion to build a 3,000 megawatt nuclear power generating station. In order to meet the growing demand for electricity, the company plans to construct a 4,000 megawatt nuclear station. Estimate the cost of the new station, if the value of the cost-capacity factor is 0.8.

Substituting the given data into Equation (17.27) yields

$$C_n = 1.5(4,000/3,000)^{0.8}$$

$$= \$1.89 \text{ billion}$$

Thus, the new station will cost $1.89 billion to construct.

17.8.2 Motor Operation Cost Estimation Model

This model is concerned with estimating operating cost of an alternating current (AC) motor. Thus, the cost to operate a motor is expressed by [26]

$$C_{om} = (0.746)\,(hp)\,T_{om}\,K_e/\beta \tag{17.28}$$

where

- C_{om} is the cost to operate the motor per year.
- hp is the motor size in horsepower.
- T_{om} is the motor annual operating time in hours.
- K_e is the electricity cost expressed in dollars per kilowatt hour ($/KWH).
- β is the motor efficiency.

Example 17.7

A 40-hp AC motor is operated for 3500 h per year and the cost of electricity is $0.08/kWh. Calculate the annual operating cost of the motor, if the motor efficiency is 90%.

Substituting the given data into Equation (17.28) yields

$$C_{om} = (0.746)\,(40)\,(3500)\,(0.08)/0.90$$

$$\simeq \$9,284$$

Thus, the annual operating cost of the motor is $9,284.

17.8.3 Corrective Maintenance Labor Cost Estimation Model

This model is concerned with estimating the annual labor cost of corrective maintenance associated with an item when its (item) mean time between failures (MTBF) and mean time to repair (MTTR) are known. Thus, the annual labor cost of the corrective maintenance is expressed by

$$C_{cm} = (SOH)\,C_{m\ell}\,(MTTR/MTBF) \tag{17.29}$$

where

- C_{cm} is the annual labor cost of corrective maintenance.
- SOH is the scheduled operating hours of the item per year.
- MTTR is the mean time to repair of the item.
- MTBF is the mean time between failures of the item.
- $C_{m\ell}$ is the cost of the corrective maintenance labor per hour.

Example 17.8

An electric motor is scheduled for 4500 h of operation in one year and its estimated MTBF is 1500 h. Whenever it fails, it takes on the average 15 h to repair. Calculate the annual cost for the motor corrective maintenance labor, if the maintenance labor cost is $30 per hour.

Inserting the given data into Equation (17.29) we get

$$C_{cm} = (4,500)\,(30)\,(15/1500)$$

$$= \$1350$$

Thus, the annual cost for the motor corrective maintenance labor is $1350.

17.9 LIFE CYCLE COSTING DATA

In order to perform an effective life cycle costing study, the availability of good cost data are essential. This can only be achieved through the existence of good cost data banks. Thus, whenever a cost data bank is being developed, careful attention must be paid to factors such as flexibility, uniformity, ready accessibility, orientation, expansion or contraction capability, size, responsiveness, and comprehensiveness [27]. Furthermore, at a minimum, a life cycle costing database should include information such as cost records, user pattern records, procedural records: operation and maintenance, and descriptive records: hardware and site.

Nevertheless, prior to embarking on a life cycle costing study, factors such as data availability, data applicability, data obsolescence, data bias, data comparability to other existing data, and data orientation towards the problem must be carefully examined.

17.9.1 Cost Data Sources

Even though there are many sources for obtaining cost related information for use in life cycle cost analysis, their quality and amount may vary quite considerably. Some of the sources for obtaining cost related data are as follows [8]:

- American Building Owners and Mangers Association (BOMA) handbook
- Unit price manuals: Marshall and Swift, Means, Dodge, Richardson, and Building Cost File
- Programmed Review of Information for Costing and Evaluation (PRICE) Model [28]
- Cost Analysis Cost Estimating (CACE) Model [28, 29]
- Budgeting Annual Cost Estimating (BACE) Model [28, 29]
- Cost for Pressure Vessels [30]
- Costs for Motors, Storage Tanks, Centrifugal Pumps, etc. [31, 32]
- Costs for Heat Exchangers [33-35]
- Costs for Varieties of Process Equipment [36-39]
- Costs for Solid Waste Shredders [40].

17.10 LIFE CYCLE COSTING ADVANTAGES AND DISADVANTAGES, RESISTANCE FACTORS AND ASSOCIATED IMPORTANT POINTS

There are many advantages to using the life cycle costing concept, including useful to control programs, making effective equipment replacement decisions, making a selection among the competing contractors, useful in reducing the total cost, and comparing the cost of competing projects.

In contrast, some of the disadvantages of life cycle costing are time consuming, expensive, and doubtful data accuracy.

Despite many advantages of the life cycle costing concept, there are many factors that still resist its usage. These factors include failure of some past procurement policies resembling life cycle costing, doubts about the life cycle costing methodology, data accuracy, data reliability, and contractor reluctance to guarantee concerned estimates [9]. Nonetheless, some of the important points associated with life cycle costing are as follows [8, 9]:

- Life cycle costing objective is to obtain maximum advantages from limited resources.
- Expect some surprises, irrespective of the excellence of the life cycle cost estimator.
- Life cycle costing is most beneficial if it is started early in the program.
- The group performing life cycle costing should be part of the finance or logistics organization.
- Ensure that all costs associated with the program are included in the life cycle cost model.
- The risk management is the essence of life cycle costing.
- Both the manufacturer and the user must organize effectively to control life cycle cost.
- Various data related difficulties may be overcome by a highly competent cost analyst.
- Perform trade-offs between life cycle cost, performance, and design to cost throughout the program.
- For good life cycle cost estimates, good data are indispensable.
- Management is key to making the life cycle costing effort worthwhile.
- In the event of life cycle costing based acquisition, ensure factors such as existence of written performance requirements, capable buyer and seller management to use life cycle cost results, sufficient time to conduct analysis, and availability of good historical cost data.

17.11 PROBLEMS

1. What are the advantages and disadvantages of life cycle costing?
2. Describe the steps associated with life cycle costing.
3. What are the reasons and uses of the life cycle costing concept?

TABLE 17.2
Data for Engineering Systems Offered by Two Different Manufacturers

No.	Description	Manufacturer A's system	Manufacturer B's system
1	Purchasing price	$80,000	$60,000
2	Cost of failure	$10,000	$5,000
3	Annual interest rate	6%	6%
4	Annual operating cost	$4,000	$5,500
5	Annual failure rate	0.004	0.004
6	Expected useful life in years	15	15

4. What are the inputs required to conduct life cycle costing?
5. What are the life cycle cost associated activities?
6. The salvage value of a computer system after a 10-year period will be $1500. Calculate the present value of this salvage value if the estimated annual interest rate is 5%.
7. A company plans to purchase an engineering system and two manufacturers are bidding for the system. Table 17.2 presents data associated with each manufacturer's system. Determine the system to be procured with respect to life cycle cost.
8. A 50-hp AC motor is operated for 2500 h per year and the electricity cost is $0.07/kWh. Calculate the annual operating cost of the motor if its efficiency is 85%.
9. List at least 10 important points associated with life cycle costing.
10. What are the important factors to be considered in developing a cost data bank for use in life cycle costing studies?

17.12 REFERENCES

1. Ryan, W.J., Procurement views of life cycle costing, *Proc. Annu. Symp. Reliability*, 164-168, 1968.
2. Wienecke-Louis, E. and Feltus, E.E., Predictive Operations and Maintenance Cost Model, Report No. ADA078052, 1979. Available from the National Technical Information Service (NTIS), Springfield, VA.
3. Life Cycle Costing in Equipment Procurement, Report No. LMI Task 4C-5, prepared by Logistics Management Institute (LMI), Washington, D.C., April 1965.
4. Life Cycle Costing Procurement Guide (interim), Guide No. LCC1, Department of Defense, Washington, D.C., July 1970.
5. Life Cycle Costing in Equipment Procurement-Casebook, Guide No. LCC2, Department of Defense, Washington, D.C., July 1970.
6. Life Cycle Costing for System Acquisitions (interim), Guide No. LCC3, Department of Defense, Washington, D.C., January 1973.
7. Dhillon, B.S., Life cycle cost: A survey, *Microelectronics and Reliability*, 21, 495-511, 1981.
8. Dhillon, B.S., *Life Cycle Cost: Techniques, Models, and Applications*, Gordon and Breach Science Publishers, New York, 1989.

9. Robert-Seldon, M., *Life Cycle Costing: A Better Method of Government Procurement,* Westview Press, Boulder, CO, 1979.
10. Lamar, W.E., Technical Evaluation Report on Design to Cost and Life Cycle Cost, North Atlantic Treaty Organization (NATO) Advisory Group for Aerospace Research and Development (AGARD) Advisory Report No. 165, May 1981. Available from the National Technical Information Service (NTIS), Springfield, VA.
11. Reiche, H., Life cycle cost, in *Reliability and Maintainability of Electronic Systems,* Arsenault, J.E. and Roberts, J.A., Eds., Computer Science Press, Potomac, MD, 1980, pp. 3–23.
12. Brown, R.J., A new market tool: Life cycle costing, *Ind. Market. Manage.*, 8, 109-113, 1979.
13. Coe, C.K., Life cycle costing by state governments, *Public Administration Rev.*, September/October 1981, pp. 564-569.
14. Eddins-Earles, M., *Factors, Formulas, and Structures for Life Cycle Costing,* Eddins-Earles, Concord, MA, 1981.
15. Bidwell, R.L., Checklist for Evaluating LCC Program Effectiveness, Product Engineering Services Office, Department of Defense, Washington, D.C., 1977.
16. Paul-Degarmo, E., Canada, J.R., and Sullivan, W.G., *Engineering Economy,* Macmillan Publishing Co., New York, 1979.
17. Dhillon, B.S., *Reliability Engineering in Systems Design and Operation,* Van Nostrand Reinhold Company, New York, 1983.
18. Newman, D.G., *Engineering Economic Analysis,* Engineering Press, San Jose, CA, 1988.
19. Siewiorek, D.P. and Swarz, R.S., *The Theory and Practice of Reliable System Design,* Digital Press, Digital Equipment Corporation, Bedford, MA, 1982.
20. Dickinson, D.B. and Sessen, L., Life cycle cost procurement of systems and spares, *Proc. Annu. Reliability Maintainability Symp.*, 282-286, 1975.
21. Monteith, D. and Shaw, B., Improved R, M, and LCC for switching power supplies, *Proc. Annu. Reliability Maintainability Symp.*, 262-265, 1979.
22. Turiel, I., Estrada, H., and Levine, M., Life cycle cost analysis of major appliances, *Energy,* 6, 945-970, 1981.
23. Desai, M.B., Preliminary cost estimating of process plants, *Chem. Eng.*, July 1981, pp. 65-70.
24. Dieter, G.E., *Engineering Design,* McGraw-Hill, New York, 1983.
25. Jelen, F.C. and Black, J.H., Eds., *Cost and Optimization Engineering,* McGraw-Hill, New York, 1983.
26. Brown, R.J. and Yanuck, R.R., *Life Cycle Costing: A Practical Guide for Energy Managers,* Fairmont Press, Atlanta, GA, 1980.
27. Bowen, B. and Williams, J., Life cycle costing and the problems of data, *Industrial. Forum,* 6, 21-24, 1975.
28. Marks, K.E., Garrison-Massey, H., and Bradley, B.D., An Appraisal of Models Used in Life Cycle Cost-Estimation for U.S. Air Force (USAF) Aircraft Systems, Report No. R-2287-AF, October 1978. Prepared by the Rand Corporation, Santa Monica, CA.
29. USAF Cost and Planning Factors, Report No. AFR 173-10, Department of the Air Force, Washington, D.C., February 6, 1975.
30. Mulet, A., Corripio, A.B., and Evans, L.B., Estimate cost of pressure vessels via correlations, *Chem. Eng.*, 88(20, 456), 1981.
31. Corripio, A.B., Chrien, K.S., and Evans, L.B., Estimate costs of heat exchangers and storage tanks via correlations, *Chem. Eng.*, 89(2, 125), 1982.

32. Corripio, A.B., Chrien, K.S., and Evans, L.B., Estimate costs of centrifugal pumps and electric motors, *Chem. Eng.*, 89(4), 115, 1982.
33. Purohit, G.P., Cost of double pipe and multi-tube heat exchangers, *Chem. Eng.*, 92, 96, 1985.
34. Woods, D.R., Anderson, S.J., and Norman, S.L., Evaluation of capital cost data: Heat exchangers, *Can. J. Chem. Eng.*, 54, 469, 1976.
35. Kumana, J.D., Cost update on specialty heat exchangers, *Chem. Eng.*, 91(13), 164, 1984.
36. Hall, R.S., Mately, J., and McNaughton, K.J., Current costs of process equipment, *Chem. Eng.*, 89(7), 80, 1982.
37. Klumpar, I.V. and Slavsky, S.T., Updated cost factors: Process equipment, commodity materials, and installation labor, *Chem. Eng.*, 92(15), 73, 1985.
38. Humphreys, K.K. and Katell, S., *Basic Cost Engineering*, Marcel Dekker, New York, 1981.
39. Peters, M.S. and Timmerhaus, K.D., *Plant Design and Economics for Chemical Engineers*, McGraw-Hill, New York, 1980.
40. Fang, C.S., The cost of shredding municipal solid waste, *Chem. Eng.*, 87(7), 151, 1980.

Index

Printed in the United States
by Baker & Taylor Publisher Services